T0360523

WEIGHTED INEQUALITIES OF HARDY TYPE

Second Edition

WEIGHTED INEQUALITIES OF HARDY TYPE

Second Edition

Alois Kufner

Academy of Sciences, Prague, Czech Republic
University of West Bohemia, Pilsen, Czech Republic
RUDN University, Moscow, Russia

Lars-Erik Persson

Luleå University of Technology, Sweden
UiT The Artic University of Norway, Norway
RUDN University, Moscow, Russia

Natasha Samko

Luleå University of Technology, Sweden

World Scientific

EW JERSEY · LONDON · SINGAPORE · BEIJING · SHANGHAI · HONG KONG · TAIPEI · CHENNAI · TOKYO

Published by

World Scientific Publishing Co. Pte. Ltd.
5 Toh Tuck Link, Singapore 596224
USA office: 27 Warren Street, Suite 401-402, Hackensack, NJ 07601
UK office: 57 Shelton Street, Covent Garden, London WC2H 9HE

Library of Congress Cataloging-in-Publication Data
Names: Kufner, Alois. | Persson, Lars-Erik, 1944– | Samko, Natasha.
Title: Weighted inequalities of Hardy type.
Description: 2nd edition / by Alois Kufner (Academy of Sciences, Czech Republic),
    Lars-Erik Persson (Luleå University of Technology, Sweden & UiT,
    The Arctic University of Norway, Norway), Natasha Samko
    (Luleå University of Technology, Sweden). | New Jersey : World Scientific, 2017. |
    Includes bibliographical references and index.
Identifiers: LCCN 2016055408 | ISBN 9789813140646 (hardcover : alk. paper)
Subjects: LCSH: Integral inequalities. | Inequalities (Mathematics)
Classification: LCC QA295 .K883 2017 | DDC 515/.243--dc23
LC record available at https://lccn.loc.gov/2016055408

British Library Cataloguing-in-Publication Data
A catalogue record for this book is available from the British Library.

Typeset by Stallion Press
Email: enquiries@stallionpress.com

Printed in Singapore

This book is dedicated
to OUR FAMILIES
for their understanding, patience
and long-lasting support

Contents

Preface

We hope that this book will help many analysts and others in their inequality hunting described above by G.H. Hardy himself. It is clear that inequalities are an essential part of virtually all areas of mathematics and there is no doubt about their importance and usefulness in various applications. This is of course reflected in the vast literature that exists on the subject. For many readers inequalities exhibit a certain elegance and beauty, and since they are also of great independent interest, they may be viewed as the *evergreen* of mathematics.

The first edition of this book was mainly devoted to a particular field in the big picture of inequalities namely Hardy-type integral inequalities mainly in weighted Lebesgue spaces. In this second edition various complements to this picture have been done, which will be described below.

These weighted integral inequalities are generalizations of those given in the fundamental work of G.H. Hardy and his contemporaries in the early 1920's. During the last four decades, the study of Hardy or Hardy-type operators focused on the characterizations of weights, for which such operators are bounded on weighted Lebesgue spaces. These results are of interest and importance — not only because the mappings are optimal in the sense that the size of the weight classes cannot be improved — but also because the weight conditions themselves are of intrinsic interest.

This intensively investigated area of mathematical analysis resulted in the publication of numerous research papers and also some books. We mention the monograph of B. Opic and A. Kufner, *Hardy-type inequalities* (quoted here frequently as [OK] in the bibliography) because its theme is precisely the weight characterizations of such operators and their relationship to the study of weighted Sobolev spaces. Hence it may be considered to be the precursor of the present book, or perhaps the present book may be seen as a continuation of that work on Hardy's inequality.

We start in Chap. 1 by giving a brief survey of various characterizations of Hardy-type inequalities in weighted Lebesgue spaces. Some natural extensions of the Hardy operator, called here Hardy-type operators, are investigated in Chap. 2. Special cases of such operators are the Riemann-Liouville fractional integral operator of order greater than or equal to one, and its conjugate, the Weyl fractional operator. The weight characterizations for which these operators are bounded on weighted Lebesgue spaces can be deduced from these general results. In Chap. 3 we consider the Hardy-Steklov operator, a variant of the Hardy operator. Here various applications and weight characterizations are given, and in particular it is observed that the weight classes for which this operator is bounded is strictly larger than those for which the Hardy operator is bounded. Chapter 4 deals with the weighted Hardy inequality in differential form even with higher order derivatives involved. Here the boundary conditions are of crucial importance. In fact they determine the weight classes in a fundamental way. In Chap. 5 the fractional order Hardy inequality is considered in a weighted setting.

The results given here relate to those given in Chaps. 3 and 4, and in addition exhibit some interesting connections with the theory of interpolation in Banach function spaces. In order to study mapping properties of classical operators on Lorentz spaces it is necessary to consider Hardy-type operators defined on the cone of decreasing functions. Chapter 6 is concerned with weight characterizations, where the operator is defined on the cone of monotone functions. Here the key result is a duality theorem from which the characterizations follow. In this new edition all these chapters have been updated with new information. Those readers who wish to pursue the subject further may consult the Comments and Remarks at the end of each of these chapters.

The research in this fascinating area has continued intensively after publishing of the first edition in 2003. In this second edition we report on some of these new developments, which complement and extend the previously described theory. These results are mainly reported on in the new Chap. 7. In particular, in Chap. 7 some results concerning Hardy-type inequalities in other function spaces than weighted Lebesgue spaces are presented and discussed (e.g. in Orlicz, Lorentz, Morrey and Hölder type spaces and also in rearrangement invariant and general Banach spaces). We also describe a powerful convexity approach to prove Hardy-type inequalities, which obviously was not discovered by Hardy himself. We also point out that the usual Muckenhoupt-Bradley and Maz'ja-Rosin conditions to characterize Hardy-type inequalities can be replaced by other (equivalent) conditions, even by some scales of (infinite many) conditions. Moreover, we include a section, where we present characterizations of Hardy's inequality for "all" parameters. In another section we describe several recent results concerning Hardy-type inequalities with Hardy operators involving various kernels. Some new multidimensional results are also included. Some questions are also raised. Finally, a new brief overview of the prehistory (before Hardy proved his original inequality in 1925) and history is included in Chap. 7. Further information concerning parts of this historical description can be found in the related monograph of A. Kufner, L. Maligranda and L.E. Persson, *The Hardy Inequality. About its*

History and Related Results (quoted here frequently as [KMP] in the bibliography).

How to read the book?

Each chapter is divided into sections, similarly as the numbers of formulas. Hence, for example **4.5** means the fifth section of Chap. 4 and (4.20) is the twentieth formula of Chap. 4. For the convenience of the reader, we added a list of symbols, notations and conventions, although, not all notions, in particular for the operators, are used systematically and consequently throughout the book.

The aim of the authors has been to collect and present *certain* of the new results in this area without any attempt to give an exhaustive picture of this huge area (which, in fact, seems to be even impossible in one book and under our limiting capacity). Naturally, many results of other authors have not been included and, thus, concerning certain parts of the book, the authors feel to be more editors than *authors*. Therefore, it is our duty to thank our colleagues who contributed in the preparation and finalization of this text. First of all, we mention the important contributions of Prof. H.P. Heinig (*McMaster University, Hamilton, Canada*), which were very substantial. His big influence of this text is particularly obvious in Chap. 3 and in parts of Chap. 6. Among other researchers who substantially have contributed with essential ideas in the text we want to mention Professors L. Maligranda, A. Gogitashvili, V. Stepanov and L. Pick.

It is also a pleasant duty to thank our colleagues who have read parts of the manuscript carefully and made comments and suggestions which have improved the material and presentation. In particular, we want to express our gratitude to Prof. S. Barza, Prof. P. Wall, Dr. P. Tomiczek and Dr. A. Wedestig. The patience and good cheer of Ms E. Ritterova, who typed the original manuscript, was invaluable and we are endebted to her. We are also deeply greatful for the help we got from the staff at World Scientific for helping us in many ways to produce the final version of this second edition from the original and complementary material we provided them with. Finally, we thank Czech Academy of Sciences, Swedish Royal Academy of Sciences and Luleå University of Technology for their financial support.

Luleå, March 2017, *burning* under the sky with the famous northern light,

A. Kufner, L.E. Persson and N. Samko

Conventions and Notation

I Conventions

- All functions are assumed to be *real-valued.*
- *Weight functions* are functions measurable and positive almost everywhere (a.e.).
- $C, C_0, C_1, \ldots$ denote *positive* constants whose exact value is not important.
- For two positive (non-negative) quantities $A, B,$
  $A \lesssim B$ means that $A \leq cB$ with some $c > 0,$
  $A \gtrsim B$ means that $A \geq cB$ with some $c > 0,$
  $A \approx B$ means that simultaneously $A \lesssim B$ and $A \gtrsim B.$
- Expressions of the form

$$0 \cdot \infty, \ 0/0, \ \infty/\infty$$

  are taken equal to zero.
- χ_E is the characteristic function of the set E, i.e.

$$\chi_E(x) = 1 \quad \text{if } x \in E, \ \chi_E(x) = 0 \quad \text{if } x \notin E.$$

- *Decreasing* means "non-increasing";
  Increasing means "non-decreasing"

(otherwise, we use the expression *strictly decreasing* and *strictly increasing*).

- $T : X \to Y$ means that the operator T maps the set (space) X into the set (space) Y.
- $X \hookrightarrow Y$ means that the imbedding of the normed linear space X into the normed linear space (i.e., the identity operator $I : X \to Y$) is *continuous* (bounded), i.e., that $\|u\|_Y \lesssim \|u\|_X$ for every $u \in X$.

II Notation

- **Sets**

$\mathbb{R}$	the real line $(-\infty, \infty)$
$\mathbb{R}^+$	the half-line $(0, \infty)$
$\mathbb{N}$	the set of positive integers
$\mathbb{N}_k$	the k-tuple $(0, 1, \ldots, k-1)$, $k \in \mathbb{N}$
$\mathbb{Z}$	the set of integers
(a, b)	the open interval
$[a, b]$	the closed interval
$[a, b),(a, b]$	half-open intervals
AC^{k-1}	the set of functions with absolutely continuous derivatives of order $k-1$
AC	the set of absolutely continuous functions
$AC^{k-1}(M_0, M_1)$	functions from AC^{k-1} satisfying some special boundary conditions
BVP	boundary value problem
E	the incidence matrix
$E(f, t)$	$= \{x \in (0, \infty) : f(x) > t\}$
M_0, M_1	subsets of $\mathbb{N}_k$
N	the set of functions with zero mean value

- **Constants and Parameters**

$A = A(a, b; u, v)$	
$\widetilde{A} = A(a, b; u, v)$	
$A^*, \widetilde{A}^*$	
a, b, c, d, A, B, C, D	some special integers

$p, q, r, \ldots$ some parameters from $(0, \infty)$

$p' = \frac{p}{p-1}$ for $0 < p < \infty$, $p \neq 1$

$p' = \infty$ for $p = 1$ (p' is the *conjugate* parameter of p)

$k(p, q)$ constant related to best constant in the Hardy inequalilty

• Functions and Measures

$f \uparrow$	a *decreasing* function
$f \downarrow$	an *increasing* function
f°	level function of f
f^{*}	decreasing rearrangement of f
f^{**}	average decreasing rearrangement, $f^{**}(x) = \frac{1}{x} \int_0^x f^*(t)dt$
λ_f	distribution function of f
$d\sigma_j^x(y)$	a positive measure on $(0, \infty)$
δ_x	the Dirac function

• Function Spaces

$C^1(0, \infty)$	the space of functions with continuous derivative
$C_0^\infty(0, \infty)$	the space of infinitely differentiable functions with compact support in $(0, \infty)$
$L^s(\mu)$	a Lebesgue space with general measure μ
$L^s(w)$	a weighted Lebesgue space with weight w
$(L^s(w))^*$	the dual space of $L^s(w)$
$\| \cdot \|_{s,w}$	the norm in $L^s(w)$
L^s	$= L^s(1)$ the "usual" unweighted Lebesgue space
$\mathcal{L}^s(\mu)$	a Lebesgue space with general measure μ
$\mathcal{L}^s(w)$	$= L^s(w^s)$
$L^{p,q}$	the Lorentz space
$L^{p,q}(w)$	a weighted Lorentz space
$\wedge^p(w)$	a weighted Lorentz space
$W_L^{1,p}(v)$	$\Big\{$ some special weighted Sobolev spaces
$W_R^{1,p}(v)$	with the norm $\|g\| = \|g'\|_{p,v}$

L_w^ϕ weighted Orlicz space

$L^{p,\lambda}(R_+)$ classical Morrey space

$\mathcal{L}^{p,\varphi}(R^n, w)$ generalized Morrey space

$V\mathcal{L}^{p,\varphi}(\Omega)$ global vanishing Morrey space

$V\mathcal{L}_{\mathrm{loc};x_0}^{p,\varphi}(\Omega)$ local vanishing Morrey space

$C^\lambda(\Omega)$ Hölder space

- **Operators**

G the geometric mean operator

     $(Gf)(x) = \exp(\frac{1}{x} \int_0^x \ln f(t)dt)$

H the Hardy operator $(Hf)(x) = \int_a^x f(t)dt$

$\widetilde{H}$ the conjugate Hardy operator $(\widetilde{H}f)(x) = \int_x^b f(t)dt$

H_a the Hardy averaging operator $(H_a f)(x) = \frac{1}{x} \int_0^x f(t)dt$

$\mathscr{H}$ a modified Hardy operator

$\widehat{\mathscr{H}}$ a modified Hardy operator

H_w a generalized Hardy averaging operator

$\widetilde{H}_w$ the conjugate operator to H_w

K the generalized Hardy-type operator

     $(Kf)(x) = \int_0^x k(x,t)f(t)dt$

T_g $= \int_0^x g(x-t)f(t)dt$, the convolution operator

$\widetilde{K}$ the conjugate operator to K

K_s $= \int_0^x k^s(x,t)f(t)dt$

$\widetilde{K}_s$ the conjugate operator to K_s

M the maximal function

S_a^b the moving averaging operator

S_γ the Steklov operator

T the Hardy-Steklov operator $(Tf)(x) = \int_{a(x)}^{b(x)} f(t)dt$

- **Duality**

$\langle g, f \rangle = \int_a^b f(t)g(t)dt$ duality

$\langle g, f \rangle_w = \int_a^b f(t)g(t)w(t)dt$ another duality

Introduction

Some Motivation

Later on in this book (and also in several books in applied mathematics) a number of examples which motivate the study of Hardy-type inequalities can be found. Just as one such example we describe the following spectral problem:

The non-linear ordinary differential equation

$$-\frac{d}{dx}\left(v(x)\left|\frac{dy}{dx}\right|^{p-2}\frac{dy}{dx}\right) + \lambda u(x)|y(x)|^{q-2}y(x) = 0 \text{ on } (a,b) \quad (0.1)$$

together with the homogeneous boundary conditions

$$y(a) := \lim_{x \to a+} y(x) = 0, \quad y(b) := \lim_{x \to b-} y(x) = 0. \quad (0.2)$$

Here, $-\infty \le a < b \le +\infty, p > 1, q > 1, u = u(x), v = v(x)$ are *weight functions*, i.e., functions which are measurable and positive a.e. in (a,b).

1

A function $y = y(x)$ is a *weak* solution of (0.1)–(0.2), if the identity

$$\int_a^b |y'(x)|^{p-2}y'(x)z'(x)v(x)dx = \lambda \int_a^b |y(x)|^{q-2}y(x)z(x)u(x)dx$$

(0.3)

holds for every function $z = z(x) \in C_0^\infty(a,b)$. [Notice that (0.3) can be obtained, multiplying (0.1) by $z(x)$ and integrating by parts.]

Putting in (0.3) $z = y$, we obtain that

$$\int_a^b |y'(x)|^p v(x)dx = \lambda \int_a^b |y(x)|^q u(x)dx.$$

(0.4)

If we introduce the weighted Lebesgue space $L^r(w) = L^r(w; a, b)$ with $r > 1$ and the weight $w = w(x)$ as

$$L^r(w) := \left\{ y = y(x), x \in (a, b), \right.$$

$$\left. \|y\|_{r,w} := \left(\int_a^b |y(x)|^r w(x)dx \right)^{1/r} < \infty \right\},$$

then we can rewrite (0.4) as

$$\|y'\|_{p,v}^p = \lambda \|y\|_{q,u}^q.$$

(0.5)

Suppose that we have an inequality of the form

$$\left(\int_a^b |f(x)|^q u(x)dx \right)^{1/q} \le C_{p,q} \left(\int_a^b |f'(x)|^p v(x)dx \right)^{1/p}$$

(0.6)

or shortly

$$\|f\|_{q,u} \le C_{p,q}\|f'\|_{p,v},$$

(0.7)

which should hold for all functions f on (a, b) such that $\|f'\|_{p,v} < \infty$ and satisfying some additional conditions (like, e.g., $f(a) = 0$ or $f(b) = 0$).

Inequality (0.6) is called *the Hardy inequality* or *Hardy's inequality in differential form* (since the function f is estimated by its derivative f').

Comparing (0.5) and (0.7) with $f = y$ we obtain after normalization (i.e., taking $\|y'\|_{p,v} = 1$) that

$$\lambda \geq \frac{1}{C_{p,q}^q}.$$

Consequently, the Hardy inequality (0.6) [more precisely, the (optimal) constant $C_{p,q}$ in this inequality] provides us with an estimate from below of the possible eigenvalues of the problem (0.1)–(0.2). Moreover, the Hardy inequality can give more information about the spectrum of differential operators like that in (0.1), and also about more-dimensional operators like the weighted p-Laplacean

$$div\ (v(x)|\nabla y(x)|^{p-2}\nabla y(x)), x \in \Omega \subset \mathbb{R}^n.$$

This is just one of the classical motivations on why to investigate the Hardy inequality.

A simple exercise

(i) If we consider the Hardy inequality (0.6) for a function $f = f(x)$ satisfying the condition $f(a) = 0$, it is easy to show that (0.6) is equivalent with the inequality

$$\left(\int_a^b \left(\int_a^x g(t)dt\right)^q u(x)dx\right)^{1/q} \leq C_{p,q} \left(\int_a^b |g(x)|^p v(x)dx\right)^{1/p} \tag{0.8}$$

for non-negative functions $g \in L^p(v; a, b)$.

Indeed, it suffices to write the function f in (0.6) in the form

$$f(x) = \int_a^x g(t)dt. \tag{0.9}$$

(ii) If we consider (0.6) for a function $f = f(x)$ satisfying the condition $f(b) = 0$ we arrive at the inequality

$$\left(\int_a^b \left(\int_x^b g(t)dt\right)^q u(x)dx\right)^{1/q} \leq C_{p,q} \left(\int_a^b |g(x)|^p v(x)dx\right)^{1/p}. \tag{0.10}$$

In this case it suffices to write f in the form

$$f(x) = \int_x^b g(t)dt. \tag{0.11}$$

Any of the inequalities (0.8) and (0.10) is again called *the Hardy inequality* and can be written shortly as

$$\|Tg\|_{q,u} \le C_{p,q}\|g\|_{p,v}, \tag{0.12}$$

where T is *the Hardy operator H* from (0.9),

$$(Hg)(x) := \int_a^x g(t)dt$$

or *the conjugate Hardy operator $\widetilde{H}$* from (0.11),

$$(\widetilde{H}g)(x) := \int_x^b g(t)dt.$$

Now, let us describe the properties of the Hardy inequality (in the form of (0.7) or (0.12)) in more detail.

Classical forms of the Hardy inequality

In a note published in 1920, G.H. Hardy [Ha2] stated (without proof) that if $a > 0$, $f(x) \ge 0$, $p > 1$ and $\int_a^\infty f^p(x)dx$ is convergent, then

$$\int_a^\infty \left(\frac{1}{x}\int_0^x f(t)dt\right)^p dx \le \left(\frac{p}{p-1}\right)^p \int_a^\infty f^p(x)dx. \tag{0.13}$$

In the 1925 note [Ha3] he wrote: "I did not give a proof, being occupied primarily with the corresponding theorem for infinite series." His main aim was to find a new, more elementary proof of Hilbert's inequality for double series, and he showed in [Ha3] that in fact this inequality follows from the discrete version of (0.13):

$$\sum_{n=1}^\infty \left(\frac{1}{n}\sum_{k=1}^n a_k\right)^p \le \left(\frac{p}{p-1}\right)^p \sum_{n=1}^\infty a_n^p \qquad (a_n \ge 0, p > 1). \tag{0.14}$$

In fact, G.H. Hardy had dealt with inequality (0.14) earlier (see [Ha1]) after a communication with M. Riesz, but without the best constant (in his estimate, the constant $[p^2/(p-1)]^p$ appears).

Finally, in his 1925 note [Ha3] G.H. Hardy wrote: "In a letter dated 21 June 1921, Prof. Landau communicated to me a direct proof of the theorem [i.e., of (0.14)] which gives the correct value $[p/(p-1)]^p$. He also pointed out that, if the integral theorem were extended to the case $a = 0$, then the theorem for series, with the correct constant, may be deduced at once by taking $f(x) = a_1$ ($0 \leq x < 1$), $f(x) = a_2$ ($1 \leq x < 2$),....." Hence G.H. Hardy stated and proved the following inequality:

Suppose that $f(x) \geq 0$, $p > 1$; that f is integrable over any finite interval $(0, X)$ and f^p is integrable over $(0, \infty)$. Then

$$\int_0^\infty \left(\frac{1}{x} \int_0^x f(t)dt \right)^p dx \leq \left(\frac{p}{p-1} \right)^p \int_0^\infty f^p(x)dx. \qquad (0.15)$$

This is the original form of Hardy's *integral* inequality, which later on has been extensively studied and used as a model example for the investigation of more general inequalities.

Very shortly after G.H. Hardy's proof of (0.15), the first "weighted" modification appeared, namely the famous inequality

$$\int_0^\infty \left(\frac{1}{x} \int_0^x f(t)dt \right)^p x^\varepsilon dx \leq \left(\frac{p}{p-1-\varepsilon} \right)^p \int_0^\infty f^p(x)x^\varepsilon dx \qquad (0.16)$$

valid — with $p > 1$ and $\varepsilon < p - 1$ — for all measurable non-negative functions f (see [HLP, Theorem 330]), where the constant $[p/(p-1-\varepsilon)]^p$ is the best possible. Let us also mention the following "dual" inequality (which can be easily derived from (0.16)):

$$\int_0^\infty \left(\frac{1}{x} \int_x^\infty f(t)dt \right)^p x^\varepsilon dx \leq \left(\frac{p}{\varepsilon+1-p} \right)^p \int_0^\infty f^p(x)x^\varepsilon dx; \qquad (0.17)$$

it holds — with $p > 1$ and $\varepsilon > p - 1$ — for all measurable non-negative functions f and the constant $[p/(\varepsilon + 1 - p)]^p$ is sharp (see again [HLP, Theorem 330]).

Remark 0.1. Concerning the dramatic prehistory before G.H. Hardy discovered his inequalities (0.15)–(0.17) we refer to [KMPe1], see also [KMP] and Sec. 7.1.

"Modern" forms of the Hardy inequality

During the last decades, inequality (0.16) was extended to the form

$$\left(\int_a^b \left(\int_a^x f(t)dt \right)^q u(x)dx \right)^{1/q} \leq C \left(\int_a^b f^p(x)v(x)dx \right)^{1/p} \quad (0.18)$$

with

- a, b real numbers satisfying

$$-\infty \leq a < b \leq \infty,$$

- u, v weight functions, i.e. measurable functions positive a.e. in the interval (a, b),
- p, q real parameters satisfying

$$0 < q \leq \infty, \quad 1 \leq p \leq \infty.$$

It is known (see [OK, Sec. 1]) that (0.18) holds for all measurable functions $f \geq 0$ *if and only if*

$$A < \infty, \quad (0.19)$$

where

$$A := \sup_{a < x < b} \left(\int_x^b u(t)dt \right)^{1/q} \left(\int_a^x v^{1-p'}(t)dt \right)^{1/p'}, \quad p' = \frac{p}{p-1}, \quad (0.20)$$

for the case

$$1 < p \leq q < \infty,$$

and

$$A := \left(\int_a^b \left(\int_x^b u(t)dt \right)^{r/q} \left(\int_a^x v^{1-p'}(t)dt \right)^{r/q'} v^{1-p'}(x)dx \right)^{1/r} \quad (0.21)$$

for the case

$$0 < q < p < \infty, \, q \neq 1, \, 1 < p < \infty$$

with $\frac{1}{r} = \frac{1}{q} - \frac{1}{p}$.

The "dual" inequality which is an extension of (0.17) has the form

$$\left(\int_a^b \left(\int_x^b f(t)dt \right)^q u(x)dx \right)^{1/q} \le C \left(\int_a^b f^p(x)v(x)dx \right)^{1/p} ; \quad (0.22)$$

it is known that (0.22) holds for all measurable functions $f \ge 0$ *if and only if*

$$\widetilde{A} < \infty, \tag{0.23}$$

where

$$\widetilde{A} := \sup_{a<x<b} \left(\int_a^x u(t)dt \right)^{1/q} \left(\int_x^b v^{1-p'}(t)dt \right)^{1/p'} \tag{0.24}$$

for the case $1 < p \le q < \infty$, and

$$\widetilde{A} := \left(\int_a^b \left(\int_a^x u(t)dt \right)^{r/q} \left(\int_x^b v^{1-p'}(t)dt \right)^{r/q'} v^{1-p'}(x)dx \right)^{1/r} \tag{0.25}$$

for the case $0 < q < p < \infty$, $q \ne 1$, $1 < p < \infty$, with $\frac{1}{r} = \frac{1}{q} - \frac{1}{p}$.

For details and proofs, as well as for the corresponding conditions in the cases p, q equal to 1 and/or ∞, see again [OK, Sec. 1].

Best constants

The best constant C in (0.18) satisfies

$$A \le C \le k(p,q)A \quad \text{for} \quad p \le q \tag{0.26}$$

and

$$q^{1/q} \left(\frac{p'q}{r} \right)^{1/q'} A \le C \le q^{1/q}(p')^{1/q'} A \quad \text{for} \quad q < p. \tag{0.27}$$

The constant $k(p,q)$ in (0.26) appears in various forms. For example,

$$k(p,q) = p^{1/q}(p')^{1/p'}$$

or

$$k(p,q) = q^{1/q}(q')^{1/p'}$$

or

$$k(p,q) = \left(1 + \frac{q}{p'}\right)^{1/q} \left(1 + \frac{p'}{q}\right)^{1/p'}$$

(see [OK, Sec. 1]) or, for $p < q$,

$$k(p,q) = \left[\frac{\Gamma(\frac{q}{s})}{\Gamma(1 + \frac{1}{s})\Gamma(\frac{q-1}{s})}\right]^{s/q}$$

with $s = \frac{q}{p} - 1$ (see V.M. Manakov [Ma1]).

These estimates remain true also for the constant C in (0.22) with A replaced by $\widetilde{A}$.

Remark 0.2. In this book, we primarily consider the cases $1 < p < \infty$, $0 < q < \infty$, omitting the limiting values $p = 1, \infty$, $q = \infty$. The reader can easily reformulate the corresponding results, using appropriately the definition of the corresponding Lebesgue spaces (and their duals).

We are not considering systematically the problem of the *best* constants in these inequalities. All constants C are assumed positive and finite and the only information given is that C is comparable to A ($C \approx A$), where A is given by (0.20) or (0.21), or that C is comparable to $\widetilde{A}$ ($C \approx \widetilde{A}$), where $\widetilde{A}$ is given by (0.24) or (0.25).

Shortly about the history

Since the Hardy inequality was proved by G.H. Hardy, a vast amount of literature exists on the inequality, its variations and generalizations. Therefore a complete list on the subject cannot be given. We refer to the books [HLP], [KoMP], [KMRS], [KMP], [MPF] and [OK] and mention only some results.

The characterization of the weights u, v for which the Hardy inequalities (0.18) and (0.22) hold in the range

$$p = q > 1$$

goes back to 1969 (G. Tomaselli [To1], G. Talenti [T1]) and to 1972 (B. Muckenhoupt [Mu1] who gave a nice and simple direct proof); many authors also refer to the "untitled and unpublished manuscript" of A. Artola. The case

$$1 < p \le q < \infty$$

was investigated independently in 1978 by J.S. Bradley [Br1] and in 1979 by V.M. Kokilashvili [Ko1] and (maybe even earlier) by V.G. Maz'ja and A.L. Rozin (see V.G. Maz'ja [Maz1]). These last two authors also characterized the case

$$1 < q < p < \infty.$$

Necessary and sufficient conditions on the weights in the case

$$0 < q < 1, \quad p > 1$$

go back to the 1987 thesis by G. Sinnamon [Si1].

Remark 0.3. More information concerning the history of Hardy-type inequalities can be found in Sec. 7.1.

Remark 0.4. In applications, most frequently *power weights* (i.e., weights of the form $x^\lambda, \lambda \in R$) appear. For the convenience of the reader, we include now some examples. The given necessary and sufficient conditions of the validity of these inequalities can be easily verified by checking whether the corresponding numbers A from (0.20) and (0.21) or $\tilde{A}$ from (0.24) and (0.25) are finite.

Example 0.5. (i) The inequality

$$\left(\int_0^\infty \left(\int_0^x f(t)dt \right)^q x^\alpha dx \right)^{1/q} \le C \left(\int_0^\infty f^p(x) x^\beta dx \right)^{1/p} \quad (0.28)$$

holds for $1 < p \le q < \infty$ if and only if

$$\beta < p - 1, \quad \alpha = \beta \frac{q}{p} - \frac{q}{p'} - 1. \quad (0.29)$$

(ii) The inequality

$$\left(\int_0^\infty \left(\int_x^\infty f(t)dt\right)^q x^\alpha dx\right)^{1/q} \le C \left(\int_0^\infty f^p(x)x^\beta dx\right)^{1/p} \quad (0.30)$$

holds for $1 < p \le q < \infty$ if and only if

$$\beta > p - 1, \quad \alpha = \beta\frac{q}{p} - \frac{q}{p'} - 1. \quad (0.31)$$

(iii) For $p > q$ the inequalities (0.28) and (0.30) *do not hold*.

(See also [OK, Example 6.7]. Compare the results for $p = q$ with the inequalities (0.16) and (0.17).)

Remark 0.6. For the case $p = q$ the sharp constants in (0.28) and (0.30) are $\frac{p}{p-\beta-1}$ and $\frac{p}{1+\beta-p}$, respectively. Concerning the case $1 < p < q < \infty$ the sharp constant in (0.28) was pointed out in 1930 by G.A. Bliss [BL1] for the case $\beta = 0$. In 2015, L.E. Persson and S. Samko [PSa1] proved that

(a) the sharp constant C in (0.28) is equal to

$$C_{pq} = \left(\frac{p-1}{p-1-\beta}\right)^{\frac{1}{p'}+\frac{1}{q}} \left(\frac{p'}{q}\right)^{\frac{1}{p}} \left(\frac{\frac{q-p}{p}\Gamma\left(\frac{pq}{q-p}\right)}{\Gamma\left(\frac{p}{q-p}\right)\Gamma\left(\frac{p(q-1)}{q-p}\right)}\right)^{\frac{1}{p}-\frac{1}{q}},$$

(b) equality in (0.28) occurs exactly when

$$f(x) = \frac{cx^{-\frac{\beta}{p-1}}}{\left(dx^{\frac{p-1-\beta}{p-1}\cdot\left(\frac{q}{p}-1\right)}+1\right)^{\frac{q}{q-p}}} \quad \text{a.e.,}$$

where c and d are positive constants, and

(c) $C_{pq} \to \frac{p}{p-1-\beta}$ as $q \to p$.

It was also proved in [PSa1] that (for fixed p and q) the inequalities (0.28) and (0.30) are in a sense equivalent. By using in particular this fact it was proved in [PSa1] that (a$'$) the sharp constant C in (0.30) is equal to

$$C^{\sharp}_{p,q} = \left(\frac{p-1}{\beta+1-p}\right)^{\frac{1}{p'}+\frac{1}{q}} \left(\frac{p'}{q}\right)^{\frac{1}{p}} \left(\frac{\frac{q-p}{p}\Gamma\left(\frac{pq}{q-p}\right)}{\Gamma\left(\frac{p}{q-p}\right)\Gamma\left(\frac{p(q-1)}{q-p}\right)}\right)^{\frac{1}{p}-\frac{1}{q}},$$

(b$'$) equality in (0.30) occurs if and only if $f(x)$ is in the form

$$f(x) = \frac{cx^{\frac{\beta}{p-1}}}{(dx^{\left(\frac{\beta+1-p}{p-1}\right)\left(\frac{q}{p}-1\right)} + 1)^{\frac{q}{q-p}}} \quad \text{a.e.,}$$

where c and d are positive constants, and

(c$'$) $C^{\sharp}_{p,q} \to \frac{p}{\beta+1-p}$ as $q \to p$.

In [PSa1] also the sharp constants in some multidimensional versions of (0.28) and (0.30) were derived.

Example 0.7. (i) The inequality

$$\left(\int_0^b \left(\int_0^x f(t)dt\right)^q x^\alpha dx\right)^{1/q} \le C \left(\int_0^b f^p(x)x^\beta dx\right)^{1/p} \qquad (0.32)$$

with $0 < b < \infty$ holds

(i-1) for $1 < p \le q < \infty$ if and only if

$$\beta < p-1, \ \alpha \ge \beta\frac{q}{p} - \frac{q}{p'} - 1; \qquad (0.33)$$

(i-2) for $1 < q < p < \infty$ if and only if

$$\beta > p-1, \ \alpha > \beta\frac{q}{p} - \frac{q}{p'} - 1. \qquad (0.34)$$

(ii) The inequality

$$\left(\int_0^b \left(\int_x^b f(t)dt\right)^q x^\alpha dx\right)^{1/q} \le C \left(\int_0^b f^p(x)x^\beta dx\right)^{1/p} \qquad (0.35)$$

with $0 < b < \infty$ holds

(ii-1) for $1 < p \le q < \infty$ if and only if

$$\beta > p - 1, \ \alpha \ge \beta\frac{q}{p} - \frac{q}{p'} - 1 \qquad (0.36)$$

or

$$\beta \le p - 1, \ \alpha > -1; \qquad (0.37)$$

(ii-2) for $1 < q < p < \infty$ if and only if

$$\beta > p - 1, \ \alpha > \beta\frac{q}{p} - \frac{q}{p'} - 1 \qquad (0.38)$$

or

$$\beta \le p - 1, \ \alpha > -1. \qquad (0.39)$$

(See also [OK, Example 6.8]. Let us emphasize the influence of the finiteness of the right endpoint b: in comparison with Example 0.7, the set of admissible values of α, β is substantially richer.)

Example 0.8. (i) The inequality

$$\left(\int_a^\infty \left(\int_a^x f(t)dt \right)^q x^\alpha dx \right)^{1/q} \le C \left(\int_a^\infty f^p(x)x^\beta dx \right)^{1/p} \qquad (0.40)$$

with $0 < a < \infty$ holds

(i-1) for $1 < p \le q < \infty$ if and only if

$$\beta < p - 1, \ \alpha \le \beta\frac{q}{p} - \frac{q}{p'} - 1 \qquad (0.41)$$

or

$$\beta \ge p - 1, \ \alpha < -1; \qquad (0.42)$$

(i-2) for $1 < p < q < \infty$ if and only if

$$\beta < p - 1, \ \alpha < \beta\frac{q}{p} - \frac{q}{p'} - 1 \qquad (0.43)$$

or

$$\beta \ge p - 1, \ \alpha < -1. \qquad (0.44)$$

(ii) The inequality

$$\left(\int_a^\infty \left(\int_x^\infty f(t)dt \right)^q x^\alpha dx \right)^{1/q} \leq C \left(\int_a^\infty f^p(x)x^\beta dx \right)^{1/p} \quad (0.45)$$

with $0 < a < \infty$ holds

(ii-1) for $1 < p \leq q < \infty$ if and only if

$$\beta > p - 1, \quad \alpha \leq \beta\frac{a}{p} - \frac{a}{p'} - 1, \quad (0.46)$$

(ii-2) for $1 < q < p < \infty$ if and only if

$$\beta > p - 1, \quad \alpha < \beta\frac{q}{p} - \frac{q}{p'} - 1. \quad (0.47)$$

(See also [OK, Example 6.9].)

Example 0.9. In inequalities (0.16) and (0.17), the value $\varepsilon = p - 1$ is excluded. In this case, we can improve the inequality by adding some logarithmic terms on both sides and considering the interval $(0, 1)$. More precisely,

(i) the inequality

$$\left(\int_0^1 \left(\int_0^x f(t)dt \right)^q \frac{1}{x}|\ln x|^\alpha dx \right)^{1/q}$$

$$\leq C \left(\int_0^1 f^p(x)x^{p-1}|\ln x|^\beta dx \right)^{1/p} \quad (0.48)$$

holds for $1 < p \leq q < \infty$ if and only if

$$\beta > p - 1, \quad \alpha = \beta\frac{q}{p} - \frac{q}{p'} - 1, \quad (0.49)$$

(ii) the inequality

$$\left(\int_0^1 \left(\int_x^1 f(t)dt \right)^q \frac{1}{x}|\ln x|^\alpha dx \right)^{1/q}$$

$$\leq C \left(\int_0^1 f^p(x)x^{p-1}|\ln x|^\beta dx \right)^{1/p} \quad (0.50)$$

holds for $1 < p \le q < \infty$ if and only if

$$\beta < p - 1, \quad \alpha = \beta \frac{q}{p} - \frac{q}{p'} - 1. \tag{0.51}$$

In the case $p > q$ the corresponding numbers A and $\widetilde{A}$ are *infinite* and inequalities (0.48) and (0.49) do not hold. See also [OK, Example 6.10 and Remark 6.11].

Remark 0.10. The progress in the theory of Hardy-type inequalities and their more general extensions has continued also after 1990 (when the book [OK] appeared). Here, we will report on some of these developments, and we will deal mainly with the following topics:

(a) Some limiting cases of the Hardy inequality for $p \to \infty$ (see Secs. 1.7, 1.8).
(b) Some weighted norm inequalities (see mainly Chap. 2).
(c) Inequalities with the Hardy-Steklov operator (Chap. 3).
(d) Higher order Hardy inequalities (Chap. 4).
(e) Fractional order Hardy inequalities and their connection with interpolation theory (Chap. 5).
(f) Integral inequalities on the cone of monotone functions (Chap. 6).

Remark 0.11. In this new edition we have included some new information in the text of all six chapters and added also a new Chap. 7 with mainly new and complementary results:

(a) On the Prehistory and History of the Hardy Inequality
(b) A Convexity Approach to Prove Hardy-type Inequalities
(c) Scales of Conditions to Characterize Hardy-type Inequalities
(d) Hardy's Inequalities for "all" Parameters
(e) More on Hardy-type Inequalities for Hardy Operators with Kernel
(f) More on Hardy-type inequalities in Other Function Spaces
(g) More on Multidimensional Hardy-type Inequalities

1

Hardy's Inequality and Related Topics

1.1. Weighted Lebesgue Spaces and the Hardy Operator

Weighted norm inequalities

The weighted Lebesgue space

$$L^s(a, b; w) = L^s(w), \tag{1.1}$$

where $0 < s \leq \infty$ and w is a weight function on (a, b), consists of all measurable functions $f = f(x)$ on (a, b) such that

$$\|f\|_{s,w} := \left(\int_a^b |f(x)|^s w(x) dx \right)^{1/s} < \infty, \quad 0 < s < \infty,$$

$$\|f\|_{\infty,w} := \operatorname*{ess\,sup}_{a<x<b} |f(x)| < \infty. \tag{1.2}$$

If H denotes the *Hardy operator*,

$$(Hf)(x) := \int_a^x f(t) dt, \tag{1.3}$$

then the Hardy inequality (0.18) can be rewritten as

$$\|Hf\|_{q,u} \le C\|f\|_{p,v} \tag{1.4}$$

and it holds for functions $f \ge 0$ if and only if

$$A < \infty \tag{1.5}$$

where A is given by (0.20) or (0.21), respectively.

Together with the Hardy operator H, we can immediately define the *conjugate Hardy operator* $\widetilde{H}$:

$$(\widetilde{H}f)(x) := \int_x^b f(t)dt. \tag{1.6}$$

The corresponding (conjugate) Hardy inequality

$$\|\widetilde{H}f\|_{q,u} \le C\|f\|_{p,v} \tag{1.7}$$

then holds for functions $f \ge 0$ if and only if

$$\widetilde{A} < \infty \tag{1.8}$$

where $\widetilde{A}$ is given by (0.24) or (0.25), respectively.

Obviously, the step from H to $\widetilde{H}$ and from the condition (1.5) to the condition (1.8) can be made by a simple substitution.

What is more important is the fact that the inequalities (1.4) and (1.7) are prototypes of a general *weighted norm inequality* of the form

$$\|Tf\|_{q,u} \le C\|f\|_{p,v} \tag{1.9}$$

where T is a *general* integral operator. Thus the inequality (1.9) asserts that T maps $L^p(v)$ (continuously) into $L^q(u)$:

$$T : L^p(v) \to L^q(u).$$

In the subsequent chapters we will obtain inequalities of the form (1.9) for several particular operators T. Sometimes, these operators are defined on certain special classes of functions f such as e.g. monotone functions.

Duality

Define the duality on the weighted Lebesgue space $L^s(w)$ with $1 < s < \infty$ by the inner product

$$\langle g, f \rangle = \int_a^b g(x)f(x)dx, \quad f \in L^s(w). \tag{1.10}$$

Then one easily sees that the dual space to $L^s(w)$ can be identified with the space $L^{s'}(\widehat{w})$ where

$$s' = \frac{s}{s-1}, \quad \widehat{w} = w^{1-s'}.$$

Specifically, $\|g\|_{s',w^{1-s'}} = \sup_{\|f\|_{s,w}=1} \langle g, f \rangle$ so that

$$(L^s(w))^* = L^{s'}(w^{1-s'}). \tag{1.11}$$

Indeed, Hölder's inequality shows that

$$|\langle g, f \rangle| \leq \int_a^b |g(x)|w^{-1/s}(x)|f(x)|w^{1/s}(x)dx$$

$$\leq \left(\int_a^b |f(x)|^s w(x)dx \right)^{1/s} \left(\int_a^b |g(x)|^{s'} w^{-s'/s}(x)dx \right)^{1/s'}$$

$$= \|f\|_{s,w} \cdot \|g\|_{s',w^{1-s'}}$$

since $s/s' = s - 1 = 1/(s'-1)$, and if

$$f = |g|^{s'-1}\operatorname{sgn} g \, w^{1-s'}/\|g\|_{s',w^{1-s'}}^{s'/s},$$

then $\|f\|_{s,w} = 1$ and $\langle g, f \rangle = \|g\|_{s',w^{1-s'}}$.

Moreover, the Hardy operators H and $\widetilde{H}$ are mutually conjugate; more precisely, if

$$H : L^p(v) \to L^q(u), \quad 1 < p, q < \infty,$$

then $(H)^* = \widetilde{H}$ and

$$\widetilde{H} : L^{q'}(u^{1-q'}) \to L^{p'}(v^{1-p'}).$$

Indeed, by Fubini's theorem,

$$\langle g, Hf \rangle = \int_a^b g(x) \int_a^x f(t)dtdx$$

$$= \int_a^b f(t) \int_t^b g(x)dxdt = \langle f, \widetilde{H}g \rangle,$$

where the first (last) brackets denote the duality in $L^q(u)$ (in $L^p(v)$).

Some other criteria

As we know, the Hardy inequality (0.18) holds if and only if $A < \infty$ where A is given by (0.20) for $p \leq q$ and by (0.21) for $p > q$. Now we will give some alternative criteria. But first, let us introduce the notation

$$V(x) := \int_a^x v^{1-p'}(t)dt. \tag{1.12}$$

Then we can rewrite the number A as

$$A = \sup_{a<x<b} \left(\int_x^b u(t)dt \right)^{1/q} V^{1/p'}(x).$$

Theorem 1.1. *Let $1 < p \leq q < \infty$. Then the Hardy inequality*

$$\left(\int_a^b \left(\int_a^x f(t)dt \right)^q u(x)dx \right)^{1/q} \leq C \left(\int_a^b f^p(x)v(x)dx \right)^{1/p} \tag{1.13}$$

holds for all functions $f \geq 0$ if and only if

$$B < \infty \tag{1.14}$$

where

$$B := \sup_{a<x<b} \left(\int_a^x v^{1-p'}(t)dt \right)^{-1/p}$$

$$\times \left(\int_a^x u(t) \left(\int_a^t v^{1-p'}(s)ds \right)^q dt \right)^{1/q}, \tag{1.15}$$

i.e.,

$$B = \sup_{a<x<b} V^{-1/p}(x) \left(\int_a^x u(t)V^q(t)dt \right)^{1/q}. \qquad (1.15^*)$$

Moreover, the constant C in (1.13) *satisfies*

$$B \leq C \leq p'B. \qquad (1.16)$$

Proof. (i) *Necessity* of condition (1.14). Suppose that the inequality (1.13) holds for all $f \geq 0$ with a constant $C < \infty$ and choose for f the function

$$f_t(x) = \chi_{(a,t)}(x)v^{1-p'}(x)$$

with $t \in (a,b)$ arbitrary but fixed. Then (1.13) implies

$$\left(\int_a^t \left(\int_a^x v^{1-p'}(s)ds \right)^q u(x)dx \right)^{1/q} \leq C \left(\int_a^t v^{1-p'}(x)dx \right)^{1/p},$$

i.e.,

$$\left(\int_a^t V^q(x)u(x)dx \right)^{1/q} \leq C(V(t))^{1/p}$$

and consequently $B \leq C < \infty$.

(ii) *Sufficiency* of condition (1.14). Due to duality, inequality (1.13) is equivalent to the inequality

$$\left(\int_a^b \left(\int_x^b g(t)dt \right)^{p'} v^{1-p'}(x)dx \right)^{1/p'} \leq C \left(\int_a^b g^{q'}(x)u^{1-q'}(x)dx \right)^{1/q'}$$

$$(1.17)$$

for $g \geq 0$ with the same constant C. For g with support in (a,b) we have by integration by parts and by Hölder's inequality

$$J := \int_a^b \left(\int_x^b g(t)dt \right)^{p'} v^{1-p'}(x)dx$$

$$= p' \int_a^b \left(\int_x^b g(t)dt \right)^{p'-1} g(x) \left(\int_a^x v^{1-p'}(t)dt \right) dx$$

$$= p' \int_a^b g(x) u^{(1-q')/q'}(x) \left(\int_x^b g(t)dt \right)^{p'-1}$$

$$\times \left(\int_a^x v^{1-p'}(t)dt \right) u^{(q'-1)/q'}(x)dx$$

$$\leq p' \left(\int_a^b g^{q'}(x) u^{1-q'}(x)dx \right)^{1/q'} J_1^{1/q}, \tag{1.18}$$

where

$$J_1 := \int_a^b \left(\int_x^b g(t)dt \right)^{q(p'-1)} \left(\int_a^x v^{1-p'}(t)dt \right)^q u(x)dx.$$

Denote

$$h(x) := \left(\int_x^b g(t)dt \right)^{q(p'-1)}.$$

Applying Fubini's theorem, we obtain

$$J_1 = \int_a^b h(x) \left(\int_a^x v^{1-p'}(t)dt \right)^q u(x)dx$$

$$= \int_a^b h(x) V^q(x) u(x)dx = \int_a^b \int_x^b [-h'(t)]dt V^q(x)u(x)dx$$

$$= \int_a^b \int_x^b [-h'(t)V^q(x)u(x)]dtdx$$

$$= \int_a^b \int_a^t [-h'(t)V^q(x)u(x)]dxdt$$

$$= \int_a^b [-h'(t)] \int_a^t u(x)V^q(x)dxdt$$

and due to (1.15*) we have

$$J_1 \leq B^q \int_a^b [-h'(t)] \left(\int_a^t v^{1-p'}(x)dx \right)^{q/p} dt.$$

Estimating the integral on the right-hand side by Minkowski's integral inequality, we obtain

$$J_1 \leq B^q \left(\int_a^b \left(\int_x^b [-h'(t)]dt \right)^{p/q} v^{1-p'}(x)dx \right)^{q/p}$$

$$= B^q \left(\int_a^b h^{p/q}(x) v^{1-p'}(x)dx \right)^{q/p}$$

$$= B^q \left(\int_a^b \left(\int_x^b g(t)dt \right)^{(p'-1)p} v^{1-p'}(x)dx \right)^{q/p}$$

$$= B^q \left(\int_a^b \left(\int_x^b g(t)dt \right)^{p'} v^{1-p'}(x)dx \right)^{q/p} = B^q J^{q/p}.$$

This together with (1.18) implies that

$$J^{1/p'} \leq p'B \left(\int_a^b g^{q'}(x) u^{1-q'}(x)dx \right)^{1/q'},$$

i.e., inequality (1.17), and thus (1.13), holds and, moreover, $C \leq p'B$. $\qquad \square$

Theorem 1.2. *Let* $0 < q < p < \infty$, $p > 1$, $1/r = 1/q - 1/p$. *Let* V *be given by* (1.12). *Then the Hardy inequality* (1.13) *holds for all functions* $f \geq 0$ *if and only if*

$$B < \infty, \tag{1.19}$$

where

$$B := \left(\int_a^b \left(\int_a^t u(s)V^q(s)ds \right)^{r/q} V^{-r/q}(t)dV(t) \right)^{1/r}. \tag{1.20}$$

Moreover, the constant C *in* (1.13) *satisfies*

$$qp^{-1/r}(p')^{1/q'} r^{-1/r'} 2^{-1/q} B \leq C \leq q^{1/q} p'B. \tag{1.21}$$

Proof.   We write

$$J := \int_a^b \left(\int_a^x f(t)dt \right)^q u(x)dx$$

$$= \int_a^b \left(\int_a^x f(t)dt \right)^q u(x)V^q(x)V^{-q}(x)dx$$

$$= q \int_a^b \left(\int_a^x f(t)dt \right)^q u(x)V^q(x) \int_x^b V^{-q-1}(s)dV(s)dx$$

$$= q \int_a^b V^{-q-1}(s) \left(\int_a^s \left(\int_a^x f(t)dt \right)^q u(x)V^q(x)dx \right) dV(s)$$

$$\leq q \int_a^b \left[\left(\int_a^s f(t)dt \right)^q V^{-q}(s) \right]$$

$$\times \left[\left(\int_a^s u(x)V^q(x)dx \right) V^{-1}(s) \right] dV(s);$$

here we have used Fubini's theorem and the fact that $\int_a^x \leq \int_a^s$ for $x \leq s$. Applying now Hölder's inequality with the parameters $\frac{p}{q}$ and $(\frac{p}{q})' = \frac{r}{q}$, we obtain

$$J \leq q \left(\int_a^b \left(\int_a^s f(t)dt \right)^p \frac{dV(s)}{V^q(s)} \right)^{q/p} B^q.$$

It is easy to see that by Theorem 1.1

$$\left(\int_a^b \left(\int_a^s f(t)dt \right)^p \frac{dV(s)}{V^p(s)} \right)^{1/p} \leq p' \left(\int_a^b f^p(s)v(s)ds \right)^{1/p}.$$

Thus, we have shown that inequality (1.13) holds with $C \leq q^{1/q}p'B < \infty$ provided $B < \infty$.

To show that the condition (1.19) is necessary, let us suppose that inequality (1.13) holds with $C < \infty$ and derive the lower bound in (1.21). Let us define

$$B_0 := \left(\int_a^b \left(\int_x^b u(t)dt \right)^{r/q} V^{r/q'}(x)dV(x) \right)^{1/r}.$$

We will show that

$$C \geq q^{1/q}(p')^{1/q'}\frac{q}{r}B_0. \tag{1.22}$$

For this purpose, denote $w(x) := v^{1-p'}(x)$ and let u_1 and v_1 be such that $0 \leq u_1 \leq u$ and $0 \leq v \leq v_1$. Suppose that the functions u_1 and w_1, $w_1(x) := v_1^{1-p'}(x)$, are integrable. Furthermore, denote

$$B_1 := \left(\int_a^b \left(\int_x^b u_1(t)dt \right)^{r/q} V_1^{r/q'}(x)dV_1(x) \right)^{1/r}$$

with

$$V_1(x) = \int_a^x v_1^{1-p'}(t)dt = \int_a^t w_1(t)dt.$$

If we choose for f the function

$$f(x) := \left(\int_x^b u_1(t)dt \right)^{r/pq} \left(\int_a^x w_1(t)dt \right)^{r/pq'} w_1(x),$$

then we have

$$\int_a^x f(t)dt \geq \left(\int_x^b u_1(t)dt \right)^{r/pq} \int_a^x \left(\int_a^t w_1(s)ds \right)^{r/pq'} w_1(t)dt$$

$$= \frac{p'q}{r} \left(\int_x^b u_1(t)dt \right)^{r/pq} \left(\int_a^x w_1(s)ds \right)^{r/p'q}.$$

Integration by parts, the last estimate and (1.13) lead to relations

$$\frac{p'q}{r} \left(\frac{q}{p'} \right)^{1/q} B_1^{r/q} = \left(\int_a^b \left(\frac{p'q}{r} \right)^q \left(\int_x^b u_1(t)dt \right)^{r/p} \right.$$

$$\times \left. \left(\int_a^x w_1(t)dt \right)^{r/p'} u_1(x)dx \right)^{1/q}$$

$$\leq \left(\int_a^b \left(\int_a^x f(t)dt \right)^q u_1(x)dx \right)^{1/q}$$

$$\leq \left(\int_a^b \left(\int_a^x f(t)dt \right)^q u(x)dx \right)^{1/q}$$

$$\leq C \left(\int_a^b f^p(x)v(x)dx \right)^{1/p}$$

$$= C \left(\int_a^b \left(\int_x^b u_1(t)dt \right)^{r/q} \right.$$

$$\times \left. \left(\int_a^x w_1(t)dt \right)^{r/q'} w_1^p(x)w^{1-p}(x)dx \right)^{1/p}$$

$$\leq C \left(\int_a^b \left(\int_x^b u_1(t)dt \right)^{r/q} \right.$$

$$\times \left. \left(\int_a^x w_1(t)dt \right)^{r/q'} w_1(x)dx \right)^{1/p}$$

$$= C \left(\int_a^b \left(\int_x^b u_1(t)dt \right)^{r/q} V_1^{r/q'}(x)dV_1(x) \right)^{1/p}$$

$$= C B_1^{r/p}.$$

Hence

$$C \geq q^{1/q}(p')^{1/q'} \frac{q}{r} B_1.$$

By approximating u and w by increasing sequences of integrable functions and applying the Monotone Convergence Theorem we derive the foregoing estimate with B_1 replaced by B_0, i.e., the estimate (1.22).

If we show that

$$B \leq (2q)^{1/q} \left(\frac{p}{r} \right)^{1/r} B_0, \tag{1.23}$$

then the lower bound in (1.21) will follow. For this purpose, write

$$B^r = \int_a^b \left(\int_a^x V^q(t)d \left(-\int_t^x u(s)ds \right) \right)^{r/q} V^{-r/q}(x)dV(x) \tag{1.24}$$

and

$$J_1 := \int_a^x V^q(t)d\left(-\int_t^x u(s)ds\right) = q\int_a^x \left(\int_t^x u(s)ds\right)V^{q-1}(t)dV(t)$$

$$= q\int_a^x \left[\left(\int_t^x u(s)ds\right)V^{q-1+q/2p}(t)\right]V^{-q/2p}(t)dV(t).$$

Hölder's inequality with parameters $\frac{r}{q}$ and $(\frac{r}{q})' = \frac{p}{q}$ leads to

$$J_1 \le q\left(\int_a^x \left(\int_t^x u(s)ds\right)^{r/q} V^{(q-1+q/2p)(r/q)}(t)dV(t)\right)^{q/r}$$

$$\times \left(\int_a^x V^{-1/2}(t)dV(t)\right)^{q/p}$$

and from (1.24) and Fubini's theorem we then have

$$B^r \le q^{r/q}2^{r/p}\int_a^b \left(\int_a^x \left(\int_t^b u(s)ds\right)^{r/q} V^{(q-1+q/2p)(r/q)}dV(t)\right)$$

$$\times V^{r/2p-r/q}(x)dV(x)$$

$$= q^{r/q}2^{r/p}\int_a^b \left(\int_t^b u(s)ds\right)^{r/q} V^{r/q'+r/2p}(t)$$

$$\times \int_t^b V^{r/2p-r/q}(x)dV(x)dV(t)$$

$$= \frac{q^{r/q}p2^{r/p+1}}{r}\int_a^b \left(\int_t^b u(s)ds\right)^{r/q} V^{r/q'}(t)dV(t) = \frac{(2q)^{r/q}p}{r}B_0^r$$

and (1.23) follows. The proof is complete. $\square$

Remark 1.3. In Sec. 7.3 we present some other alternative conditions to characterize Hardy-type inequalities. Especially we present the fact that these conditions can even be replaced by infinite many conditions, namely (at least) four scales of conditions. Each scale is described by a parameter t, $0 < t < \infty$, and the usual (e.g. Muckenhoupt-Bradley, Maz'ja-Rosin and Persson-Stepanov presented above) conditions are just points on these scales.

1.2. Hardy's Inequality with Derivatives

The "differential" form of Hardy's inequality

Let us denote by

$$AC^{k-1}(a,b) = AC^{k-1}$$

with $k \in \mathbb{N}$ the set of all absolutely continuous functions on (a,b) with absolutely continuous derivatives up to the order $k - 1$, and consider the inequality

$$\left(\int_a^b |g(x)|^q u(x) dx \right)^{1/q} \leq C \left(\int_a^b |g'(x)|^p v(x) dx \right)^{1/p} \qquad (1.25)$$

for functions $g \in AC^0$ which satisfy either

$$g(a) = 0 \qquad (1.26)$$

or

$$g(b) = 0. \qquad (1.27)$$

Further, denote by

$$W_L^{1,p}(v) \quad \text{and} \quad W_R^{1,p}(v)$$

the sets of all functions $g \in AC^0$ which satisfy respectively the "initial condition" (1.26) or (1.27)[1] and for which the right-hand side of (1.25) is finite:

$$\|g'\|_{p,v} < \infty.$$

Then the sets $W_L^{1,p}(v)$ and $W_R^{1,p}(v)$ are normed linear spaces with the same norm $\|g'\|_{p,v}$ and the inequality (1.25) is equivalent to the Hardy

---

[1]The subscript L [R] indicates that the function vanishes at the left [right] endpoint of the interval (a,b). Of course, for $a = -\infty$, $g(a) = 0$ means that $\lim_{x \to -\infty} g(x) = 0$, and analogously for $b = \infty$.

inequality (0.18) (for $g \in W_L^{1,p}(v)$) and/or to the Hardy inequality (0.22) (for $g \in W_R^{1,p}(v)$). Indeed, it suffices to take

$$g(x) = \int_a^x f(t)dt \ (= (Hf)(x))$$

and/or

$$g(x) = \int_x^b f(t)dt \ (= (\widetilde{H}f)(x))$$

(see also [OK, Lemma 1.10]).

Consequently, the Hardy inequality (1.25), which can be rewritten also in the form

$$\|g\|_{q,u} \le C\|g'\|_{p,v}, \tag{1.28}$$

expresses the *continuous imbedding*

$$W_L^{1,p}(v) \hookrightarrow L^q(u) \tag{1.29}$$

and/or

$$W_R^{1,p}(v) \hookrightarrow L^q(u) \tag{1.30}$$

and the condition (1.5) is necessary and sufficient for (1.29) while the condition (1.8) is necessary and sufficient for (1.30).

In Chaps. 4 and 5, generalizations and modifications of the inequality (1.28) are studied. Specifically, in Chap. 4 we investigate, for $k \in \mathbb{N}$, $k > 1$, the *higher order (k-th order) Hardy inequality*

$$\|g\|_{q,u} \le C\|g^{(k)}\|_{p,v} \tag{1.31}$$

for functions $g \in AC^{k-1}$ subject to certain additional "boundary conditions" which extend the conditions (1.26) or (1.27). More precisely, we look for (necessary and sufficient) conditions on the parameters p, q and on the weight functions u, v, under which inequality (1.31) holds for some particular subsets of AC^{k-1}.

Furthermore, in Chap. 5 we will investigate also *fractional order Hardy inequalities*, i.e., inequalities of the form

$$\|g\|_{q,u} \le C\|g^{(\lambda)}\|_{p,v}, \quad 0 < \lambda < 1,$$

and

$$\|g^{(\lambda)}\|_{q,u} \leq C\|g'\|_{p,v}, \quad 0 < \lambda < 1$$

where the expression $\|g^{(\lambda)}\|_{r,w}$, in accordance with the commonly used notion of fractional order derivatives, is interpreted as

$$\left(\int_a^b \int_a^b \frac{|g(x) - g(y)|^r}{|x - y|^{1+\lambda r}} w(x,y) dx dy \right)^{1/r}.$$

Here $w(x,y)$ is a *two-dimensional* measurable weight function defined and positive a.e. in $(a,b) \times (a,b)$.

Remark 1.4 (two-sided conditions). The Hardy inequality (1.25) can also be considered for functions

$$g \in W_L^{1,p}(v) \cap W_R^{1,p}(v),$$

i.e., functions satisfying *both* the conditions (1.26) and (1.27):

$$g(a) = g(b) = 0. \tag{1.32}$$

In this case, a necessary and sufficient condition for (1.25) to hold is that for parameters p, q satisfying $1 < p \leq q < \infty$ we have

$$\sup_{(c,d) \subset (a,b)} \left[\left(\int_c^d u(t) dt \right)^{1/q} \right.$$
$$\left. \times \left(\min \left\{ \int_a^c v^{1-p'}(t) dt, \int_d^b v^{1-p'}(t) dt \right\} \right)^{1/p'} \right] < \infty \tag{1.33}$$

(see [OK, Sec. 8]; this result is due to P. Gurka [Gu1]). A modification of this condition as well as a condition for the case $p > q$ will be given later.

1.3. Some Notations and Modifications

Remark 1.5. (i) Necessary and sufficient conditions for the validity of Hardy's inequality — either in the form (0.18) or (0.22) or in its "differential" form (1.25) — are expressed in terms of the constants A and $\widetilde{A}$ (cf. (0.20), (0.21) and (0.24), (0.25), respectively). In order to emphasize the role of the weight functions u, v, and also the interval (a, b), we will sometimes use the notation

$$A = A(a, b; u, v), \quad \widetilde{A} = \widetilde{A}(a, b; u, v). \tag{1.34}$$

(ii) Several authors have used the *modified* weighted Lebesgue space

$$\mathscr{L}^s(a, b; w) = \mathscr{L}^s(w) \tag{1.35}$$

defined as the set of all measurable functions f such that

$$fw \in L^s,$$

i.e., normed (for $1 \le s < \infty$) by

$$\|f\|_{s,w} := \left(\int_a^b |f(x)w(x)|^s dx \right)^{1/s}.$$

The connection with the weighted Lebesgue spaces $L^s(w)$ introduced by (1.2) is evident:

$$\mathscr{L}^s(w) = L^s(w^s).$$

Of course, in this case the weight conditions involve the finiteness of

$$A := A(a, b; u^q, v^p) < \infty$$

etc. For instance, inequality (1.25) with u and v replaced by u^q and v^p, respectively, for functions g satisfying (1.26), i.e., the inequality

$$\left(\int_a^b |g(x)u(x)|^q dx \right)^{1/q} \le C \left(\int_a^b |g'(x)v(x)|^p dx \right)^{1/p},$$

holds if and only if

$$\sup_{a < x < b} \left(\int_x^b u^q(t) dt \right)^{1/q} \left(\int_a^x v^{-p'}(t) dt \right)^{1/p'} < \infty \tag{1.36}$$

for $1 < p \le q < \infty$.

(iii) Sometimes, instead of investigating the Hardy operator H as an operator acting between *weighted* Lebesgue spaces $L^p(v)$ and

$L^q(u)$, several authors work with a *modified* operator $\mathscr{H}$,

$$(\mathscr{H}f)(x) := u(x) \int_a^x f(t)v(t)dt, \qquad (1.37)$$

as an operator between unweighted Lebesgue spaces L^s:

$$\mathscr{H} : L^p \to L^q. \qquad (1.38)$$

An easy calculation shows that the corresponding necessary and sufficient conditions for $\|\mathscr{H}f\|_q \le C\|f\|_p$ to hold have the (more symmetric) form

$$\sup_{a<x<b} \left(\int_x^b u^q(t)dt \right)^{1/q} \left(\int_a^x v^{p'}(t)dt \right)^{1/p'} < \infty$$

for $1 < p \le q < \infty$ and

$$\left(\int_a^b \left(\int_x^b u^q(t)dt \right)^{r/q} \left(\int_a^x v^{p'}(t)dt \right)^{r/q'} v^{p'}(x)dx \right)^{1/r} < \infty$$

for $p > q$ with $\frac{1}{r} = \frac{1}{q} - \frac{1}{p}$.

Similarly we could proceed with the operator $\widehat{\mathscr{H}}$:

$$(\widehat{\mathscr{H}}f)(x) := u(x) \int_x^b f(t)v(t)dt.$$

Clearly, these notions are equivalent. But since we will be interested in the "differential" form of Hardy's inequality (1.25), which is connected to the inequalities (0.18) (or (1.4)) for H and (1.7) for $\widetilde{H}$ due to the fact that

$$g'(x) = f(x), \ g(a) = 0 \quad \text{for} \quad g(x) = (Hf)(x),$$
$$g'(x) = -f(x), \ g(b) = 0 \quad \text{for} \quad g(x) = (\widetilde{H}f)(x),$$

it is more convenient to deal with the Hardy operators H and $\widetilde{H}$ acting between *weighted* Lebesgue spaces.

(iv) Formulas for A and $\widetilde{A}$ in the case $p > q$ [see (0.21) and (0.25)] can also be rewritten as

$$A^* := \left(\int_a^b \left(\int_a^x v^{1-p'}(t)dt \right)^{r/p'} \left(\int_x^b u(t)dt \right)^{r/p} u(x)dx \right)^{1/r}$$

and

$$\widetilde{A}^* := \left(\int_a^b \left(\int_x^b v^{1-p'}(t)dt \right)^{r/p'} \left(\int_a^x u(t)dt \right)^{r/p} u(x)dx \right)^{1/r}$$

since integration by parts shows that

$$A = \left(\frac{p'}{q} \right)^{1/r} A^* \quad \text{and} \quad \widetilde{A} = \left(\frac{p'}{q} \right)^{1/r} \widetilde{A}^*.$$

Now we can replace the conditions $A < \infty$ and $\widetilde{A} < \infty$ by $A^* < \infty$ and $\widetilde{A}^* < \infty$, respectively. The advantage is that these conditions are also meaningful for $q = 1$ and useful for $0 < q < 1$ where $q' < 0$, see also Sec. 7.4.

1.4. Hardy's Inequality for Some Special Classes of Functions

"Inner point" and "mixed" conditions

The "differential" Hardy inequality

$$\left(\int_a^b |g(x)|^q u(x)dx \right)^{1/q} \leq C \left(\int_a^b |g'(x)|^p v(x)dx \right)^{1/p} \tag{1.39}$$

was considered for functions $g \in AC^0$ satisfying one of the following conditions:

$$g(a) = 0 \quad \text{or} \quad g(b) = 0 \quad \text{or} \quad g(a) = g(b) = 0. \tag{1.40}$$

These additional conditions are natural since they avoid the meaningless case when g is a (nonzero) constant.

In what follows we will consider classes of functions $g \in AC^0$ satisfying either an *inner point condition*

$$g(c) = 0 \quad \text{with} \quad c \in (a, b), \tag{1.41}$$

c fixed, or a *mixed condition*

$$\lambda_1 g(a) + \lambda_2 g(b) = 0 \tag{1.42}$$

with λ_1, λ_2 nonzero constants.

First, let us note that condition (1.42) can be rewritten in the form

$$g(a) + \lambda g(b) = 0, \quad \lambda \neq 0. \tag{1.43}$$

However, now $\lambda \neq -1$ since for $\lambda = -1$, condition (1.43) is satisfied by any function $g(x) \equiv \text{const.} (\neq 0)$, for which inequality (1.39) is meaningless.

Theorem 1.6. *Let c be fixed, $a < c < b$. Then inequality* (1.39) *holds for every $g \in AC^0$ satisfying* (1.41) *if and only if*

$$\widetilde{A}(a, c; u, v) < \infty \tag{1.44}$$

and

$$A(c, b; u, v) < \infty. \tag{1.45}$$

Let $\lambda \in \mathbb{R}$ be fixed, $\lambda \neq 0$, $\lambda \neq -1$. Then inequality (1.39) *holds for every $g \in AC^0$ satisfying* (1.43) *if and only if*

$$A(a, b; u, v) < \infty \tag{1.46}$$

$$\widetilde{A}(a, b; u, v) < \infty. \tag{1.47}$$

Proof.   (i) *Necessity of* (1.44) *and* (1.45). Suppose that (1.39) holds for every g satisfying (1.41) and choose g such that $g(x) \equiv 0$ for $x \in [c, b)$. Then inequality (1.39) attains the form

$$\left(\int_a^c |g(x)|^q u(x) dx \right)^{1/q} \leq C \left(\int_a^c |g'(x)|^p v(x) dx \right)^{1/p}, \tag{1.48}$$

which is the Hardy inequality on the interval (a, c), and consequently, the condition (1.44) is satisfied. Similarly, for g such that $g(x) \equiv 0$ for $x \in (a, c]$, (1.39) attains the form

$$\left(\int_c^b |g(x)|^q u(x) dx \right)^{1/q} \leq C \left(\int_c^b |g'(x)|^p v(x) dx \right)^{1/p}, \qquad (1.49)$$

which is the Hardy inequality on the interval (c, b) and consequently, the condition (1.45) is satisfied.

(ii) *Sufficiency of* (1.44) *and* (1.45). Suppose that conditions (1.44) and (1.45) are satisfied. Then inequality (1.48) holds for $g_1 \in AC^0(a, c)$ with $g_1(c) = 0$, and inequality (1.49) holds for $g_2 \in AC^0(c, b)$ with $g_2(c) = 0$. Taking $g \in AC^0(a, b)$ such that

$$g(x) = \begin{cases} g_1(x) \text{ for } x \in (a, c], \\ g_2(x) \text{ for } x \in [c, b), \end{cases}$$

we find that

$$\int_a^b |g(x)|^q u(x) dx = \int_a^c |g_1(x)|^q u(x) dx + \int_c^b |g_2(x)|^q u(x) dx$$

$$\leq C^q \left[\left(\int_a^c |g_1'(x)|^p v(x) dx \right)^{q/p} \right.$$

$$\left. + \left(\int_c^b |g_2'(x)|^p v(x) dx \right)^{q/p} \right]$$

$$\leq C^q \left[\left(\int_a^b |g'(x)|^p v(x) dx \right)^{q/p} \right.$$

$$\left. + \left(\int_a^b |g'(x)|^p v(x) dx \right)^{q/p} \right]$$

$$= 2C^q \left(\int_a^b |g'(x)|^p v(x) dx \right)^{q/p}.$$

(iii) If we define the operator T as $\frac{1}{\lambda+1}H - \frac{\lambda}{\lambda+1}\widetilde{H}$, i.e.

$$(Tf)(x) = \frac{1}{\lambda+1}\int_a^x f(t)dt - \frac{\lambda}{\lambda+1}\int_x^b f(t)dt, \qquad (1.50)$$

then $g(x) = (Tf)(x)$ satisfies condition (1.43) and $g' = f$. Consequently, instead of the Hardy inequality (1.39), we can consider the inequality

$$\|Tf\|_{q,u} \le C\|f\|_{p,v}. \qquad (1.51)$$

Since

$$\|Tf\|_{q,u} \le \left|\frac{1}{\lambda+1}\right|\|Hf\|_{q,u} + \left|\frac{\lambda}{\lambda+1}\right|\|\widetilde{H}f\|_{q,u},$$

conditions (1.46) and (1.47) are obviously *sufficient* for (1.51).

For $\lambda \in (-1,0)$ it is $\frac{1}{\lambda+1} > 0$ and $-\frac{\lambda}{\lambda+1} > 0$, and consequently, for $f \ge 0$ we have

$$Hf \le (\lambda+1)Tf, \quad \widetilde{H}f \le \frac{\lambda+1}{|\lambda|}Tf.$$

Thus, for $f \ge 0$, the conditions (1.46) and (1.47) are also *necessary* since if (1.51) holds, then the inequalities

$$\|Hf\|_{q,u} \le C(\lambda+1)\|f\|_{p,v} \text{ and } \|\widetilde{H}f\|_{q,u} \le C(\lambda+1)|\lambda|^{-1}\|f\|_{p,v}$$

also hold, which implies (1.46) and (1.47).

For $\lambda < -1$ or $\lambda > 0$, T is a combination of H and $\widetilde{H}$ with one positive and one negative coefficient. Therefore, the foregoing argument cannot be used and the necessity follows from the more general case considered in Chap. 2 (see Theorem 2.3). □

1.5. The Role of the Interval

The interval (a, b)

In the previous considerations, the interval (a,b) was arbitrary, $-\infty \le a < b \le +\infty$. In the classical case (0.16), G.H. Hardy considered the particular interval

$$(a, b) = (0, \infty). \tag{1.52}$$

For this interval, the necessary and sufficient condition for the validity of the general Hardy inequality (0.18) or (1.25), i.e., of the inequalities

$$\left(\int_0^\infty \left(\int_0^x f(t)dt \right)^q u(x)dx \right)^{1/q} \le C \left(\int_0^\infty f^p(x)v(x)dx \right)^{1/p} \tag{1.53}$$

with $f \ge 0$ and

$$\left(\int_0^\infty |g(x)|^q u(x)dx \right)^{1/q} \le C \left(\int_0^\infty |g'(x)|^p v(x)dx \right)^{1/p}$$

with $g(0) = 0$ reads, for $1 < p \le q < \infty$, as

$$A = A(0, \infty; u, v) := \sup_{x>0} \left(\int_x^\infty u(t)dt \right)^{1/q} \left(\int_0^x v^{1-p'}(t)dt \right)^{1/p'} < \infty. \tag{1.54}$$

Let us show that all other possible choices of the interval (a, b) can be reduced to this special case.

(i) Consider the case

$$(a, b) = (-\infty, \infty).$$

Then the inequality

$$\left(\int_{-\infty}^\infty \left(\int_{-\infty}^x f(t)dt \right)^q u(x)dx \right)^{1/q} \le C \left(\int_{-\infty}^\infty f^p(x)v(x)dx \right)^{1/p} \tag{1.55}$$

holds for all $f \ge 0$ if and only if

$$A(-\infty, \infty; u, v) = \sup_{x \in \mathbb{R}} \left(\int_x^\infty u(t)dt \right)^{1/q} \left(\int_{-\infty}^x v^{1-p'}(t)dt \right)^{1/p'} < \infty. \tag{1.56}$$

Indeed: Let $t = \ln s$ and $x = \ln y$. Then $dt = \frac{1}{s}ds$, $dx = \frac{1}{y}dy$, and under these substitutions, inequality (1.55) becomes

$$\left(\int_0^\infty \left(\int_0^y \frac{f(\ln s)}{s} ds \right)^q \frac{u(\ln y)}{y} dy \right)^{1/q}$$
$$\leq C \left(\int_0^\infty \left(\frac{f(\ln y)}{y} \right)^p y^{p-1} v(\ln y) dy \right)^{1/p}. \qquad (1.57)$$

Denote

$$\widetilde{u}(y) = \frac{u(\ln y)}{y}, \quad \widetilde{f}(y) = \frac{f(\ln y)}{y}, \quad \widetilde{v}(y) = y^{p-1} v(\ln y).$$

Then (1.57) can be rewritten as

$$\left(\int_0^\infty \left(\int_0^y \widetilde{f}(s) ds \right)^q \widetilde{u}(y) dy \right)^{1/q} \leq C \left(\int_0^\infty \widetilde{f}^p(y) \widetilde{v}(y) dy \right)^{1/p},$$

which is (1.53) for $\widetilde{f}, \widetilde{u}, \widetilde{v}$, and it holds, according to (1.54), if and only if

$$\sup_{r>0} \left(\int_r^\infty \widetilde{u}(s) ds \right)^{1/q} \left(\int_0^r \widetilde{v}^{1-p'}(s) ds \right)^{1/p'} < \infty,$$

i.e., if and only if

$$\sup_{r>0} \left(\int_r^\infty \frac{u(\ln s)}{s} ds \right)^{1/q} \left(\int_0^r \frac{v^{1-p'}(\ln s)}{s} ds \right)^{1/p'} < \infty.$$

Making the above substitution, we can write this condition as

$$\sup_{r>0} \left(\int_{\ln r}^\infty u(t) dt \right)^{1/q} \left(\int_{-\infty}^{\ln r} v^{1-p'}(t) dt \right)^{1/p'} < \infty,$$

and this is (1.56) with $x = \ln r$, since $x \in \mathbb{R}$ for $r > 0$.

(ii) Consider the case

$$a = 0, \quad 0 < b < \infty.$$

Then the inequality

$$\left(\int_0^b \left(\int_0^x f(t)dt\right)^q u(x)dx\right)^{1/q} \leq C \left(\int_0^b f^p(x)v(x)dx\right)^{1/p} \quad (1.58)$$

holds for all $f \geq 0$ if and only if

$$A(0,b;u,v) = \sup_{0<x<b} \left(\int_x^b u(t)dt\right)^{1/q} \left(\int_0^x v^{1-p'}(t)dt\right)^{1/p'} < \infty.$$

$$(1.59)$$

Again, let $t = \frac{b}{s+1}$ and $x = \frac{b}{y+1}$. Then $dt = -\frac{b}{(s+1)^2}ds$ and $dx = -\frac{b}{(y+1)^2}dy$, and under these substitutions, inequality (1.58) becomes

$$\left(\int_\infty^0 \left(\int_\infty^y f\left(\frac{b}{s+1}\right) \frac{-b}{(s+1)^2}ds\right)^{1/q} u\left(\frac{b}{y+1}\right) \frac{-b}{(y+1)^2}dy\right)^{1/q}$$

$$\leq C \left(\int_\infty^0 f^p\left(\frac{b}{y+1}\right) v\left(\frac{b}{y+1}\right) \frac{-b}{(y+1)^2}dy\right)^{1/p}. \quad (1.60)$$

If

$$\widetilde{u}(y) = u\left(\frac{b}{y+1}\right) \frac{b}{(y+1)^2},$$

$$\widetilde{f}(y) = f\left(\frac{b}{y+1}\right) \frac{b}{(y+1)^2},$$

$$\widetilde{v}(y) = v\left(\frac{b}{y+1}\right) \left(\frac{b}{(y+1)^2}\right)^{1-p},$$

then (1.60) can be rewritten as

$$\left(\int_0^\infty \left(\int_y^\infty \widetilde{f}(s)ds\right)^q \widetilde{u}(y)dy\right)^{1/q} \leq C \left(\int_0^\infty \widetilde{f}^p(y)\widetilde{v}(y)dy\right)^{1/p}.$$

But this is the conjugate Hardy inequality for $\widetilde{f}, \widetilde{u}, \widetilde{v}$, which is satisfied if and only if

$$\widetilde{A}(0,\infty;\widetilde{u},\widetilde{v}) = \sup_{r>0} \left(\int_0^r \widetilde{u}(s)ds\right)^{1/q} \left(\int_r^\infty \widetilde{v}^{1-p'}(s)ds\right)^{1/p'} < \infty,$$

i.e., if and only if

$$\sup_{r>0} \left(\int_0^r u \left(\frac{b}{s+1} \right) \frac{b}{(s+1)^2} ds \right)^{1/q}$$

$$\times \left(\int_r^\infty v^{1-p'} \left(\frac{b}{s+1} \right) \left(\frac{b}{(s+1)^2} \right)^{(1-p)(1-p')} ds \right)^{1/p'} < \infty.$$

Making the above substitution, this condition becomes

$$\sup_{r>0} \left(\int_{\frac{b}{r+1}}^b u(t)dt \right)^{1/q} \left(\int_0^{\frac{b}{r+1}} v^{1-p'}(t)dt \right)^{1/p'} < \infty,$$

which is (1.59) with $x = \frac{b}{r+1}$, since for $r > 0$, it is $x \in (0,b)$.

(iii) Finally, consider the case

$$-\infty < a < b \leq \infty.$$

Again we can deduce from (1.53) and (1.54) that the inequality

$$\left(\int_a^b \left(\int_a^x f(t)dt \right)^q u(x)dx \right)^{1/q} \leq C \left(\int_a^b f^p(x)u(x)dx \right)^{1/p} \quad (1.61)$$

holds for all $f \geq 0$ if and only if

$$A(a,b;u,v) = \sup_{a<x<b} \left(\int_x^b u(t)dt \right)^{1/q} \left(\int_a^x v^{1-p'}(t)dt \right)^{1/p'} < \infty. \quad (1.62)$$

For this purpose, write

$$h_c(x) := h(x+c).$$

The substitutions $t = s + a$ and then $x = y + a$ show that (1.61) can be rewritten as

$$\left(\int_0^{b-a} \left(\int_0^y f_a(s)ds \right)^q u_a(y)dy \right)^{1/q} \leq C \left(\int_0^{b-a} f_a^p(y)v_a(y)dy \right)^{1/p}.$$

But this is case (ii) with b replaced by $b - a$ and f, u, v by f_a, u_a, v_a, respectively. Thus this inequality is satisfied if and only if

$$\sup_{0<r<b-a} \left(\int_r^{b-a} u_a(s)ds \right)^{1/q} \left(\int_0^r v_a^{1-p'}(s)ds \right)^{1/p'} < \infty$$

or

$$\sup_{0<r<b-a} \left(\int_r^{b-a} u(s+a)ds \right)^{1/q} \left(\int_0^r v^{1-p'}(s+a)ds \right)^{1/p'} < \infty,$$

i.e.,

$$\sup_{0<r+a<b} \left(\int_{r+a}^{b} u(t)dt \right)^{1/q} \left(\int_0^{r+a} v^{1-p'}(t)dt \right)^{1/p'} < \infty,$$

and this is (1.62) with $x = r + a$.

Remark 1.7. The "universality" of the interval $(0, \infty)$ was shown in Sec. 1.5 only for the case $p \le q$. By the same approach one can also prove it for $p > q$ $(0 < q < \infty, 1 < p < \infty)$.

1.6. Compactness of the Hardy Operator

Motivation

For Ω a domain in $\mathbb{R}^N$ with (smooth) boundary $\partial\Omega$, let $W_0^{1,p}(\Omega)$ be the *Sobolev space,*

$$W_0^{1,p}(\Omega) = \{g : |\nabla g| \in L^p(\Omega), \; g|_{\partial\Omega} = 0\}.$$

It is well-known that the imbedding

$$W_0^{1,p}(\Omega) \hookrightarrow L^q(\Omega) \tag{1.63}$$

is *continuous* if $1 < q \le \frac{Np}{N-p}$ and *compact* if $1 < q < \frac{Np}{N-p}$ $(1 < p < N)$. Hence, the value

$$p^* = \frac{Np}{N-p} \tag{1.64}$$

is the so-called *critical exponent* of the imbedding (1.63), i.e., the imbedding is compact for $q < p^*$, continuous for $q \le p^*$ and *does not take place* for $q > p^*$.

If Ω is the ball $\{x \in \mathbb{R}^N \, |x| < R\}$ and if we consider *radial functions,* $g(x) = g(|x|) = g(r), \; 0 < r < R$, then we can rewrite the imbedding (1.63) in the form of the Hardy inequality

$$\left(\int_0^R |g(r)|^q r^{N-1} dr \right)^{1/q} \leq C \left(\int_0^R |g'(r)|^p r^{N-1} dr \right)^{1/p} \qquad (1.65)$$

with very special weights: $u(r) = v(r) = r^{N-1}$.

Hence, the following natural question occurs: For a fixed p, $p > 1$, and for given (general) weight functions u and v, does there exist a parameter $p^* = p^*(p, u, v)$, $p^* \geq p$, such that the imbedding described for differentiable functions g with $g' \in L^p(v)$ and with $g(R) = 0$ by the inequality

$$\left(\int_0^R |g(r)|^q u(r) dr \right)^{1/q} \leq C \left(\int_0^R |g'(r)|^p v(r) dr \right)^{1/p} \qquad (1.66)$$

is continuous for $q \leq p^*$, compact for $q < p^*$ and does not take place for $q > p^*$?

And since inequality (1.66) is equivalent to the Hardy inequality

$$\|\widetilde{H} f\|_{q,u} \leq C \|f\|_{p,v} \qquad (1.67)$$

for the conjugate Hardy operator

$$(\widetilde{H} f)(x) = \int_x^R f(t) dt,$$

we can reformulate the problem just mentioned for the operator $\widetilde{H}$: Does there exist, for $p > 1$ fixed, a *critical exponent* $p^* = p^*$ (p, u, v) such that $\widetilde{H}$ maps $L^p(v)$ into $L^q(u)$ compactly for $q < p^*$, continuously for $q \leq p^*$ and that the Hardy inequality (1.67) does not hold for $q > p^*$?

Notation

Let p be fixed, $p > 1$. We shall consider functions defined on the interval $(0, R)$, $R \leq \infty$, and for simplicity, we will assume that the weight functions u, v satisfy

$$u \in L^1(0, R), \ v^{1-p'} \in L^1(r, R) \text{ for every } r < R,$$
$$\text{but } v^{1-p'} \notin L^1(0, R). \tag{1.68}$$

Using the notation

$$U(r) := \int_0^r u(t)dt, \quad V(r) := \int_r^R v^{1-p'}(t)dt, \tag{1.69}$$

assumption (1.68) means that

$$U(0) = 0, \ U(R) < \infty, \ V(0) = \infty, \ V(R) = 0. \tag{1.70}$$

If we further denote for $1 < p \le q < \infty$

$$B_q(r) = U^{1/q}(r)V^{1/p'}(r)$$

then we know that the inequality (1.67) holds — i.e., that the mapping $\widetilde{H} : L^p(v) \to L^q(u)$ is *continuous* — if and only if

$$\sup_{0 < r < R} B_q(r) < \infty. \tag{1.71}$$

If we want the mapping $\widetilde{H}$ to be *compact*, this condition has to be *strengthened:* the corresponding necessary and sufficient condition then reads

$$\lim_{r \to 0} B_q(r) = \lim_{r \to R} B_q(r) = 0. \tag{1.72}$$

(For details, see [OK, Sec. 7].)

Remark 1.8. (i) Due to assumption (1.70), we immediately have $B_q(R) = 0$. Hence, we need only to take care of the first condition in (1.72), namely, the behaviour of $B_q(r)$ for r small, $r > 0$.

(ii) We also immediately see that *if the mapping* $\widetilde{H} : L^p(u) \to L^q(v)$ *is continuous for some* $\widehat{q} > p$, *then it is compact for every* $q < \widehat{q}$, $1 < p \le q$.

Indeed, we have

$$B_q(r) = B_{\widehat{q}}(r)U^{1/q - 1/\widehat{q}}(r)$$

and since $B_{\widehat{q}}(r)$ is bounded in $(0, R)$ due to the continuity of $\widetilde{H}$ (see (1.71)) and $U(r) \to 0$ for $r \to 0$, we obtain that $B_q(r) \to 0$ for $r \to 0$ (we have $1/q - 1/\widehat{q} > 0$).

Assumption 1.9. *We will assume throughout this section that there exists at least one $\widetilde{q} > p$ such that*

$$\sup_{0 < r < R} B_{\widetilde{q}}(r) < \infty,$$

i.e., that the (conjugate) Hardy inequality (1.67) holds for $q = \widetilde{q}$.

Definition 1.10. Let us consider the set

$$S := \{s > p : \sup_{0 < r < R} B_s(r) < \infty\}. \tag{1.73}$$

Due to Assumption 1.9, the set S is *non-empty*: $\widetilde{q} \in S$. Consequently, we can define the number (possibly ∞)

$$p^* = \sup S. \tag{1.74}$$

Remark 1.11. (i) Assumption 1.9 is reasonable since for general weights satisfying (1.70) it might happen that $\sup_{0 < r < R} B_q(r) = \infty$ for every $q \geq p$. The couple

$$u(r) = r^\alpha, \alpha > 1; \ v(r) = r^{2(p-1)} \exp\frac{1 - p}{r}.$$

may serve, for $R < \infty$, as an example.

(ii) It seems that the number p^* from (1.74) could serve as the critical exponent for the mapping $\widetilde{H}$. Indeed: For $q > p^*$, the Hardy inequality (1.67) does not hold due to the definition of the set S – we have $\sup_{0 < r < R} B_q(r) = \infty$. For $q < p^*$, there is a $\widehat{q} \in S$ with $\widehat{q} > q$ according to the definition of S, and the mapping $\widetilde{H}$ is compact due to Remark 1.8 (ii).

It is still not clear what can happen if $q = p^*$. We will see later (see Examples 1.14 and 1.15 below) that for an appropriate choice of the weights u, v, all possibilities can occur: the mapping

$$\widetilde{H} : L^p(v) \to L^{p^*}(u)$$

can be compact, or continuous but not compact, or not continuous at all.

But first, let us show that for p^* we have the formula

$$p^* = p' \liminf_{r \to 0} \frac{|\log U(r)|}{\log V(r)}. \tag{1.75}$$

For this purpose, we introduce the notation

$$W(r) := \frac{|\log U(r)|}{\log V(r)}, \tag{1.76}$$

$$W_0 := \liminf_{r \to 0} W(r). \tag{1.77}$$

Then obviously

$$B_q^q(r) = U(r)V^{q/p'}(r) = (V(r))^{(q-p'W(r))/p'} \tag{1.78}$$

and (if W_0 is finite)

$$\limsup_{r \to 0}(q - p'W(r)) = q - p'W_0. \tag{1.79}$$

Lemma 1.12. *Let $p > 1$ and suppose that the weight functions u, v satisfy (1.68). Then Assumption 1.9 holds if and only if $W_0 > p - 1$.*

Proof. Assume first that $p - 1 < W_0 < \infty$, and let q be such that $p \leq q < p'W_0$. Then it follows from (1.78) and (1.79) that $\lim_{r \to 0} B_q^q(r) = 0$ and thus, $q \in S$. Hence, the whole interval $(p, p'W_0)$ belongs to S, and Assumption 1.9 follows. The proof for the case $W_0 = \infty$ is similar.

Conversely, assume that Assumption 1.9 holds, i.e., that $S \neq \emptyset$. Then there exists a $q > p$ such that $\sup_{0<r<R} B_q^q(r) < \infty$, which implies, in view of (1.68) and (1.70), that

$$q - p'W(r) \leq 0 \quad \text{for} \quad r > 0 \quad \text{sufficiently small,}$$

and hence

$$W_0 = \liminf_{r \to 0} W(r) \geq \frac{q}{p'} > \frac{p}{p'} = p - 1. \qquad \square$$

Now, we are able to prove the main result of this section.

Theorem 1.13. *Let $p > 1$ and suppose that the weight functions u, v satisfy (1.68) and Assumption 1.9. Let p^* be given by (1.74). Then formula (1.75) holds and the mapping $\widetilde{H} : L^p(v) \to L^q(u)$ is compact for $1 \leq q < p^*$.*

If $p^* < \infty$ and $q > p^*$, then the Hardy inequality (1.67) does not hold.

Moreover, for $q = p^*$, the mapping $\widetilde{H} : L^p(v) \to L^{p^*}(u)$ is continuous if and only if $U(r)V^{p^*/p'}(r)$ is bounded for r near zero, and it is compact if and only if $\lim_{r \to 0} U(r)V^{p^*/p'}(r) = 0$.

Proof.   Let W_0 be given by (1.77). From the proof of Lemma 1.12 we know that $(p, p'W_0) \subseteq S$, which implies that $p^* \geq p'W_0$. Thus, the result follows if $W_0 = \infty$.

Let now $W_0 < \infty$. If $q > p'W_0$, then $q \notin S$, i.e., $\limsup_{r \to 0} B_q^q(r) = \infty$, since choosing a sequence $\{r_n\}$, $r_n \to 0$, such that $W(r_n) \to W_0$ as $n \to \infty$, we find that

$$\lim_{n \to \infty} (q - p'(W(r_n))) = q - p'W_0 > 0$$

and thus, due to (1.78), $\lim_{n \to \infty} B_q^q(r_n) = \infty$. Consequently, $p^* = p'W_0$ must hold.

The remaining part of the proof follows from our previous considerations (see Remark 1.8 (ii)) and from [OK, Sec. 7]. $\square$

Example 1.14. (i) If we choose, for $R < \infty$,

$$u(r) = r^\alpha, \ \alpha > 1 \quad \text{and} \quad v(r) = r^{p-1},$$

we have $B_q(r) = (r^{\alpha+1}/(\alpha + 1))^{1/q} (\log \frac{R}{r})^{r/p'}$, and hence $B_q(r) \to 0$ as $r \to 0$. Thus $S = (p, \infty)$ and $p^* = \infty$.

(ii) If we choose, for $R < \infty$,

$$u(r) = r^\alpha, \alpha > -1 \text{ and } v(r) = r^{\beta(1-p)} \text{ with } \beta < -1,$$

we find that

$$B_q(r) = \text{const}[r^{(\alpha+1)p'/q}(r^{\beta+1} - R^{\beta+1})]^{1/p'}$$

and

$$p^* = \frac{(\alpha+1)p'}{-\beta - 1}.$$

Consequently, the operator $\widetilde{H} : L^p(v) \to L^q(u)$ is continuous for $q \le p^*$ and compact for $q < p^*$.

(iii) For the particular choice $u(r) = v(r) = r^{N-1}$, i.e. $\alpha = N - 1$ and $\beta = (N-1)/(1-p)$ in the foregoing example, we obtain for $p^* = Np/(N-p)$ the critical exponent from (1.64).

The next example shows that the operator $\widetilde{H}$ need not necessarily be continuous in the *critical* case as a mapping from $L^p(v)$ into $L^{p^*}(u)$.

Example 1.15. Let $v(r) = r^{N-1}$, $1 < p < N$, and let k be any number such that $k > p - 1$. Assume that

$$U(r) = \int_0^r u(s)ds := r^{k(N-p)/(p-1)}m(r)$$

for r near 0, where m is a positive smooth function such that

$$\lim_{r \to 0} \frac{\log m(r)}{\log r} = 0.$$

Conditions (1.70) are satisfied, and since we can show that

$$\lim_{r \to 0} \frac{\log V(r)}{|\log r|} = \frac{N-p}{p-1},$$

we find that

$$W_0 = \lim_{r \to 0} W(r) = \lim_{r \to 0} \frac{|\log U(r)|}{\log V(r)} = k > p - 1,$$

i.e., Assumption 1.9 is satisfied due to Lemma 1.12.

For $p^* = p'k$ we find that

$$B_{p^*}^{p^*}(r) = V^k(r)U(r) = \text{const}\left(1 - \left(\frac{r}{R}\right)^{(N-p)/(p-1)}\right)^k m(r).$$

Hence

- for $m(r) = |\log r|$, the Hardy inequality (1.67) – with $q = p^*$ – does not hold,
- for $m(r) = 1$ the mapping $\widetilde{H} : L^p(v) \to L^{p^*}(u)$ is continuous but not compact,
- for $m(r) = |\log r|^{-1}$ it is compact.

Remark 1.16. As a consequence of our motivating example on pp. 39–40, we have considered in (1.66) functions g vanishing at the right end — $g(R) = 0$, which leads to the conjugate Hardy operator $\widetilde{H}$. Clearly, it is possible to consider the "classical" Hardy operator $H : (Hf)(x) = \int_0^x f(t)dt$, and also an arbitrary interval (a, b).

1.7. Some Limiting Inequalities — Preliminary Results

A limiting case of Hardy's inequality (0.15)

The well-known *Knopp inequality*

$$\int_0^\infty \exp\left(\frac{1}{x}\int_0^x \ln f(t)dt\right) dx \le e \int_0^\infty f(x)dx \qquad (1.80)$$

where e is the best constant (see K. Knopp [Kn1] or [HLP, p. 250]) can be considered as a limit, for p tending to infinity, of the classical Hardy inequality (0.15) used for the function $f^{1/p}$,

$$\int_0^\infty \left(\frac{1}{x}\int_0^x f^{1/p}(t)dt\right)^p dx \le \left(\frac{p}{p-1}\right)^p \int_0^\infty f(x)dx.$$

Indeed, the *geometric mean of f*,

$$\exp\left(\frac{1}{x}\int_0^x \ln f(t)dt\right),$$

satisfies

$$\lim_{p \to \infty} \left(\frac{1}{x} \int_0^x f^{1/p}(t)dt \right)^p = \exp \left(\frac{1}{x} \int_0^x \ln f(t)dt \right)$$

(see [HLP, p. 139]) while $[p/(p-1)]^p \to e$ for $p \to \infty$.

Moreover, we even have strict inequality in (1.80) whenever the integral on the right hand side converges. We also note that inequality (1.80) was known before the Knopp paper even by G.H. Hardy himself; see [Ha3, p. 156] and cf. Remark 1.17 below.

A limiting case of the discrete inequality (0.14)

A similar limiting procedure in (0.14) leads to the *Carleman inequality*

$$\sum_{n=1}^{\infty} \sqrt[n]{a_1 a_2 \ldots a_n} \le e \sum_{n=1}^{\infty} a_n \qquad (a_n \ge 0), \qquad (1.81)$$

again with strict inequality sign whenever the sum on the right hand side converges.

Remark 1.17. Carleman's original proof of (1.81) was based on the Lagrange multiplier method (see T. Carleman [C1]). According to G. H. Hardy [Ha3, p. 156], the above limiting argument was observed for the first time probably by G. Pólya.

Several authors have given other proofs, generalizations or applications of (1.80)–(1.81); let us mention G.H. Hardy [Ha3], [Ha4], K. Knopp [Kn1], L. Carleson [Ca1], J.A. Cochran and C.S. Lee [CL1], H.P. Heinig [He1], E.R. Love [Lov1–Lov3], B. Yang and L. Debnath [YD1], H. Alzer [Al1], P. Yan and G. Sun [YS1] and S. Kaijser *et al.* [KP01]. For more references and information see [MPF, Chap. 4], [KoMP] and the review articles by J. Pečarić and K.B. Stolarsky [PS1] and M. Johansson *et al.* [JPW1].

Carleson's result

In his 1954 paper, L. Carleson [Ca1] was obviously interested in finding an elementary proof of (1.81) but he formulated his result as an integral inequality which also proves (1.80). Namely, he proved:

Let $m(x)$ be a convex function on $(0, \infty)$ satisfying $m(0) = 0$. Then, for $p > -1$,

$$\int_0^\infty x^p e^{-m(x)/x} dx \leq e^{p+1} \int_0^\infty x^p e^{-m'(x)} dx. \qquad (1.82)$$

Let f^* be the decreasing rearrangement of f (see (6.2)) and let us apply (1.82) with the convex function

$$m(x) = -\int_0^x \ln f^*(t) dt.$$

Then we obtain

$$\int_0^\infty x^p \exp\left(\frac{1}{x} \int_0^x \ln f^*(t) dt\right) dx \leq e^{p+1} \int_0^\infty x^p f^*(x) dx,$$

which for $p = 0$ leads to (1.80).

Similarly, applying (1.82) with the convex function $m(x)$ defined as the polygon connecting the points $(0, 0)$ and $(n, \sum_{i=1}^n \ln(1/a_i^*))$, $n = 1, 2, \dots$ [where $\{a_i^*\}_{i=1}^\infty$ denotes the decreasing rearrangement of $\{a_i\}_{i=1}^\infty$], we obtain for $p = 0$ that

$$\sum_{n=1}^\infty \sqrt[n]{a_1 a_2 \dots a_n} \leq \sum_{n=1}^\infty \sqrt[n]{a_1^* a_2^* \dots a_n^*} \leq e \sum_{n=1}^\infty a_n^* = e \sum_{n=1}^\infty a_n$$

because on $(n-1, n)$, we have $m'(x) = \ln(1/a_n^*)$ and

$$\frac{m(x)}{x} \leq \frac{m(n)}{n} = \frac{1}{n} \sum_{i=1}^n \ln(1/a_i^*).$$

Proof of (1.82)

Using Hölder's inequality and the convexity of m:

$$m(kx) \geq m(x) + (k-1)x m'(x), \quad k > 1,$$

we conclude that, for any $A > 0$,

$$\frac{1}{k^{p+1}} \int_0^A x^p e^{-m(x)/x} dx \leq \int_0^A x^p e^{-m(kx)/(kx)} dx$$

$$\leq \int_0^A x^p e^{-m(x)/(kx)-((k-1)/k)m'(x)} dx$$

$$\leq \left(\int_0^A x^p e^{-m(x)/x} dx \right)^{1/k}$$

$$\times \left(\int_0^A x^p e^{-m'(x)} dx \right)^{(k-1)/k}$$

and hence

$$\int_0^A x^p e^{-m(x)/x} \leq k^{(p+1)k/(k-1)} \int_0^A x^p e^{-m'(x)} dx.$$

Now, inequality (1.82) follows by letting $A \to \infty$ and $k \to 1$, i.e., $k^{k/(k-1)} \to e$. $\qquad\qquad\qquad\qquad\qquad\qquad\qquad\qquad\qquad\qquad\square$

Remark 1.18. (i) Another generalization which covers both (1.80) and (1.81) was proved in S. Kaijser *et al.* [KPO1]. In particular, this result implies the inequality

$$\int_0^\infty \exp \left(\frac{1}{M(x)} \int_0^x \ln f(t) dM(t) \right) dM(x) \leq e \int_0^\infty f(x) dM(x)$$

with $M(x)$ a right-continuous and increasing function on $[0, \infty)$.

(ii) It is easy to see that Knopp's inequality (1.80) implies Carleman's inequality (1.81). In fact, applying (1.80) with $f(x) = a_n$, $x \in [n-1, n)$, $n = 1, 2, \ldots$, and making some straightforward calculations we obtain (1.81).

1.8. Limiting Inequalities — General Results

A generalization of inequality (1.80)

Besides the *Hardy averaging operator* H_a:

$$(H_a f)(x) := \frac{1}{x} \int_0^x f(t)dt \qquad (1.83)$$

which can also be called the arithmetic mean operator, let us introduce the *geometric mean operator* G by

$$(Gf)(x) := \exp\left(\frac{1}{x} \int_0^x \ln f(t)dt\right), \quad f \geq 0. \qquad (1.84)$$

In this section we shall consider the weighted integral inequality

$$\left(\int_0^\infty (Gf)^q(x)u(x)dx\right)^{1/q} \leq C \left(\int_0^\infty f^p(x)v(x)dx\right)^{1/p} \qquad (1.85)$$

for $u(x) \geq 0$, $v(x) \geq 0$ and $0 < p, q < \infty$, as a limit of some Hardy inequalities.

Recall that u, v are weight functions and that Knopp's inequality (1.80) is a special case of (1.85) with $p = q = 1$, $u(x) = v(x) \equiv 1$, $C = e$.

For later purposes we also introduce the notation

$$w(x) := \left[G\left(\frac{1}{v(x)}\right)\right]^{q/p} u(x). \qquad (1.86)$$

Some preliminaries

(i) As already mentioned in Sec. 1.7,

$$(Gf)(x) = \lim_{\alpha \to 0+} \{H_a(f^\alpha(x))\}^{1/\alpha} = \lim_{\alpha \to 0+} \left(\frac{1}{x} \int_0^x f^\alpha(t)dt\right)^{1/\alpha}. \qquad (1.87)$$

Moreover,

$$G(f^s) = (G(f))^s, \quad s > 0, \qquad (1.88)$$

$$G(fg) = G(f)G(g), \qquad (1.89)$$

the scale of power means $(H_a(f^\alpha)(x))^{1/\alpha} = (\frac{1}{x}\int_0^x f^\alpha(t)dt)^{1/\alpha}$ is increasing in α by Hölder's inequality and, in particular,

$$G(f)(x) \le (H_a f)(x). \tag{1.90}$$

By replacing $f(x)$ by $f(x)v^{-1/p}(x)$ in (1.85) and using (1.88) and (1.89) we see that (1.85) is equivalent to the inequality

$$\left(\int_0^\infty (Gf)^q(x)w(x)dx\right)^{1/q} \le C \left(\int_0^\infty f^p(x)dx\right)^{1/p} \tag{1.91}$$

or even equivalent to

$$\left(\int_0^\infty (Gf)^{qs/p}(x)w(x)dx\right)^{p/qs} \le C^{p/s} \left(\int_0^\infty f^s(x)dx\right)^{1/s} \tag{1.92}$$

for any $s > 0$.

(ii) For an operator $T : X \to Y$ where X, Y are normed linear spaces, the operator norm is defined by

$$\|T\|_{X \to Y} := \sup_{f \ne 0} \frac{\|Tf\|_Y}{\|f\|_X}$$

and $\|T\|_{X \to Y}$ is the least constant C in the estimate

$$\|Tf\|_X \le C\|f\|_Y.$$

We will use this operator norm notation in the sequel; notice that the Hardy inequality (1.12) can be rewritten as $\|Hf\|_{q,u} \le C\|f\|_{p,v}$ so that

$$H : L^p(v) \to L^q(u) \quad \text{and} \quad C = \|H\|_{L^p(v) \to L^q(u)},$$

and similarly (1.91) means that

$$G : L^p \to L^q(w) \quad \text{and} \quad C = \|G\|_{L^p \to L^q(w)}.$$

(iii) The equivalence of (1.85), (1.91) and (1.92) thus implies that

$$\|G\|_{L^p(v) \to L^q(u)} = \|G\|_{L^p \to L^p(w)} = \|G\|_{L^s \to L^{qs/p}(w)}^{s/p}. \tag{1.93}$$

Moreover, (1.90) implies that we obtain sufficient conditions for (1.85) to hold by just using (1.90) and known Hardy-type inequalities for the case

$$H_a : L^s \to L^{qs/p}(w)$$

with a suitable s (cf. (1.92)). In this way we also obtain an upper bound for the operator norm $\|G\|_{L^p(v) \to L^q(u)}$ (see (1.93)). Consequently, it only remains to find a corresponding lower bound and to show that (1.85) can be obtained as a limiting case of the corresponding Hardy inequalities.

The limiting procedure

In view of our considerations in Sec. 1.7 and due to (1.87), we study inequality (1.91) — and thus inequality (1.85) — as a limiting inequality (for $\alpha \to 0+$) of the following scale of inequalities:

$$\left(\int_0^\infty (H_a f^\alpha)^{q/\alpha}(x) w(x) dx \right)^{1/q} \leq C_\alpha \left(\int_0^\infty f^p(x) dx \right)^{1/p}, \ \alpha > 0,$$
(1.94)

which of course is equivalent to the Hardy inequalities

$$\left(\int_0^\infty (H_a f)^{q/\alpha}(x) w(x) dx \right)^{\alpha/q} \leq C_\alpha^\alpha \left(\int_0^\infty f^{p/\alpha}(x) dx \right)^{\alpha/p}, \ \alpha > 0.$$
(1.95)

We conclude that (1.91) — and thus (1.85) — holds for some $C < \infty$ if $\lim_{\alpha \to 0} C_\alpha < \infty$, and, moreover,

$$\|G\|_{L^p(v) \to L^q(u)} = \|G\|_{L^p \to L^q(w)} = \lim_{\alpha \to 0} \|H_a\|_{L^{p/\alpha} \to L^{q/\alpha}(w)}^{1/\alpha}.$$
(1.96)

Now, we are ready to formulate the main result for the case $0 < p \leq q < \infty$.

Theorem 1.19. *Let* $0 < p \leq q < \infty$. *Then the inequality* (1.85), *i.e.,*

$$\left(\int_0^\infty (Gf)^q(x) u(x) dx \right)^{1/q} \leq C \left(\int_0^\infty f^p(x) v(x) dx \right)^{1/p},$$

holds for $f \geq 0$ if and only if

$$B := \sup_{t>0} t^{-1/p} \left(\int_0^t w(x)dx \right)^{1/q} < \infty \tag{1.97}$$

with w given by (1.86).

Moreover, if C is the least constant in (1.85) (i.e., $C = \|G\|_{L^p(v) \to L^q(u)}$), then

$$B \leq C \leq e^{1/p}B. \tag{1.98}$$

Proof. Assume that (1.97) holds. Apply the alternative form of Hardy's inequality described in Theorem 1.1 with $a = 0$, $b = \infty$, $v(x) \equiv 1$, $u(x) = w(x)x^{-q/\alpha}$ and p and q replaced by p/α and q/α, respectively, where $0 < \alpha < p$. Then (1.95) and thus (1.94) holds and, moreover,

$$C_\alpha^a \leq \frac{p/\alpha}{p/\alpha - 1} B^\alpha.$$

By letting $\alpha \to 0+$ we find that (1.91) and thus (1.85) holds with

$$C \leq e^{1/p}B$$

since $\lim_{\alpha \to 0+} \left(\frac{p/\alpha}{p/\alpha - 1} \right)^{1/\alpha} = e^{1/p}$.

Conversely, suppose that (1.85) holds and apply the equivalent inequality (1.91) to the test function

$$f_t(x) = \chi_{[0,t]}(x) + h_\varepsilon(x)\chi_{(t,\infty)}(x)$$

with a fixed $t > 0$ and with $h_\varepsilon(x) > 0$, ε arbitrary positive, chosen so that

$$\int_t^\infty h_\varepsilon^p(x)dx < \varepsilon.$$

Then it follows that

$$\left(\int_0^t w(x)dx \right)^{1/q} (t + \varepsilon)^{-1/p} \leq C < \infty$$

and, since ε is arbitrary,

$$B = \sup_{t>0} \left(\int_0^t w(x)dx \right)^{1/q} t^{-1/p} \leq C < \infty.$$

Hence, the proof is complete. $\qquad\qquad\qquad\qquad\qquad\qquad\qquad$ $\square$

Remark 1.20. By arguing as in the proof of Theorem 1.19 but using Theorem 1.2 (instead of Theorem 1.1) with $a = 0$, $b = \infty$, $v(x) \equiv 1$, $u(x) = w(x)x^{-q/\alpha}$ and p, q replaced by p/α, q/α, respectively, together with (1.96) and the known estimates of $\|H_\alpha\|_{L^{p/\alpha} \to L^{q/\alpha}(w)}$, we obtain the main result for the case $0 < q < p < \infty$. For details, see L.E. Persson and V. Stepanov [PeSt1].

Theorem 1.21. *Let $0 < q < p < \infty$, $1/r = 1/q - 1/p$. Then the inequality (1.85), i.e., the inequality*

$$\|Gf\|_{q,u} \leq C\|f\|_{p,v},$$

holds for $f \geq 0$ if and only if

$$B_0 := \left(\int_0^\infty \left(\frac{1}{x} \int_0^x w(t)dt \right)^{r/p} w(x)dx \right)^{1/r} < \infty \qquad (1.99)$$

with w given by (1.86).
    Moreover,

$$\|G\|_{L^p(v) \to L^q(u)} \approx B_0$$

with factors of equivalence depending only on p and q.

Remark 1.22. The results obtained in this section can be extended to some more general geometric mean operators. Just as an example, let us investigate operators G_s, $s > 0$:

$$(G_sf)(x) = \exp \left(\frac{s}{x^s} \int_0^x t^{s-1}\ln f(t)dt \right) \qquad (1.100)$$

(see e.g., J.A. Cochran and C.-S. Lee [CL1]). Then it can be shown by some obvious substitutions and by Theorem 1.19 that the inequality

$$\|G_s f\|_{q,u} \leq C \|f\|_{p,v}$$

holds for $f \geq 0$ in the case $0 < p \leq q < \infty$ if and only if

$$D_s := \sup_{t>0} t^{-1/p} \left(\int_0^t w_s(x) dx \right)^{1/q} < \infty,$$

where

$$w_s(x) = \left[G \left(\frac{1}{v_s(x)} \right) \right]^{q/p} u_s(x)$$

with $u_s(x) = \frac{1}{s} x^{1/s-1} u(x^{1/s})$, $v_s(x) = \frac{1}{s} x^{1/s-1} v(x^{1/s})$. Moreover,

$$D_s \leq \|G_s\|_{L^p(v) \to L^q(u)} \leq e^{1/p} D_s.$$

Similarly, also Theorem 1.21 can be generalized. For details, see L.E. Persson and V. Stepanov [PeSt1].

Example 1.23. (i) Applying the result just mentioned to $u(t) = t^\alpha$, $v(t) = t^\beta$, $\alpha, \beta \in \mathbb{R}$, we obtain that the inequality

$$\left(\int_0^\infty x^\alpha \left(\exp sx^{-s} \int_0^x t^{s-1} \ln f(t) dt \right)^q dx \right)^{1/q}$$

$$\leq C \left(\int_0^\infty f^p(x) x^\beta dx \right)^{1/p}$$

holds for $f \geq 0$ with $0 < p \leq q < \infty$ and $s > 0$ if and only if

$$\left(\frac{1+\alpha}{s} \right) \frac{1}{q} - \left(\frac{1+\beta}{s} \right) \frac{1}{p} = 0.$$

And that, moreover,

$$C \leq s^{1/q-1/p} e^{(1+\beta)/(\alpha p)}.$$

(ii) In particular, for $p = q = 1$, $\alpha = \beta$ we obtain the well-known inequality

$$\int_0^\infty x^\alpha \exp\left(sx^{-s}\int_0^x t^{s-1}\ln f(t)dt\right) dx \le e^{(\alpha+1)/s}\int_0^\infty x^\alpha f(x)dx$$

due to J.A. Cochran and C.-S. Lee [CL1] (cf. also H.P. Heinig [He1] and E.R. Love [Lov1]).

Remark 1.24. Some new information concerning such limit Pólya-Knopp type inequalities can be found in the Ph.D. theses by A. Wedestig [We1] and E. Ushakova [U1]. For the multidimensional case see also Sec. 7.8.

1.9. Miscellanea

The discrete case

Inequality (0.14), i.e.,

$$\sum_{n=1}^\infty \left(\frac{1}{n}\sum_{k=1}^n a_k\right)^p \le \left(\frac{p}{p-1}\right)^p \sum_{n=1}^\infty a_n^p, \quad 1 < p < \infty, \qquad (1.101)$$

is the discrete analogue of the classical Hardy inequality (0.15). The general weighted extension of (1.101) (i.e., the discrete analogue of inequality (0.18) with *sequence* weights $\{u_n\}$, $\{v_n\}$) has the form

$$\left[\sum_{n=1}^\infty u_n \left(\sum_{k=1}^n a_k\right)^q\right]^{1/q} \le C \left[\sum_{n=1}^\infty v_n a_n^p\right]^{1/p} \qquad (1.102)$$

for the whole index range $0 < q < \infty$, $1 < p < \infty$. This inequality can be derived from the general "continuous" (i.e., integral) case using step functions or using the "measure" characterization (i.e., the case where $u(x)dx$ and $v(x)dx$ are replaced by rather general measures). E.g., for $1 < p \le q < \infty$, inequality (1.102) holds for every non-negative sequence $\{a_n\}$ if and only if

$$\sup_{k>0} \left(\sum_{n=k}^\infty u_n\right)^{1/q} \left(\sum_{n=0}^k v_n^{1-p'}\right)^{1/p'} < \infty,$$

which is a discrete analogue of the condition $A < \infty$ from (0.19) with A given by (0.20).

But not all results can be derived via the continuous case. For some characterizations as well as for direct proofs see G. Bennett [Be1], [Be2]. See also L. Leindler [Le1].

Amalgams

In the case of discrete Hardy inequalities, the spaces under consideration are weighted *sequence spaces* ℓ^p. For amalgams, we need a combination of L^p- and ℓ^p-spaces. More precisely, a function f on $(0, \infty)$ is said to belong to the *weighted amalgam* of $L^p(\ell^{\tilde{p}}; w)$ if

$$\|f\|_{p,\tilde{p},w} = \left\{ \sum_{n=0}^{\infty} \left[\int_n^{n+1} |f(x)|^p w(x)dx \right]^{\tilde{p}/p} \right\}^{1/\tilde{p}} < \infty$$

with $1 < p, \tilde{p} < \infty$ and w a weight function. Characterizations of the weights u, v for which the Hardy operator (and some similar operators) are bounded from $L^p(\ell^{\tilde{p}}; v)$ to $L^q(\ell^{\tilde{q}}; u)$, $1 < p \le q < \infty$, $1 < \tilde{p} \le \tilde{q} < \infty$, were proved by C. Carton-Lebrun *et al.* [CHH1]. They are expressed as combinations of the conditions for the "continuous" case (on the intervals $[n, n+1]$, $n = 0, 1, 2, \ldots$) and of the conditions in the discrete case.

For the remaining indices $p, \tilde{p}, q, \tilde{q}$, the problem seems to be still open.

Corresponding results for the differential form of the Hardy inequality can be found in H.P. Heinig and A. Kufner [HK1].

A special two-dimensional Hardy operator

Let us consider the operator H_2 defined by

$$(H_2 f)(x, y) = \int_0^x \int_0^y f(s, t)dtds$$

and the corresponding inequality

$$\left(\int_0^\infty \int_0^\infty (H_2 f)^q (x, y) u(x, y) dx dy \right)^{1/q}$$

$$\leq C \left(\int_0^\infty \int_0^\infty f^p (x, y) v(x, y) dx dy \right)^{1/p} \qquad (1.103)$$

for $f \geq 0$ with u, v weight functions on $(0, \infty) \times (0, \infty)$. For the case $1 < p \leq q < \infty$, it was shown by B. Muckenhoupt [Mu3] that the condition

$$\sup_{x>0, y>0} \left(\int_x^\infty \int_y^\infty u(s, t) dt ds \right)^{1/q} \left(\int_0^x \int_0^y v^{1-p'}(s, t) dt ds \right)^{1/p'} < \infty$$
$$(1.104)$$

which corresponds to the condition $A < \infty$ from (0.19) with A given by (0.20), is only necessary for (1.103) to hold. Then E. Sawyer [Sa2] has found two additional conditions which — together with (1.104) — are necessary and sufficient.

The corresponding full characterization of the weights for the case $q < p$ or for N-dimensional analogues of H_2 with $N > 2$ seems not to be fully investigated.

For the special case $u(x, y) = u_1(x) u_2(y)$ and $v(x, y) = v_1(x) v_2(y)$, the full answer can be given by a successive use of two one-dimensional Hardy inequalities. See J. Appell and A. Kufner [AK1].

Remark 1.25. Some new information concerning this case can be found in the Ph.D. theses by A. Wedestig [We1] and E. Ushakova [U1] (see also the references therein). See also Sec 7.7.

An N-dimensional Hardy operator

The classical Hardy inequality

$$\int_0^\infty \left(\frac{1}{x} \int_0^x f(t) dt \right)^p dx \leq \left(\frac{p}{p-1} \right)^p \int_0^\infty f^p(x) dx, \ 1 < p < \infty,$$

for functions $f \geq 0$ defined on $(0, \infty)$ is equivalent to the inequality

$$\int_{-\infty}^{\infty} \left(\frac{1}{2|x|} \int_{-|x|}^{|x|} f(t)dt \right)^p dx \leq \left(\frac{p}{p-1} \right)^p \int_{-\infty}^{\infty} f^p(x)dx$$

for functions $f \geq 0$ defined on $(-\infty, \infty)$.

Now, for $x \in \mathbb{R}^N$, let us denote by $B(x)$ the ball $\{y \in \mathbb{R}^N : |y| \leq |x|\}$ and by $|B(x)|$ its volume. In M. Christ and L. Grafakos [CG1], it is shown that the *N-dimensional Hardy operator* $\mathscr{H}_N$ defined by

$$(\mathscr{H}_N f)(x) = \frac{1}{|B(x)|} \int_{B(x)} f(y)dy, \; x \in \mathbb{R}^N,$$

satisfies

$$\int_{\mathbb{R}^N} |\mathscr{H}_N f(x)|^p dx \leq \left(\frac{p}{p-1} \right)^p \int_{\mathbb{R}^N} |f(x)|^p dx, \; 1 < p < \infty,$$

the constant $(\frac{p}{p-1})^p$ being again the best possible. In P. Drábek *et al.* [DHK1], this *Hardy inequality* was extended to general *N*-dimensional weights u, v and to the whole range of the parameters $p, q, 1 < p < \infty, 0 < q < \infty$. The (necessary and sufficient) conditions for the validity of the inequality

$$\left(\int_{\mathbb{R}^N} |\mathscr{H}_N f(x)|^q u(x)dx \right)^{1/q} \leq C \left(\int_{\mathbb{R}^N} |f(x)|^p v(x)dx \right)^{1/p} \quad (1.105)$$

are exactly the analogues of the corresponding conditions for dimension one; hence, for $1 < p \leq q < \infty$ this condition reads

$$\sup_{x \in \mathbb{R}^N} \left(\int_{\mathbb{R}^N \setminus B(x)} u(y)dy \right)^{1/q} \left(\int_{B(x)} v^{1-p'}(y)dy \right)^{1/p'} < \infty.$$

Remark 1.26. Some new information concerning the best constant in the powerweighted multidimensional case with $1 < p < q < \infty$ can be found in the paper [PSa1] by L.E. Persson and S. Samko (cf. also Remark 0.6).

The "differential" form of (1.105)

The *differential* form of the N-dimensional Hardy inequality (1.105), corresponding to the one-dimensional inequality (1.25), can be written as

$$\left(\int_\Omega |g(x)|^q u(x) dx \right)^{1/q} \leq C \left(\int_\Omega |\nabla u(x)|^p v(x) dx \right)^{1/p}. \qquad (1.106)$$

Here Ω is a (usually bounded) domain in $\mathbb{R}^N$ with reasonable (e.g. locally Lipschitz) boundary $\partial\Omega$ or the whole $\mathbb{R}^N$ and g is considered to be from $C_0^\infty(\Omega)$ or from an appropriate subspace of the weighted Sobolev space $W^{1,p}(\Omega; v)$ where the right hand side of (1.106) is a norm.

We will not deal here in detail with general inequalities of the form (1.106) which are often called *N-dimensional Hardy inequalities* or *weighted Friedrichs* (Poincaré, Sobolev) *inequalities*. These inequalities are dealt with extensively in [OK]; see also A. Kufner [Ku1] or books devoted to Sobolev spaces (R.A. Adams [A1], V. Maz'ja [Maz1]). Here, let us only mention that inequalities of the form (1.106), mostly with power type weights but under rather weak conditions on the domain, have been investigated — among others — by A. Wannebo [W1], [W2].

Not only Lebesgue spaces

Characterizations of weights (or measures) for which the Hardy operator is bounded between weighted Orlicz spaces are *in general* not available. Partial results in this direction were given by a number of authors. We mention here only G. Palmieri [P1] and also N. Levinson [Lev1] who has shown in particular that if the non-negative function Φ defined on $(0, \infty)$ satisfies

$$\Phi(t)\Phi''(t) \geq \left(1 - \frac{1}{p} \right) (\Phi'(t))^2, \ p > 1, \qquad (1.107)$$

then

$$\int_0^\infty \Phi \left(\frac{1}{x} \int_0^x f(t) dt \right) dx \leq \left(\frac{p}{p-1} \right)^p \int_0^\infty \Phi(f(x)) dx \qquad (1.108)$$

for $f \geq 0$ unless $\Phi(f) \equiv 0$. The *proof* is easy: If we denote $\Psi(t) = (\Phi(t))^{1/p}$, then by (1.107) $\Psi''(t) \geq 0$ and hence $\Psi(t)$ is convex. Thus by Jensen's inequality

$$\Psi\left(\frac{1}{x}\int_0^x f(t)dt\right) \leq \frac{1}{x}\int_0^x \Psi(f(t))dt$$

and by Hardy's inequality (0.15) applied to $\Psi(f(x))$

$$\int_0^\infty \left(\frac{1}{x}\int_0^x \Psi(f(t))dt\right)^p dx \leq \left(\frac{p}{p-1}\right)^p \int_0^\infty (\Psi(f(x)))^p dx.$$

Now, (1.108) follows from the last two inequalities since $\Phi = \Psi^p$.

This result was modified by M.J. Carro and H.P. Heinig [CH1] as follows: If P, Q are non-negative increasing functions on $[0, \infty)$ satisfying $P(0) = Q(0) = 0$, then the inequality

$$\int_0^\infty P\left(\frac{1}{x}\int_0^x f(t)dt\right) dx \leq C_1 \int_0^\infty Q(C_2 f(x))dx$$

for $f \geq 0$ with positive constants C_1, C_2 holds if and only if there exist positive constants C_3, C_4 such that for all $t > 0$

$$P(t) + t\int_0^t \frac{P(y)}{y^2}dy \leq C_3 Q(C_4 t).$$

Example 1.27. A particular choice of the function Φ or P, Q in the foregoing subsection can lead to interesting modifications of the Hardy inequality. Let us consider three special cases.

(i) The function $\Phi(t) = t^p$, $p > 1$, satisfies (1.107). In this case, (1.108) is the classical Hardy inequality (0.15).

(ii) The function $\Phi(t) = t^{-q}$ with $q > 0$ satisfies (1.107) for every $p > 1$. Hence (1.108) has the form

$$\int_0^\infty \left(\frac{1}{x}\int_0^x f(t)dt\right)^{-q} dx \leq \left(\frac{p}{p-1}\right)^p \int_0^\infty f^{-q}(x)dx,$$

i.e., we have the classical Hardy inequality also for *negative* powers.

(iii) The function $\Phi(t) = e^{t^a}$ with $a \geq 1$ or $a < 0$ again satisfies (1.107), and consequently, for such a we have

$$\int_0^\infty \exp\left[\frac{1}{x}\int_0^x f(t)dt\right]^a dx \leq \left(\frac{p}{p-1}\right)^p \int_0^\infty \exp f^a(x)dx.$$

Compare with the Knopp inequality (1.80).

Remark 1.28. Some new information concerning Hardy-type inequalities in Morrey-type and Hölder-type spaces are proved and discussed in Sec. 7.6. Moreover, we refer to the books [KMP] and [KMRS] concerning complementary information in this connection. In particular, some of this information concerning Orlicz, Lorentz, rearrangement invariant, general Banach and variable exponent $L^{p(\cdot)}$ spaces are also summarized and discussed in Sec. 7.6.

A Hardy-Knopp inequality

Let us state an inequality related to inequality (1.108):

Proposition 1.29. *Let Φ be a positive, convex and strictly increasing function on $[0, \infty)$. Then for every function $f \geq 0$ we have*

$$\int_0^\infty \Phi\left(\frac{1}{x}\int_0^x f(t)dt\right)\frac{dx}{x} \leq \int_0^\infty \Phi(f(x))\frac{dx}{x}. \tag{1.109}$$

Proof. By Jensen's inequality and Fubini's theorem,

$$\int_0^\infty \Phi\left(\frac{1}{x}\int_0^x \Phi^{-1}(f(t))dt\right)\frac{dx}{x} \leq \int_0^\infty \frac{1}{x^2}\left(\int_0^x f(t)dt\right)dx$$

$$= \int_0^\infty f(t)\left(\int_t^\infty \frac{1}{x^2}dx\right)dt$$

$$= \int_0^\infty f(t)\frac{dt}{t}.$$

Now, (1.109) follows immediately by replacing $f(t)$ by $\Phi(f(t))$. $\square$

Example 1.30. (i) If we choose $\Phi(x) = e^x$ in (1.109) and then replace $f(x)$ by $\ln f(x)$, we obtain

$$\int_0^\infty \exp\left(\frac{1}{x}\int_0^x \ln f(t)dt\right)\frac{dx}{x} \le \int_0^\infty f(x)\frac{dx}{x}.$$

Finally, replacing $f(x)$ by $xf(x)$, we have Knopp's inequality (1.80).

(ii) The choice $\Phi(x) = x^p$ with $p \ge 1$ in (1.109) yields

$$\int_0^\infty \left(\frac{1}{x}\int_0^x f(t)dt\right)^p \frac{dx}{x} \le \int_0^\infty f^p(x)\frac{dx}{x}, \qquad (1.110)$$

which is Hardy's inequality (0.16) for $\varepsilon = -1$. For the case $p > 1$, inequality (1.110) can be rewritten into the "usual" form (0.15), i.e.

$$\int_0^\infty \left(\frac{1}{x}\int_0^x g(t)dt\right)^p dx \le \left(\frac{p}{p-1}\right)^p \int_0^\infty g^p(x)dx, \qquad (1.111)$$

by some straightforward calculations; here, $g(x) = f(x^{(p-1)/p})x^{-1/p}$.

Note that Hardy's inequality (1.110) holds for $p \ge 1$, i.e., also for $p = 1$, while in (1.111) we need to have $p > 1$.

Remark 1.31. This convexity approach to prove Hardy's inequality, its limit Pólya-Knopp inequality and their power weighted versions was obviously not discovered by Hardy himself. In particular, the basic form (1.110) of Hardy's inequality shows directly via substitutions that Hardy's inequality (both in non-weighted and weighted forms) holds also for $p < 0$. More information and results connected to this convexity approach can be found in the paper [PeS3] by L.E. Persson and N. Samko, see also [BPS1], [PeO1] and Sec. 7.2.

Indefinite weights

Up to now, we have considered weight functions u, v which have been *positive* a.e. However, V.G. Maz'ja and I.E. Verbitsky [MV1] have derived necessary and sufficient conditions for the validity of the Hardy inequality for the case $p = q = 2$ with weights which *change sign.* For the particular case $v(x) \equiv 1$ their result reads as follows:

The Hardy inequality

$$\left| \int_0^\infty |g(x)|^2 u(x) dx \right| \leq C \int_0^\infty |g'(x)|^2 dx \qquad (1.112)$$

holds for every $g \in C_0^\infty(0, \infty)$ *if and only if*

$$\sup_{r>0} r \int_r^\infty |\gamma(x)|^2 dx < \infty, \qquad (1.113)$$

where $\gamma(x) = \int_x^\infty u(t) dt$.

Moreover, the corresponding imbedding is *compact* if and only if

$$r \int_r^\infty |\gamma(x)|^2 dx = O(1) \quad \text{for} \quad r \to 0 \quad \text{and} \quad r \to \infty.$$

There is also an extension with a (non-negative) weight v on the right hand side of (1.112). Then the necessary and sufficient condition reads

$$\sup_{r>0} \left(\int_0^r \frac{dx}{v(x)} \right) \left(\int_r^\infty |\gamma(x)|^2 \frac{dx}{v(x)} \right) < \infty.$$

Remark 1.32. All Hardy-type inequalities we presented so far are for the cases $1 < p \leq q < \infty$ and $0 < q < p < \infty, q \neq 1, p > 1$. However, such results can also be proved for some other parameters. In Sec. 7.4 some new results for such cases are presented and discussed.

1.10. Comments and Remarks

1.10.1. The *historical remarks* in the Introduction are of course subjective and do not claim to cover the immense amount of literature concerned with Hardy's inequality, e.g., let us mention the paper by E. Sawyer [Sa1], the book [KMP] and cf. also Sec. 7.1.

Instead of considering weighted Lebesgue spaces $L^r(w)$, it is also possible to deal with spaces $L^r(\mu)$ with μ a measure where $w(x)dx$ is replaced by $d\mu$. A characterization of measures μ and ν for which H is bounded from $L^p(\nu)$ into $L^q(\mu)$ (i.e., with $d\nu$ a $d\mu$ instead of $v(x)dx$ and $u(x)dx$, respectively) was given by B. Muckenhoupt [Mu1] in the case $p = q > 1$ and by G. Sinnamon [Si1], [Si3] in the remaining cases of the indices. See also V.G. Maz'ja [Maz1].

1.10.2. The alternative conditions of the validity of Hardy's inequality in Theorems 1.1 and 1.2 are taken from L.E. Persson and V. Stepanov [PeSt1]. The proofs given here are new in the sense, that in the paper mentioned, the proofs are given for the interval $(0, \infty)$, while here, we have a general interval (a, b). However, at least the criterion from Theorem 1.2 appeared earlier: G. Tomaselli [To1] derived the condition $B < \infty$ with B from (1.14) for $p = q$, the result for $1 < p \le q < \infty$ is due to V. Stepanov [St5], see also G. Sinnamon and V. Stepanov [SS1].

There are also other criteria for the validity of Hardy's inequality, e.g., the finiteness of the number

$$K = \frac{p'}{q} \inf_f \sup_{a<x<b} \frac{1}{f(x)} \int_a^x u(t) \left[f(t) + \int_a^t v^{1-p'}(s)ds \right]^{q/p'+1} dt,$$

where the infimum is taken over all positive measurable functions f. This result is again due to G. Tomaselli [To1] for $p = q$ and to P. Gurka [Gu1] for $1 < p \le q < \infty$ (see [OK, Lemma 4.3]). See also Sec. 7.3.

1.10.3. The compactness of the Hardy operators H and $\widetilde{\widetilde{H}}$ as well as the compactness of the corresponding imbeddings (1.29) and (1.30) are dealt with in [OK, Sec. 1.7]. Here, the results mentioned in Sec. 1.6 have been taken mainly from M. García-Huidobro *et al.* [GKMY1]. See also A. Kufner [Ku7].

1.10.4. The *differential form* analogue of the discrete Hardy inequality (1.102) is, of course,

$$\left(\sum_{n=1}^{\infty} u_n |a_n|^q \right)^{1/q} \le C \left(\sum_{n=1}^{\infty} v_n |a_n - a_{n+1}|^p \right)^{1/p}.$$

1.10.5. The two-dimensional Hardy operator H_2 on p. 57 is obviously the composition of two one-dimensional operators H — one with respect to the variable y and the second with respect to the variable x:

$$(H_2 f)(x, y) = H_x(H_y f)(x, y).$$

Clearly, analogous results can be derived also by combining H with the conjugate operator $\widetilde{H}$, i.e., for

$$(\widetilde{H}_2 f)(x, y) = \int_0^x \int_y^\infty f(s, t) dt ds,$$

$$(H^2 f)(x, y) = \int_x^\infty \int_y^\infty f(x, t) dt ds$$

etc.

1.10.6. The N-dimensional Hardy operator $\mathscr{H}_N$ from p. 59 was modified by H.P. Heinig and G. Sinnamon [HS1] replacing the ball $B(x)$ by a general domain in $\mathbb{R}^N$ "centered" at x and generated by a convex symmetric neighbourhood of the point 0. G. Sinnamon [Si5] extended these results to more general *star-shaped* domains.

Moreover, also results for the *conjugate* $\widetilde{\mathscr{H}_N}$ of $\mathscr{H}_N$, defined by

$$(\widetilde{\mathscr{H}_N} f)(x) = \frac{1}{|B(x)|} \int_{\mathbb{R}^N \setminus B(x)} f(y) dy,$$

can be derived. Some more information concerning multidimensional Hardy-type inequalities can be found in Sec. 7.7.

1.10.7. A complete characterization of the Orlicz space for which the *differential* form of Hardy's inequality provides an imbedding

$$W_L^{1,p}(v) \hookrightarrow L^\Phi$$

is given by V.G. Maz'ja [Maz1]. The function Φ is characterized in terms of capacities.

1.10.8. There are several results for the case where Orlicz norm inequalities are replaced by *modular* inequalities of the form

$$Q^{-1} \left[\int_0^\infty Q[u_1(x)(Hf)(x)] u_0(x) dx \right]$$

$$\leq P^{-1} \left[\int_0^\infty P[v_1(x) f(x)] v_0(x) dx \right].$$

Q. Lai [L1] dealt with the case that P, Q are Young's functions such that $Q^{-1} \circ P$ is convex and $u_1 = v_1 \equiv 1$, the general case is due to H.P. Heinig and L. Maligranda [HM1]. The weight characterization when P is a Young function, Q "weakly convex" and $Q^{-1} \circ P$ not necessarily convex (which corresponds to the Lebesgue space case with indices $q < p$) was proved by Q. Lai [L3].

1.10.9. For a multidimensional version of some special cases of the limiting inequality in Sec. 1.8 see also C. Sbordone and I. Wik [SW1], P. Drábek *et al.* [DHK1] and B. Gupta *et al.* [GJPW1].

1.10.10. The factor $e^{1/p}$ in Theorem 1.18, formula (1.98), is the best possible when $p = q$ and is attained when $u(x) = v(x) \equiv 1$. For $p = q = 1$ an alternative criterion as that in Theorem 1.33 was given by H.P. Heinig [He1, Theorem 1.4]. (The condition (1.97) is different!) For the case $p < q$ the factor $e^{1/p}$ can be improved. Further details can be found in L.E. Persson and V. Stepanov [PeSt1].

1.10.11. The ideas and results in Sec. 1.8 are taken from L.E. Persson and V. Stepanov [PeSt1]. Some previous results of this type have been derived by B. Opic and P. Gurka [OG1] for the case $0 < p \le q$, by L. Pick and B. Opic [PO1] for the case $q < p$ and by H.P. Heinig *et al.* [HKK1] for the whole scale $p, q > 0$. However, in these papers different approaches were used and the papers do not contain estimates of the operator norms given here.

The results derived in L.E. Persson and V. Stepanov [PeSt1] hold in fact for much more general integral operators than the Hardy operator; see M. Nasyrova *et al.* [NPS1] and also the Ph.D. thesis by M. Nasyrova [N2].

Further facts and some supplementary results can be found in P. Jain *et al.* [JPW1], [JPW2]. See also the Ph.D. thesis of A. Wedestig [We1].

1.10.12. The case of *decreasing* functions is dealt with in detail in Chap. 6. Here, let us only mention that inequality (1.1) implies, that for $p > 1$, $\varepsilon < p - 1$ and f decreasing, $f \ge 0$, we have

$$\int_0^\infty \left(\frac{1}{x} \int_0^x f(t)dt \right)^p x^\varepsilon dx \approx \int_0^\infty f^p(x)x^\varepsilon dx. \qquad (1.114)$$

In fact, this equivalence holds for *every* $p > 0$, $\varepsilon < p - 1$, $\varepsilon \neq -1$, and the best constants can be found in J. Bergh *et al.* [BBP1], [BBP2].

1.10.13. According to (1.114), the norms of f and of $H_a f$ in $L^p(x^\varepsilon)$ are equivalent, with $(H_a f)(x) = \frac{1}{x}(Hf)(x)$. A similar property is possessed also by the operator

$$I - H_a.$$

Namely,

$$\int_0^\infty \left(f(x) - \frac{1}{x} \int_0^x f(t)dt \right)^p x^\varepsilon dx \approx \int_0^\infty f^p(x)x^\varepsilon dx \qquad (1.115)$$

for $f \geq 0$ with $p > 1$, $\varepsilon < p - 1$, $\varepsilon \neq 1$. This equivalence was proved by N. Krugljak *et al.* [KrMP2] and for decreasing functions earlier by C. Bennett *et al.* [BDS1].

In fact, for $p = 2$ there is equality in (1.115) even for more general weights than x^ε, see N. Kaiblinger *et al.* [KaMP1]; see also Chap. 5.

1.10.14. The proof of the inequality (1.109) is taken from S. Kaijser *et al.* [KPO1]. Note that the same proof gives that also the inequality

$$\int_0^b \phi \left(\frac{1}{x} \int_0^x f(t)dt \right) \frac{dx}{x} \leq \int_0^b \phi(f(x)) \left(1 - \frac{x}{b} \right) \frac{dx}{x}$$

holds for all b such that $0 < b \leq \infty$ provided $f(t) \geq 0$ and ϕ is positive, convex and strictly increasing. In particular, by using this inequality for $b < \infty$ and with the special cases from Example 1.27 we obtain the following improvements of the Knopp and Hardy inequalities:

$$\int_0^b \exp \left(\frac{1}{x} \int_0^x f(t)dt \right) dx \leq e \int_0^b f(x) \left(1 - \frac{x}{b} \right) dx, \qquad (1.116)$$

and for $p > 1$,

$$\int_0^{b_0} \left(\frac{1}{x} \int_0^x g(t)dt \right)^p dx \leq \left(\frac{p}{p-1} \right)^p \int_0^{b_0} g^p(x) \left(1 - \left(\frac{x}{b_0} \right)^{(p-1)/p} \right) dx,$$

$$(1.117)$$

where $b_0 = b^{p/(p-1)}$ and $g(x) = f(x^{(p-1)/p})x^{-1/p}$ as before. The inequalities (1.116) and (1.117) have also been proved by A. Čižmešija and J.E. Pečarić [CP3] (cf. also [CP1]) by using a mixed-mean inequalities technique. For more information, results and references concerning the technique mentioned see the review article [CP2] by the same authors.

2

Some Weighted Norm Inequalities

2.1. Preliminaries

A general weighted norm inequality

As already mentioned in Chap. 1, now we investigate inequalities of the form

$$\|Tf\|_{q,u} \leq C\|f\|_{p,v} \tag{2.1}$$

or, more precisely,

$$\left(\int_a^b |(Tf)(x)|^q u(x)dx\right)^{1/q} \leq C\left(\int_a^b |f(x)|^p v(x)dx\right)^{1/p}, \tag{2.2}$$

where T is an operator of the form

$$(Tf)(x) = \int_a^b k(x,t)f(t)dt \tag{2.3}$$

with $k(x,t)$ a given kernel, u, v weight functions and $-\infty \leq a < b \leq \infty$.

We are interested in conditions on u, v and on the parameters p, q under which inequality (2.2) is satisfied for certain classes of functions f.

Example 2.1. (i) The first examples of the operator T are provided by the Hardy operator H and its conjugate $\widetilde{H}$, that is

$$(Hf)(x) = \int_a^x f(t)dt \quad \text{and} \quad (\widetilde{H}f)(x) = \int_x^b f(t)dt.$$

These operators are of the form (2.3) with the kernel $k(x,t) = \chi_{(a,x)}(t)$, $x < b$, or $k(x,t) = \chi_{(x,b)}(t)$, $x > a$, respectively. Necessary and sufficient conditions for the validity of the corresponding inequality (2.1) have been given in Chap. 1.

(ii) A simple extension provides the *modified* Hardy operator $\mathscr{H}$ defined as

$$(\mathscr{H}f)(x) := \varphi(x) \int_a^x \psi(t)f(t)dt \qquad (2.4)$$

with given (non-negative) functions φ, ψ, and similarly $\widehat{\mathscr{H}}$:

$$(\widehat{\mathscr{H}}f)(x) := \widehat{\varphi}(x) \int_x^b \widehat{\psi}(t)f(t)dt. \qquad (2.5)$$

The operator $\mathscr{H}$ is again of the form (2.3) with the kernel

$$k(x,t) = \chi_{(a,x)}(t)\varphi(x)\psi(t), \quad x < b, \qquad (2.6)$$

and an easy calculation shows that the inequality

$$\|\mathscr{H}f\|_{q,u} \le C\|f\|_{p,v}$$

holds for all $f \ge 0$ if and only if

$$A = A(a, b; u\varphi^q, v\psi^{-p}) < \infty$$

with A given by (0.8) for $p \le q$ and by (0.9) for $p > q$ (of course, with u replaced by $u\varphi^q$ and v by $v\psi^{-p}$). An analogous result holds for $\widehat{\mathscr{H}}$. (Notice that operators of the form $\mathscr{H}$ and $\widehat{\mathscr{H}}$ have already been mentioned in Chap. 1 — cf. (1.37) — where they acted between classical (i.e. non-weighted) Lebesgue spaces.)

Remark 2.2. Recall that the expression $v\psi^{-p}$ appears in A in the form $(v(x)\psi^{-p}(x))^{1-p'} = v^{1-p'}(x)\psi^{p'}(x)$ (behind an integral sign, see formulas (0.8), (0.9)) so that ψ can even vanish somewhere in (a, b).

2.2. A Special Operator

The case $T = \mathscr{H} + \widehat{\mathscr{H}}$

Now, we will deal with a particular operator T defined as

$$(Tf)(x) := \varphi_1(x) \int_a^x \psi_1(t)f(t)dt + \varphi_2(x) \int_x^b \psi_2(t)f(t)dt \qquad (2.7)$$

with φ_i, ψ_i, $i = 1, 2$, given functions defined on (a, b) (not necessarily non-negative). Thus, the kernel k in (2.3) has the special form

$$k(x, t) = \chi_{(a,x)}(t)\varphi_1(x)\psi_1(t) + \chi_{(x,b)}(t)\varphi_2(x)\psi_2(t).$$

The main result of this section is

Theorem 2.3. *Let T be defined by (2.7). Then inequality (2.1) holds if and only if*

$$A(a, b; u|\varphi_1|^q, v|\psi_1|^{-p}) < \infty \qquad (2.8)$$

and

$$\widetilde{A}(a, b; u|\varphi_2|^q, v|\psi_2|^{-p}) < \infty \qquad (2.9)$$

with A and $\widetilde{A}$ given by (0.20) and (0.24), respectively, for $1 < p \leq q < \infty$ and by (0.21) and (0.25), respectively, for $0 < q < p < \infty$, $1 < p < \infty$.

Proof. Denote

$$(S_1f)(x) := \varphi_1(x) \int_a^x \psi_1(t)f(t)dt;$$

$$(S_2f)(x) := \varphi_2(x) \int_x^b \psi_2(t)f(t)dt. \qquad (2.10)$$

Then $T = S_1 + S_2$ and consequently,

$$\|Tf\|_{q,u} \leq \|S_1f\|_{q,u} + \|S_2f\|_{q,u}. \qquad (2.11)$$

(i) *Sufficiency.* Due to Example 2.1 (ii), conditions (2.8) and (2.9) imply the validity of the inequalities

$$\|S_1 f\|_{q,u} \le C\|f\|_{p,v} \quad \text{and} \quad \|S_2 f\|_{q,u} \le C\|f\|_{p,v}. \tag{2.12}$$

Thus, inequality (2.1) follows from (2.11).

(ii) *Necessity* for non-negative functions φ_i, ψ_i. Suppose that inequality (2.1) holds and that $f \ge 0$. Then obviously

$$\|S_i f\|_{q,u} \le \|T f\|_{q,u}, \quad i = 1, 2,$$

and consequently, both the inequalities in (2.12) hold. This implies that conditions (2.8) (for $i = 1$) and (2.9) (for $i = 2$) hold.

(iii) *Necessity* for general functions φ_i, ψ_i. Suppose again that inequality (2.1) holds and define, for $\varepsilon > 0$, a new weight function v_ε as

$$v_\varepsilon(x) = \max\{v(x), |\psi_1(x)|^p \varepsilon\}.$$

We have $v(x) \le v_\varepsilon(x)$ and consequently, $\|f\|_{p,v} \le \|f\|_{p,v_\varepsilon}$. Thus, it follows from (2.1) that

$$\|T f\|_{q,u} \le C\|f\|_{p,v_\varepsilon}. \tag{2.13}$$

Let α, β be such that $a < \alpha < \beta < b$ and define

$$\tilde{f}(x) = \begin{cases} 0 & \text{for } x \in (a, \alpha] \cup [\beta, b), \\ |\psi_1(x)|^{p'-1} v_\varepsilon^{1-p'}(x) \, \text{sgn } \psi_1(x) & \text{for } x \in (\alpha, \beta). \end{cases}$$

Then $\tilde{f} \in L^p(v_\varepsilon)$ since, due to the definition of v_ε and $\tilde{f}$,

$$\|\tilde{f}\|_{p,v_\varepsilon}^p = \int_a^b |\tilde{f}(x)|^p v_\varepsilon(x) dx = \int_\alpha^\beta |\psi_1(x)|^{p(p'-1)} v_\varepsilon^{p(1-p')}(x) v_\varepsilon(x) dx$$

$$= \int_\alpha^\beta |\psi_1(x)|^{p'} v_\varepsilon^{1-p'}(x) dx \le \int_\alpha^\beta |\psi_1(x)|^{p'} (|\psi_1(x)|^p \varepsilon)^{1-p'} dx$$

$$= \varepsilon^{1-p'} (\beta - \alpha) < \infty.$$

Further,

$$\|T\widetilde{f}\|_{q,u}^q = \int_a^b |T\widetilde{f}(x)|^q u(x)dx$$

$$\geq \int_\beta^b \left| \varphi_1(x) \int_a^x \psi_1(t)\widetilde{f}(t)dt + \varphi_2(x) \int_x^b \psi_2(t)\widetilde{f}(t)dt \right|^q u(x)dx$$

$$= \int_\beta^b |\varphi_1(x)|^q \left| \int_a^x \psi_1(t)\widetilde{f}(t)dt \right|^q u(x)dx$$

$$\geq \int_\beta^b |\varphi_1(x)|^q u(x) \left(\int_\alpha^\beta |\psi_1(t)|^{p'} v_\varepsilon^{1-p'}(t)dt \right)^q dx$$

$$= \left(\int_\beta^b |\varphi_1(x)|^q u(x)dx \right) \left(\int_\alpha^\beta |\psi_1(t)|^{p'} v_\varepsilon^{1-p'}(t)dt \right)^q$$

since for $x > \beta$ we have $\int_x^b \psi_2(t)\widetilde{f}(t)dt = 0$ and

$$\left| \int_a^x \psi_1(t)\widetilde{f}(t)dt \right| \geq \int_\alpha^\beta |\psi_1(t)|^{p'} v_\varepsilon^{1-p'}(t)dt.$$

Using these estimates in (2.13) (for $f = \widetilde{f}$), we obtain

$$\left(\int_\beta^b |\varphi_1(x)|^q u(x)dx \right)^{1/q} \left(\int_\alpha^\beta |\psi_1(t)|^{p'} v_\varepsilon^{1-p'}(t)dt \right)$$

$$\leq C \left(\int_\alpha^\beta |\psi_1(x)|^{p'} v_\varepsilon^{1-p'}(x)dx \right)^{1/p},$$

i.e.,

$$\left(\int_\beta^b |\varphi_1(x)|^q u(x)dx \right)^{1/q} \left(\int_\alpha^\beta |\psi_1(x)|^{p'} v_\varepsilon^{1-p'}(x)dx \right)^{1/p'} \leq C.$$

If we denote

$$B(\beta; a, b; U, V) := \left(\int_\beta^b U(x)dx \right)^{1/q} \left(\int_a^\beta V^{1-p'}(x)dx \right)^{1/p'}, \quad (2.14)$$

then the last estimate reads

$$B(\beta; \alpha, b; u|\varphi_1|^q, v_\varepsilon|\psi_1|^{-p}) \leq C,$$

where C is independent of α and ε. Now let $\alpha \to a$ and $\varepsilon \to 0$ (via a subsequence), then $v_\varepsilon \to v$ and we obtain

$$\sup_{a<\beta<b} B(\beta; a, b; u|\varphi_1|^q, v|\psi_1|^{-p}) \leq C < \infty. \qquad (2.15)$$

But this is condition (2.8) where, of course, A is given by formula (0.20) which in fact corresponds to the case $p \leq q$. Thus, condition (2.15), i.e. condition (2.8) with $p \leq q$, is necessary for (2.1) to hold.

Next, denote

$$\widetilde{B}(\beta; a, b; U, V) := \left(\int_a^\beta U(x) dx \right)^{1/q} \left(\int_\beta^b V^{1-p'}(x) dx \right)^{1/p'}. \qquad (2.16)$$

Then it can be shown completely analogously that with the choice

$$v_\varepsilon(x) = \max\{v(x), |\psi_2(x)|^p \varepsilon\},$$

$$\widetilde{f}(x) = \begin{cases} |\psi_2(x)|^{p'-1} v_\varepsilon^{1-p'}(x) \, \mathrm{sgn}\, \psi_2(x) & \text{for } x \in (\alpha, \beta), \\ 0 & \text{otherwise} \end{cases}$$

the following estimate holds:

$$\sup_{a<\beta<b} \widetilde{B}(\beta; a, b; u|\varphi_2|^q, v|\psi_2|^{-p}) \leq C < \infty. \qquad (2.17)$$

This is condition (2.9) where, of course, $\widetilde{A}$ is now given by formula (0.24).

Thus, we have shown that conditions (2.15) and (2.17) are necessary for (2.1) to hold, and Theorem 2.3 is proved for the case $p \leq q$.

It remains to prove the necessity of conditions (2.8) and (2.9) for the case $p > q$ (and for general functions φ_i, ψ_i). This will be done below, but before we need to introduce some notation and derive some auxiliary estimates, which we will do in the following remark.

Remark 2.4. (i) Let us again emphasize that conditions (2.15) and (2.17) are in fact conditions (2.8) and (2.9) with A and $\widetilde{A}$ corresponding to the case $p \leq q$. Nevertheless, conditions (2.15) and (2.17) are *necessary for every choice* of p, q, $1 < p < \infty$, $0 < q < \infty$.

(ii) We now suppose that

$$0 < q < p < \infty, \quad 1 < p < \infty.$$

Denote

$$A_1(\varphi, \psi) = \left(\int_a^b \left(\int_x^b u(t)|\varphi(t)|^q dt \right)^{r/q} \left(\int_a^x v^{1-p'}(t)|\psi(t)|^{p'} dt \right)^{r/q'} \right.$$

$$\left. \times v^{1-p'}(x)|\psi(x)|^{p'} dx \right)^{1/r},$$

$$A_2(\varphi, \psi) = \left(\int_a^b \left(\int_a^x u(t)|\varphi(t)|^q dt \right)^{r/q} \left(\int_x^b v^{1-p'}(t)|\psi(t)|^{p'} dt \right)^{r/q'} \right.$$

$$\left. \times v^{1-p'}(x)|\psi(x)|^{p'} dx \right)^{1/r} \tag{2.18}$$

with $\frac{1}{r} = \frac{1}{q} - \frac{1}{p}$.

Then we have

$$A_1(\varphi, \psi_1 + \psi_2) \lesssim (A_1(\varphi, \psi_1) + A_1(\varphi, \psi_2)),$$

$$A_2(\varphi, \psi_1 + \psi_2) \lesssim (A_2(\varphi, \psi_1) + A_2(\varphi, \psi_2)). \tag{2.19}$$

Indeed, since $A_1(\varphi, \psi)$ is nothing else than $A(a, b; u|\varphi|^q, v|\psi|^{-p})$ from formula (0.21), we can rewrite it due to Remark 1.5 (iv) as

$$\left(\frac{p'}{q} \right)^{1/r} A_1^*(\varphi, \psi)$$

where

$$A_1^*(\varphi, \psi) = \left(\int_a^b \left(\int_x^b u(t)|\varphi(t)|^q dt \right)^{r/p} \right.$$

$$\left. \times \left(\int_a^x v^{1-p'}(t)|\psi(t)|^{p'} dt \right)^{r/p'} u(x)|\varphi(x)|^q dx \right)^{1/r}.$$

Now the first estimate in (2.19) follows from the obvious inequalities

$$\left(\int_a^x v^{1-p'}(t)|\psi_1(t) + \psi_2(t)|^{p'} dt \right)^{r/p'}$$

$$\leq \left(2^{p'-1} \int_a^x v^{1-p'}(t)(|\psi_1(t)|^{p'} + |\psi_2(t)|^{p'}) dt \right)^{r/p'}$$

$$\leq \left[\left(\int_a^x v^{1-p'}(t)|\psi_1(t)|^{p'} dt \right)^{r/p'} + \left(\int_a^x v^{1-p'}(t)|\psi_2(t)|^{p'} dt \right)^{r/p'} \right],$$

and the second estimate in (2.19) can be derived analogously.

The necessity of conditions (2.8), (2.9) for $p > q$

Suppose that, for the operator T defined by (2.7), inequality (2.1) holds. Thus, T is a *continuous* operator,

$$T : L^p(v) \to L^q(u),$$

and $\|T\| \leq C$ with C the constant from (2.1). We want to show that conditions (2.8), (2.9) are satisfied, which means, due to formula (2.18), that

$$A_1(\varphi_1, \psi_1) < \infty, \quad A_2(\varphi_2, \psi_2) < \infty. \tag{2.20}$$

Notice that the first condition in (2.20) is necessary and sufficient for the continuity of the operator S_1 from (2.10) while the second condition in (2.20) is necessary and sufficient for the continuity of the operator S_2 from (2.10),

$$S_i : L^p(v) \to L^q(u), \quad i = 1, 2.$$

We will proceed by *contradiction*. If conditions (2.20) are *not* satisfied, then we can show that necessarily

$$A_1(\varphi_1, \psi_1) = \infty \quad \text{and} \quad A_2(\varphi_2, \psi_2) = \infty. \tag{2.21}$$

Indeed, if, e.g., $A_1(\varphi_1, \psi_1) = \infty$ but $A_2(\varphi_2, \psi_2) < \infty$, then the operator S_1 is *not* continuous while S_2 is continuous. Consequently, the operator T as the sum of S_1 and S_2 *cannot be continuous*, which contradicts our assumption of its continuity.

Hence, assume (2.21). We now construct a sequence of intervals $[c_i, d_i]$, $i = 1, 2, \ldots$, by the following procedure: Take $[c_1, d_1] = [a, b]$ and let d be its interior point, $c_1 < d < d_1$. Denote

$$\Delta_1 = (c_1, d), \quad \Delta_2 = (d, d_1), \quad \psi_{ij} = \psi_i \chi_{\Delta_j}, \quad i, j = 1, 2.$$

Since $\psi_i = \psi_{i1} + \psi_{i2}$, it follows from (2.19) and (2.21) that at least one of the numbers $A_j(\varphi_j, \psi_{j1})$ and $A_j(\varphi_j, \psi_{j2})$ must be infinite. If

$$A_1(\varphi_1, \psi_{11}) = \infty, \quad A_2(\varphi_2, \psi_{21}) < \infty$$

or

$$A_1(\varphi_1, \psi_{11}) < \infty, \quad A_2(\varphi_2, \psi_{21}) = \infty,$$

then we take for $[c_2, d_2]$ the interval $[c_1, d]$; if

$$A_1(\varphi_1, \psi_{12}) = \infty, \quad A_2(\varphi_2, \psi_{22}) < \infty$$

or

$$A_1(\varphi_1, \psi_{12}) < \infty, \quad A_2(\varphi_2, \psi_{22}) = \infty,$$

then we take for $[c_2, d_2]$ the interval $[d, d_1]$, and stop the procedure.

If none of these four possibilities occurs, then at least for one i (equal to 1 or 2), we have

$$A_1(\varphi_1, \psi_{1i}) = \infty, \quad A_2(\varphi_2, \psi_{2i}) = \infty; \qquad (2.22)$$

in this case we take for $[c_2, d_2]$ the interval $\overline{\Delta_i}$ and continue the procedure just described, choosing an interior point d, $c_2 < d < d_2$, and considering the intervals $\Delta_{11} = (c_2, d)$, $\Delta_{21} = (d, d_2)$. Now, Δ_i plays the role of the interval (a, b) and conditions (2.22) play the role of conditions (2.21).

So, we have constructed a (finite or infinite) sequence of intervals such that

$$[c_1, d_1] \supset [c_2, d_2] \supset [c_3, d_3] \supset \ldots.$$

Denote $\delta_k = [c_k, d_k]$ and $\psi_{ik} = \psi_i \chi_{\delta_k}$, $i = 1, 2$.

Suppose that the sequence $\{\delta_k\}$ is *finite* and let δ_n be its last term. The operator T_n,

$$(T_n f)(x) = \varphi_1(x) \int_a^x \psi_{1n}(t)f(t)dt + \varphi_2(x) \int_x^b \psi_{2n}(t)f(t)dt$$

is the restriction of T to functions with support in δ_n and, moreover, it is *not continuous*, since it is the sum of two operators S_{1n}, S_{2n} from which one is continuous and one not continuous due to the construction of the interval δ_n. Hence, T is not continuous, either, and we have the desired contradiction.

If the sequence $\{\delta_k\}$ is *infinite*, then there is a point c,

$$c \in \bigcap_{k=1}^{\infty} [c_k, d_k].$$

We now show that, for every k,

$$\psi_1(x) = 0 \quad \text{a.e. in } (a, c_k),$$
$$\psi_2(x) = 0 \quad \text{a.e. in } (d_k, b). \tag{2.23}$$

Indeed, by (2.15) and Remark 2.4 (i), it follows that

$$\left(\int_{d_k}^b u(t)|\varphi_1(t)|^q dt \right)^{1/q} \left(\int_{c_k}^{d_k} v^{1-p'}(t)|\psi_1(t)|^{p'} dt \right)^{1/p'} \leq C < \infty,$$

and since $\varphi_1(t) \neq 0$ for a.e. t, we have

$$\left(\int_{c_k}^{d_k} v^{1-p'}(t)|\psi_1(t)|^{p'} dt \right)^{1/p'} < \infty.$$

Since ψ_{1k} is zero outside (c_k, d_k), it follows from the definition of A_1 (see (2.18)) that

$$A_1(\varphi_1, \psi_{1k}) = \left(\int_{c_k}^{d_k} \left(\int_x^b u(t)|\varphi_1(t)|^q dt \right)^{r/q} \right.$$

$$\left. \times \left(\int_{c_k}^x v^{1-p'}(t)|\psi_{1k}(t)|^{p'} dt \right)^{r/q'} v^{1-p'}(x)|\psi_{1k}(x)|^{p'} dx \right)^{1/r}$$

$$\leq \left(\int_{c_k}^b u(t)|\varphi_1(t)|^q dt \right)^{1/q}$$

$$\times \left(\int_{c_k}^{d_k} \left(\int_{c_k}^x v^{1-p'}(t)|\psi_1(t)|^{p'} dt \right)^{r/q'} v^{1-p'}(x)|\psi_1(x)|^{p'} dx \right)^{1/r}$$

$$= \left(\int_{c_k}^b u(t)|\varphi_1(t)|^q dt \right)^{1/q}$$

$$\times \left(\frac{1}{\frac{r}{q'}+1} \int_{c_k}^{d_k} \frac{d}{dx} \left(\int_{c_k}^x v^{1-p'}(t)|\psi_1(t)|^{p'} dt \right)^{r/q'+1} dx \right)^{1/r}$$

$$\leq \left(\int_{c_k}^b u(t)|\varphi_1(t)|^q dt \right)^{1/q} \left(\int_{c_k}^{d_k} v^{1-p'}(t)|\psi_1(t)|^{p'} dt \right)^{1/p'}.$$

The construction of the interval $[c_k, d_k]$ implies that

$$A_1(\varphi_1, \psi_{1k}) = \infty$$

and consequently,

$$\left(\int_{c_k}^b u(t)|\varphi_1(t)|^q dt \right)^{1/q} = \infty.$$

Now, due to (2.15),

$$\left(\int_x^b u(t)|\varphi_1(t)|^q dt \right)^{1/q} \left(\int_a^x v^{1-p'}(t)|\psi_1(t)|^{p'} dt \right)^{1/p'} \leq C < \infty$$

for every $x \in (a, c_k)$, and this implies that $\psi_1(x) = 0$ a.e. in (a, c_k). The proof that $\psi_2(x) = 0$ a.e. in (d_k, b) follows analogously.

In view of (2.23) we have

$$\psi_1(x) = 0 \quad \text{a.e. in } (a, c),$$

$$\psi_2(x) = 0 \quad \text{a.e. in } (c, b),$$

and consequently, the operator T can be rewritten as

$$(Tf)(x) = (T_1f)(x) + (T_2f)(x),$$

where

$$(T_1 f)(x) = \chi_{(c,b)}(x)\varphi_1(x) \int_c^x \psi_1(t)f(t)dt,$$

$$(T_2 f)(x) = \chi_{(a,c)}(x)\varphi_2(x) \int_x^c \psi_2(t)f(t)dt.$$

Clearly,

$$\|T_1\| \approx A_1(\varphi_1, \psi_1), \quad \|T_2\| \approx A_2(\varphi_2, \psi_2).$$

For the norm of T we have $\|T\| \leq C < \infty$, but on the other hand,

$$\|T\| \geq \frac{1}{2}(\|T_1\| + \|T_2\|) \geq C_1(A_2(\varphi_2, \psi_2) + A_1(\varphi_1, \psi_1)) = \infty$$

due to (2.21). But this is a contradiction, and thus we have completed the proof of Theorem 2.3.   □

2.3. General Hardy-type Operators. The Fundamental Lemma

Definition 2.5. We will say that K is a *general Hardy-type operator* if it has the form

$$(Kf)(x) := \int_a^x k(x,t)f(t)dt, \quad x \in (a,b).$$

For simplicity, we will deal with the case $(a,b) = (0,\infty)$, i.e., with the operator

$$(Kf)(x) := \int_0^x k(x,t)f(t)dt, \quad x > 0. \tag{2.24}$$

The results for a general interval (a,b) will be summarized at the end of this chapter (see Sec. 2.6).

We suppose that the kernel k satisfies

$$k(x,t) \geq 0, \quad 0 < t < x,$$

$$k \text{ is increasing in } x \text{ and decreasing in } t \tag{2.25}$$

and

$$k(x,t) \approx k(x,z) + k(z,t), \quad 0 < t < z < x. \tag{2.26}$$

Such kernels are sometimes called *Oinarov kernels*.

The conjugate operator $\widetilde{K}$ to K is given by

$$(\widetilde{K}g)(x) := \int_x^\infty k(t,x)g(t)dt, \quad x > 0. \tag{2.27}$$

Remark 2.6. If the kernel $k(x,t)$ satisfies the monotonicity conditions from (2.25), then it follows at once that

$$k(x,t) \geq \frac{1}{2}(k(x,z) + k(z,t)) \quad \text{for } 0 < t < z < x$$

holds, i.e., the *lower* bound implied by (2.26) is satisfied.

Example 2.7. The following kernels satisfy conditions (2.25) and (2.26):

(i) The *convolution kernel*

$$k(x,t) = \varphi(x - t),$$

where φ satisfies

$$\varphi(a + b) \approx \varphi(a) + \varphi(b), \quad 0 < a, b < \infty.$$

(ii) The kernel

$$k(x,t) = \Phi(\tfrac{t}{x}),$$

where Φ satisfies

$$\Phi(ab) \approx \Phi(a) + \Phi(b), \quad 0 < a, b < \infty.$$

(iii) The particular cases of (i), (ii)

$$k(x,t) = (x - t)^\alpha, \quad \alpha \geq 0, \quad x > t > 0,$$
$$k(x,t) = \log^\beta \frac{x}{t}, \quad \beta \geq 0, \quad x > t > 0.$$

(iv) The "integral" kernel

$$k(x,t) = \int_t^x \varphi(\tau)d\tau, \quad x > t > 0,$$

with a non-negative function φ.

Remarks and Notation 2.8. (i) In the sequel, weight functions will be characterized for which the operator K is bounded between weighted Lebesgue spaces for a wide range of indices. This problem has been extensively studied in the primary literature (see, for example, S. Bloom and R. Kerman [BK1], F.J. Martín-Reyes and E. Sawyer [MS1], R. Oinarov [O1], [O2], V.D. Stepanov [St1], [St2] and the sources cited therein).

(ii) Let us mention that an important role will be played by *duality*. If the operator K from (2.24) maps $L^p(v)$ into $L^q(u)$,

$$K : L^p(v) \to L^q(u)$$

for $1 < p,q < \infty$ boundedly with norm C, then analogously its conjugate $\widetilde{K}$ from (2.27) maps $L^{q'}(u^{1-q'})$ into $L^{p'}(v^{1-p'})$,

$$\widetilde{K} : L^{q'}(u^{1-q'}) \to L^{p'}(v^{1-p'})$$

with the same norm C. Indeed: by duality we have

$$\left(\int_0^\infty |(\widetilde{K}g)(x)|^{p'} v^{1-p'}(x)dx \right)^{1/p'} = \sup_{\|f\|_{p,v}=1} \left| \int_0^\infty (\widetilde{K}g)(x)f(x)dx \right|$$

$$= \sup_{\|f\|_{p,v}=1} \left| \int_0^\infty g(x)(Kf)(x)dx \right|$$

$$\leq \sup_{\|f\|_{p,v}=1} \|Kf\|_{q,u} \|g\|_{q',u^{1-q'}}$$

$$\leq C\|g\|_{q',u^{1-q'}},$$

where we have applied Fubini's theorem, Hölder's inequality and the assumption that

$$\|Kf\|_{q,u} \leq C\|f\|_{p,v}. \tag{2.28}$$

(See also p. 17.)

(iii) The particular operator K,

$$(Kf)(x) := \frac{1}{\Gamma(\alpha)} \int_0^x (x - t)^\alpha f(t)dt, \quad \alpha > -1$$

is called the *Riemann-Liouville operator* and its conjugate

$$(\widetilde{K}f)(x) := \frac{1}{\Gamma(\alpha)} \int_x^\infty (t - x)^\alpha f(t)dt$$

is called the *Weyl fractional integral operator*. From the results below it follows in particular that weights will be characterized for which these operators are bounded on weighted Lebesgue spaces when $\alpha \geq 0$. The case $-1 < \alpha < 0$ is not covered here because $k(x,t) = (x-t)^\alpha$, $-1 < \alpha < 0$, is not an Oinarov kernel. (Recall that in Example 2.7 (iii), this kernel appeared only with $\alpha \geq 0$!)

(iv) In order to find necessary and sufficient conditions on the weight functions u and v under which inequality (2.28), i.e., the inequality

$$\left(\int_0^\infty |(Kf)(x)|^q u(x)dx \right)^{1/q} \leq C \left(\int_0^\infty |f(x)|^p v(x)dx \right)^{1/p} \quad (2.29)$$

holds, we need a technical lemma. To simplify the notation, we denote for $s \geq 0$

$$(K_s h)(x) := \int_0^x k^s(x,t)h(t)dt,$$

$$(\widetilde{K}_s h)(x) := \int_x^\infty k^s(t,x)h(t)dt \quad (2.30)$$

with k an Oinarov kernel. Note that if $s = 0$, then (2.30) reduces to the Hardy operators H and $\widetilde{H}$, respectively.

We will write

$$K_1 h = Kh, \quad \widetilde{K}_1 h = \widetilde{K}h.$$

Furthermore, let us emphasize that without loss of generality we can assume that

$$f \geq 0. \quad (2.31)$$

Now, we are ready to formulate our fundamental lemma.

Lemma 2.9. *Suppose that $1 < q < \infty$ and that $K_q, \widetilde{K}_q$ are defined by (2.30).*

(a) *If $Kf \in L^q(u)$, then*

$$\int_0^\infty (Kf)^q(x)u(x)dx \approx \int_0^\infty f(x)(K_0f)^{q-1}(x)(\widetilde{K}_qu)(x)dx$$
$$+ \int_0^\infty f(x)(Kf)^{q-1}(x)(\widetilde{K}u)(x)dx.$$
(2.32)

(b) *If $\widetilde{K}g \in L^q(u)$, then*

$$\int_0^\infty (\widetilde{K}g)^q(x)u(x)dx \approx \int_0^\infty g(x)(\widetilde{K}_0g)^{q-1}(x)(K_qu)(x)dx$$
$$+ \int_0^\infty g(x)(\widetilde{K}g)^{q-1}(x)(Ku)(x)dx.$$

Proof. Since the proof of (b) is quite similar to that of (a), we omit it proving the first part only.

Denote the left and right hand sides of (2.32) by I and J, respectively. Fubini's theorem and obvious estimates yield

$$I = \int_0^\infty u(x) \left(\int_0^x k(x,t)f(t)dt \right)^q dx$$
$$= \int_0^\infty u(x) \left(\int_0^x k(x,z)f(z)dz \right)^{q-1} \left(\int_0^x k(x,t)f(t)dt \right) dx$$
$$= \int_0^\infty f(t) \int_t^\infty k(x,t)u(x) \left[\left(\int_0^t + \int_t^x \right) k(x,z)f(z)dz \right]^{q-1} dxdt$$
$$\approx \int_0^\infty f(t) \int_t^\infty k(x,t)u(x) \left(\int_0^t k(x,z)f(z)dz \right)^{q-1} dxdt$$
$$+ \int_0^\infty f(t) \int_t^\infty k(x,t)u(x) \left(\int_t^x k(x,z)f(z)dz \right)^{q-1} dxdt$$
$$=: I_0 + I_1.$$

Since $z < t < x$ in the integral I_0 one obtains from (2.26) that

$$k(x,z) \approx k(x,t) + k(t,z),$$

and consequently,

$$I_0 \approx \int_0^\infty f(t) \left(\int_t^\infty k^q(x,t)u(x)dx \right) \left(\int_0^t f(z)dz \right)^{q-1} dt$$

$$+ \int_0^\infty f(t) \int_t^\infty k(x,t)u(x) \left(\int_0^t k(t,z)f(z)dz \right)^{q-1} dxdt$$

$$= \int_0^\infty f(t)(\widetilde{K}_q u)(t)(K_0 f)^{q-1}(t)dt$$

$$+ \int_0^\infty f(t)(\widetilde{K} u)(t)(K f)^{q-1}(t)dt = J.$$

Thus $I \approx J + I_1$, and since $I_1 \geq 0$, the lower bound of (a) follows.

To complete the proof we must show that $I_1 \lesssim J$.

Consider first the case $q = 2$. Fubini's theorem yields

$$I_1 = \int_0^\infty f(t) \int_t^\infty k(x,t)u(x) \int_t^x k(x,z)f(z)dzdxdt$$

$$= \int_0^\infty f(t) \int_t^\infty f(z) \int_z^\infty k(x,t)k(x,z)u(x)dxdzdt.$$

Since $t < z < x$, we conclude that $k(x,t) \approx k(x,z) + k(z,t)$ and consequently, by a repeated use of Fubini's theorem,

$$I_1 \approx \int_0^\infty f(t) \int_t^\infty f(z) \int_z^\infty k^2(x,z)u(x)dxdzdt$$

$$+ \int_0^\infty f(t) \int_t^\infty f(z) \int_z^\infty k(z,t)k(x,z)u(x)dxdzdt$$

$$= \int_0^\infty f(z) \int_0^z f(t)dt \int_z^\infty k^2(x,z)u(x)dxdz$$

$$+ \int_0^\infty f(z) \left(\int_z^\infty k(x,z)u(x)dx \right) \int_0^z k(x,t)f(t)dtdz$$

$$= \int_0^\infty f(z)(K_0 f)(z)(\widetilde{K}_2 u)(z)dz + \int_0^\infty f(z)(\widetilde{K} u)(z)(K f)(z)dz$$

$$= J.$$

Thus, we have proved the result for $q = 2$.

If $q \neq 2$, let us write

$$I_1 = \int_0^\infty f(t) I_1(t) dt,$$

where

$$
\begin{aligned}
I_1(t) &:= \int_t^\infty k(x,t) u(x) \left(\int_t^x k(x,z) f(z) dz \right)^{q-1} dx \\
&= \int_t^\infty k(x,t) u(x) \left(\int_t^x k(x,s) f(s) ds \right)^{q-2} \int_t^x k(x,z) f(z) dz dx \\
&= \int_t^\infty f(z) \int_z^\infty k(x,z) k(x,t) u(x) \\
&\quad \times \left(\int_t^x k(x,s) f(s) ds \right)^{q-2} dx dz.
\end{aligned}
$$

Here we have again used Fubini's theorem. But since $t < z < x$, condition (2.26) yields $k(x,t) \approx k(x,z) + k(z,t)$ and consequently,

$$
\begin{aligned}
I_1(t) &\approx \int_t^\infty f(z) \int_z^\infty k^2(x,z) u(x) \left(\int_t^x k(x,s) f(s) ds \right)^{q-2} dx dz \\
&\quad + \int_t^\infty f(z) \int_z^\infty k(x,z) k(z,t) u(x) \\
&\quad \times \left(\int_t^x k(x,s) f(s) ds \right)^{q-2} dx dz \\
&=: I_1^0(t) + I_1^1(t).
\end{aligned}
$$

Thus,

$$I_1 \approx \int_0^\infty f(t) I_1^0(t) dt + \int_0^\infty f(t) I_1^1(t) dt =: I_1^0 + I_1^1. \tag{2.33}$$

If $q > 2$, we apply Hölder's inequality (with exponents $q-1$ and $\frac{q-1}{q-2}$) to obtain

$$
\begin{aligned}
I_1^0(t) &= \int_t^\infty f(z) \int_z^\infty k^{q/(q-1)+(q-2)/(q-1)}(x,z) \\
&\quad \times u^{1/(q-1)+(q-2)/(q-1)}(x) \left(\int_t^x k(x,s) f(s) ds \right)^{q-2} dx dz
\end{aligned}
$$

$$\leq \int_t^\infty f(z) \left(\int_z^\infty k^q(x,z) u(x) \right)^{1/(q-1)}$$

$$\times \left(\int_z^\infty k(x,z) u(x) \left(\int_t^x k(x,s) f(s) ds \right)^{q-1} dx \right)^{(q-2)/(q-1)} dz$$

$$\leq \int_t^\infty f(z) (\widetilde{K}_q u)^{1/(q-1)}(z)$$

$$\times \left(\int_z^\infty k(x,z) u(x) (Kf)^{q-1}(x) dx \right)^{(q-2)/(q-1)} dz$$

and thus, applying Fubini's theorem (twice) and again Hölder's inequality,

$$I_1^0 \leq \int_0^\infty f(t) \int_t^\infty f(z) (\widetilde{K}_q u)^{1/(q-1)}(z)$$

$$\times \left(\int_z^\infty k(x,z) u(x) (Kf)^{q-1}(x) dx \right)^{(q-2)/(q-1)} dz dt$$

$$= \int_0^\infty f(z) (\widetilde{K}_q u)^{1/(q-1)}(z)$$

$$\times \left(\int_z^\infty k(x,z) u(x) (Kf)^{q-1}(x) dx \right)^{(q-2)/(q-1)} \int_0^z f(t) dt dz$$

$$\leq \left(\int_0^\infty f(z) (\widetilde{K}_q u)(z) (K_0 f)^{q-1}(z) dz \right)^{1/(q-1)}$$

$$\times \left(\int_0^\infty f(z) \left(\int_z^\infty k(x,z) u(x) (Kf)^{q-1}(x) dx \right) dz \right)^{(q-2)/(q-1)}$$

$$= \left(\int_0^\infty f(z) (\widetilde{K}_q u)(z) (K_0 f)^{q-1}(z) dz \right)^{1/(q-1)}$$

$$\times \left(\int_0^\infty u(x) (Kf)^q(x) dx \right)^{(q-2)/(q-1)}$$

$$= \left(\int_0^\infty f(z) (\widetilde{K}_q u)(z) (K_0 f)^{q-1}(z) dz \right)^{1/(q-1)} I^{(q-2)/(q-1)}.$$

Similarly, using the fact that $q > 2$, we find that

$$
\begin{aligned}
I_1^1 &= \int_0^\infty f(t) I_1^1(t) dt \\
&= \int_0^\infty f(t) \int_t^\infty f(z) k(x,t) \int_z^\infty k(x,z) u(x) \\
&\quad \times \left(\int_t^x k(x,s) f(s) ds \right)^{q-2} dx\, dz\, dt \\
&= \int_0^\infty f(z) \int_0^z k(z,t) f(t) \int_z^\infty k(x,z) u(x) \\
&\quad \times \left(\int_t^x k(x,s) f(s) ds \right)^{q-2} dx\, dt\, dz \\
&\leq \int_0^\infty f(z)(Kf)(z) \left(\int_z^\infty k(x,z) u(x)(Kf)^{q-2}(x) dx \right) dz \\
&\leq \int_0^\infty f(z)(Kf)(z) \left(\int_z^\infty k(x,z) u(x) dx \right)^{1/(q-1)} \\
&\quad \times \left(\int_z^\infty k(x,z) u(x)(Kf)^{q-1}(x) dx \right)^{(q-2)/(q-1)} dz \\
&\leq \left(\int_0^\infty f(z)(Kf)^{q-1}(z)(\widetilde{K}u)(z) dz \right)^{1/(q-1)} \\
&\quad \times \left(\int_0^\infty f(z) \left(\int_z^\infty k(x,z) u(x)(Kf)^{q-1}(x) dx \right) dz \right)^{(q-2)/(q-1)} \\
&= \left(\int_0^\infty f(z)(Kf)^{q-1}(z)(\widetilde{K}u)(z) dz \right)^{1/(q-1)} \\
&\quad \times \left(\int_0^\infty u(x)(Kf)^{q-1}(x) \left(\int_0^z k(x,z) f(x) dz \right) dx \right)^{(q-2)/(q-1)} \\
&= \left(\int_0^\infty f(z)(Kf)^{q-1}(z)(\widetilde{K}u)(z) dz \right)^{1/(q-1)} I^{(q-2)/(q-1)}.
\end{aligned}
$$

Consequently,

$$I_1 \leq C \left(\int_0^\infty f(z)(\widetilde{K}_q u)(z)(K_0 f)^{q-1}(z)dz \right.$$

$$\left. + \int_0^\infty f(z)(Kf)^{q-1}(z)(\widetilde{K}u)(z)dz \right)^{1/(q-1)} I^{(q-2)/(q-1)}$$

$$= C J^{1/(q-1)} I^{(q-2)/(q-1)}$$

$$\leq C_1 J^{1/(q-1)} (I_0 + I_1)^{(q-2)/(q-1)}$$

since $I \approx I_0 + I_1$. The indices $q-1$ and $\frac{q-1}{q-2}$ are conjugate, and hence

$$I_1 \leq C_2 \frac{J}{q-1} + \frac{I_0 + I_1}{(q-1)/(q-2)}.$$

But we have shown that $I_0 \approx J$, and thus

$$I_1 \leq C_3 J + \frac{q-2}{q-1} I_1,$$

i.e.,

$$I_1 \leq C_3(q-1)J.$$

Hence the result follows for $q > 2$.

Finally, if $1 < q < 2$, then

$$I_1^0(t) = \int_t^\infty f(z) \int_z^\infty k^2(x,z)u(x) \left(\int_t^x k(x,s)f(s)ds \right)^{q-2} dxdz$$

$$= \int_t^\infty f(z) \int_z^\infty k^q(x,z)u(x) \left(\int_t^x k(x,s)f(s)ds \right)^{q-2}$$

$$\times k^{2-q}(x,z)dxdz$$

and since $x > z > t$,

$$\left(\int_t^x k(x,s)f(s)ds \right)^{q-2} \leq \left(\int_t^z k(x,s)f(s)ds \right)^{q-2}$$

$$\leq k^{q-2}(x,z) \left(\int_t^z f(s)ds \right)^{q-2},$$

where the second inequality follows from the fact that k is decreasing in the second variable. Hence

$$I_1^0(t) \leq \int_t^\infty f(z) \left(\int_t^z f(s)ds \right)^{q-2} \int_z^\infty k^q(x,z)u(x)dxdz,$$

and Fubini's theorem yields

$$
\begin{aligned}
I_1^0 &\leq \int_0^\infty f(t) \int_t^\infty f(z) \left(\int_t^z f(s)ds \right)^{q-2} (\tilde{K}_q u)(z)dzdt \\
&= \int_0^\infty f(z)(\tilde{K}_q u)(z) \int_0^z f(t) \left(\int_t^z f(s)ds \right)^{q-2} dtdz \\
&= \frac{1}{q-1} \int_0^\infty f(z)(\tilde{K}_q u)(z)(K_0 f)^{q-1}(z)dz,
\end{aligned}
$$

since

$$
\begin{aligned}
\int_0^z f(t) \left(\int_t^z f(s)ds \right)^{q-2} dt &= -\frac{1}{q-1} \int_0^z \frac{d}{dt} \left(\int_t^z f(s)ds \right)^{q-1} dt \\
&= \frac{1}{q-1} \left(\int_0^z f(s)ds \right)^{q-1} \\
&= \frac{1}{q-1}(K_0 f)^{q-1}(z).
\end{aligned}
$$

Similarly,

$$
\begin{aligned}
I_1^1(t) &= \int_t^\infty k(z,t)f(z) \int_z^\infty k(x,z)u(x) \left(\int_t^x k(x,s)f(s)ds \right)^{q-2} dxdz \\
&\leq \int_t^\infty k(z,t)f(z) \int_z^\infty k(x,z)u(x) \left(\int_t^z k(z,s)f(s)ds \right)^{q-2} dxdz
\end{aligned}
$$

since $t < z < x$ and k is increasing in the first variable. Hence

$$
\begin{aligned}
I_1^1 \leq \int_0^\infty f(t) \int_t^\infty k(z,t)f(z) \int_z^\infty k(x,z)u(x) \\
\times \left(\int_t^z k(z,s)f(s)ds \right)^{q-2} dxdzdt
\end{aligned}
$$

$$= \int_0^\infty f(z) \int_0^z k(z,t)f(t)$$

$$\times \left(\int_t^z k(z,s)f(s)ds \right)^{q-2} \int_z^\infty k(x,z)u(x)dxdtdz$$

$$= \frac{1}{q-1} \int_0^\infty f(z)(Kf)^{q-1}(z)(\widetilde{K}u)(z)dz$$

and therefore

$$I_1 \approx I_1^0 + I_1^1$$

$$\leq \frac{1}{q-1} \left[\int_0^\infty f(z)(\widetilde{K}_q u)(z)(K_0 f)^{q-1}(z)dz \right.$$

$$\left. + \int_0^\infty f(z)(\widetilde{K}u)(z)(Kf)^{q-1}(z)dz \right]$$

$$= \frac{1}{q-1} J.$$

This completes the proof of the lemma. □

2.4. General Hardy-type Operators. The Case $p \leq q$

Now we are able to state and prove the first main theorem of this chapter describing necessary and sufficient conditions of the validity of inequality (2.29).

Theorem 2.10. *Let $1 < p \leq q < \infty$. Let K be the general Hardy-type operator from Definition 2.5 and let K_s and $\widetilde{K}_s$ be given by (2.30). Then the inequality*

$$\left(\int_0^\infty (Kf)^q(x)u(x)dx \right)^{1/q} \leq C \left(\int_0^\infty f^p(x)v(x)dx \right)^{1/p} \quad (2.34)$$

holds for all $f \geq 0$ if and only if

$$A_0 := \sup_{t>0} (\widetilde{K}_q u)^{1/q}(t)(K_0 v^{1-p'})^{1/p'}(t) < \infty \quad (2.35)$$

and

$$A_1 := \sup_{t>0} (\widetilde{K}_0 u)^{1/q}(t)(K_{p'} v^{1-p'})^{1/p'}(t) < \infty. \quad (2.36)$$

The best constant C in (2.34) satisfies

$$C \approx \max(A_0, A_1). \tag{2.37}$$

Remark 2.11. Due to (2.30), we can write A_0 and A_1 also in the form

$$A_0 = \sup_{t>0} \left(\int_t^\infty k^q(s,t)u(s)ds \right)^{1/q} \left(\int_0^t v^{1-p'}(s)ds \right)^{1/p'}$$

$$A_1 = \sup_{t>0} \left(\int_t^\infty u(s)ds \right)^{1/q} \left(\int_0^t k^{p'}(t,s)v^{1-p'}(s)ds \right)^{1/p'}. \tag{2.38}$$

If K is the *Hardy operator* H, i.e., if we take $k(x,t) \equiv 1$, then A_0 and A_1 coincide and are nothing else than the constant A from (0.20).

Proof of Theorem 2.10

(i) *Necessity* of conditions (2.35), (2.36).

Suppose

$$(K_{p'}v^{1-p'})(t) := \int_0^t k^{p'}(t,s)v^{1-p'}(s)ds < \infty \tag{2.39}$$

for some $t > 0$. [If not, then $v \equiv 0$ a.e. But if (2.34) is satisfied then the left-hand side in (2.34) is zero and hence $u \equiv 0$ a.e., and the convention $0 \cdot \infty = 0$ implies that (2.35) and (2.36) hold.]

If (2.39) is satisfied for some $t > 0$, define a function f_t by

$$f_t(x) := \chi_{[0,t]}(x)k^{p'-1}(t,x)v^{1-p'}(x).$$

Then it follows from (2.34) that

$$\left(\int_0^\infty (Kf_t)^q(x)u(x)dx \right)^{1/q} \leq C \left(\int_0^\infty f_t^p(x)v(x)dx \right)^{1/p}$$

$$= C \left(\int_0^t k^{p'}(t,x)v^{1-p'}(x)dx \right)^{1/p}$$

$$= C(K_{p'}v^{1-p'})^{1/p}(t).$$

But since k is increasing in the first variable, we have

$$\left(\int_0^\infty (Kf_t)^q(x)u(x)dx \right)^{1/q}$$

$$\geq \left(\int_t^\infty u(x) \left(\int_0^t k(x,s)k^{p'-1}(t,s)v^{1-p'}(s)ds \right)^q dx \right)^{1/q}$$

$$\geq \left(\int_t^\infty u(x)dx \right)^{1/q} \left(\int_0^t k^{p'}(t,s)v^{1-p'}(s)ds \right)$$

$$= (\widetilde{K}_0 u)^{1/q}(t)(K_{p'}v^{1-p'})(t)$$

and it follows that

$$(\widetilde{K}_0 u)^{1/q}(t)(K_{p'}v^{1-p'})^{1/p'}(t) \leq C.$$

Hence (2.36) holds.

To see that (2.35) is also satisfied note that (2.34) is equivalent to

$$\left(\int_0^\infty (\widetilde{K}g)^{p'}(x)v^{1-p'}(x)dx \right)^{1/p'} \leq C \left(\int_0^\infty g^{q'}(x)u^{1-q'}(x)dx \right)^{1/q'}$$

$$(2.40)$$

with the same constant C as in (2.34) (see Remark 2.8 (ii)). As above we may assume that

$$(\widetilde{K}_q u)(t) := \int_t^\infty k^q(s,t)u(s)ds < \infty$$

for some $t > 0$ and for such t define a function g_t by

$$g_t(x) = \chi_{[t,\infty)}(x)k^{q-1}(x,t)u(x).$$

Then (2.40) yields

$$(K_0 v^{1-p'})^{1/p'}(t)(\widetilde{K}_q u)(t)$$

$$= \left(\int_0^t v^{1-p'}(s) \left(\int_t^\infty k^q(x,t)u(x)dx \right)^{p'} ds \right)^{1/p'}$$

$$\leq \left(\int_0^t v^{1-p'}(s) \left(\int_t^\infty k(x,s)k^{q-1}(x,t)u(x)dx \right)^{p'} ds \right)^{1/p'}$$

$$\leq \left(\int_0^\infty (\widetilde{K} g_t)^{p'}(s) v^{1-p'}(s) ds \right)^{1/p'}$$

$$\leq C \left(\int_0^\infty g_t^{q'}(x) u^{1-q'}(x) dx \right)^{1/q'}$$

$$= C \left(\int_t^\infty k^q(x,t) u(x) dx \right)^{1/q'} = C(\widetilde{K}_q u)^{1/q'}(t)$$

so that

$$(\widetilde{K}_q u)^{1/q}(t)(K_0 v^{1-p'})^{1/p'}(t) \leq C \qquad (2.41)$$

and hence (2.35) holds.

In particular,

$$\max(A_0, A_1) \leq C, \qquad (2.42)$$

i.e., the lower estimate in (2.37) holds.

(ii) *Sufficiency* of conditions (2.35), (2.36).

We may assume that f has a compact support in $(0, \infty)$ and that $0 < \|f\|_{p,v} < \infty$. The general case then follows via Fatou's lemma.

Now by the upper estimate of (2.32) from Lemma 2.9 we obtain

$$I := \int_0^\infty u(x)(Kf)^q(x) dx$$

$$\lesssim \left(\int_0^\infty f(x)(K_0 f)^{q-1}(x)(\widetilde{K}_q u)(x) dx \right.$$

$$\left. + \int_0^\infty f(x)(Kf)^{q-1}(x)(\widetilde{K} u)(x) dx \right)$$

$$=: C(J_0 + J_1).$$

By Hölder's inequality

$$J_0 := \int_0^\infty f(x)(K_0 f)^{q-1}(x)(\widetilde{K}_q u)(x) dx$$

$$\leq \left(\int_0^\infty f^p(x) v(x) dx \right)^{1/p}$$

$$\times \left(\int_0^\infty v^{1-p'}(x)(K_0 f)^{p'(q-1)}(x)(\widetilde{K}_q u)^{p'}(x) dx \right)^{1/p'}. \qquad (2.43)$$

If we denote

$$w(x) = v^{1-p'}(x)(\widetilde{K}_q u)^{p'}(x),$$

then the second factor on the right-hand side in (2.43) takes the form

$$\left(\int_0^\infty w(x) \left(\int_0^x f(t)dt\right)^{p'(q-1)} dx\right)^{1/p'}$$

and can be estimated from above by

$$C_2^{q-1} \left(\int_0^\infty v(x)f^p(x)dx\right)^{(q-1)/p}$$

via the standard Hardy inequality (cf. (0.18)) if and only if

$$C_3 := \sup_{t>0} \left(\int_t^\infty w(x)dx\right)^{1/(p'(q-1))} \left(\int_0^t v^{1-p'}(x)dx\right)^{1/p'} < \infty$$

(cf. condition (0.19); the application of the Hardy inequality is correct since $1 < p \leq q < \infty$ implies that $p'(q-1) \geq p$). Moreover, we have $C_2 \approx C_3$. But by (2.35),

$$\begin{aligned}
\int_t^\infty w(x)dx &= \int_t^\infty v^{1-p'}(x)(\widetilde{K}_q u)^{p'}(x)dx \\
&\leq A_0^{qp'} \int_t^\infty v^{1-p'}(x)(K_0 v^{1-p'})^{-q}(x)dx \\
&= A_0^{qp'} \int_t^\infty v^{1-p'}(x) \left(\int_0^x v^{1-p'}(s)ds\right)^{-q} dx \\
&= \frac{A_0^{qp'}}{1-q} \left(\int_0^x v^{1-p'}(s)ds\right)^{1-q} \Bigg|_t^\infty \\
&\leq \frac{A_0^{qp'}}{q-1} \left(\int_0^t v^{1-p'}(s)ds\right)^{1-q}
\end{aligned}$$

and hence

$$C_3 \leq A_0^{q'}(q-1)^{-1/(p'(q-1))}.$$

Therefore,

$$J_0 \le C_2^{q-1} \left(\int_0^\infty v(x) f^p(x) dx \right)^{1/p+(q-1)/p} \le C_4 A_0^q \|f\|_{p,v}^q. \quad (2.44)$$

Similarly,

$$J_1 := \int_0^\infty f(x)(Kf)^{q-1}(x)(\widetilde{K}u)(x) dx$$

$$\le \left(\int_0^\infty f^p(x) v(x) dx \right)^{1/p}$$

$$\times \left(\int_0^\infty v^{1-p'}(x)(\widetilde{K}u)^{p'}(x)(Kf)^{p'(q-1)}(x) dx \right)^{1/p'}. \quad (2.45)$$

Since the function Kf is increasing, the second factor on the right hand side in (2.45) can be written as

$$\left(\int_0^\infty v^{1-p'}(x)(\widetilde{K}u)^{p'}(x) \left(\int_0^x d(Kf)^{p'(q-1)}(s) \right) dx \right)^{1/p'}$$

$$= \left(\int_0^\infty \left(\int_s^\infty v^{1-p'}(x)(\widetilde{K}u)^{p'}(x) dx \right) d(Kf)^{p'(q-1)}(s) \right)^{1/p'}.$$

By Minkowski's integral inequality and by (2.36) we have

$$\int_s^\infty v^{1-p'}(x)(\widetilde{K}u)^{p'}(x) dx = \int_s^\infty v^{1-p'}(x) \left(\int_x^\infty k(t,x)u(t)dt \right)^{p'} dx$$

$$\le \left(\int_s^\infty u(t) \left(\int_s^t v^{1-p'}(x)k^{p'}(t,x)dx \right)^{1/p'} dt \right)^{p'}$$

$$\le \left(\int_s^\infty u(t) \left(\int_0^t v^{1-p'}(x)k^{p'}(t,x)dx \right)^{1/p'} dt \right)^{p'}$$

$$= \left(\int_s^\infty u(t)(K_{p'} v^{1-p'})^{1/p'}(t)dt \right)^{p'}$$

$$\leq A_1^{p'} \left(\int_s^\infty u(t)(\widetilde{K}_0 u)^{-1/q}(t) dt \right)^{p'}$$

$$= A_1^{p'} \left(-q' \int_s^\infty d(\widetilde{K}_0 u)^{1/q'}(t) \right)^{p'} = (q')^{p'} A_1^{p'} (\widetilde{K}_0 u)^{p'/q'}(s).$$

Hence

$$J_1 \leq q' A_1 \|f\|_{p,v} \left(\int_0^\infty (\widetilde{K}_0 u)^{p'/q'}(s) d(Kf)^{p'q/q'}(s) \right)^{1/p'},$$

and by Minkowski's integral inequality (we have $p'/q' \geq 1$ since $p \leq q$) it follows that

$$J_1 \leq A_1 q' \|f\|_{p,v} \left(\int_0^\infty u(s)(Kf)^q(s) ds \right)^{1/q'}. \tag{2.46}$$

Thus

$$I = \|Kf\|_{q,u}^q \lesssim (J_0 + J_1) \lesssim (\|f\|_{p,v}^q + \|f\|_{p,v} I^{1/q'})$$

$$= C_0 \|f\|_{p,v}^q + C_0 \|f\|_{p,v} I^{1/q'}.$$

Using Young's inequality $ab \leq \frac{a^q}{q} + \frac{b^{q'}}{q'}$, $q > 1$, with $a = C_0 \|f\|_{p,v}$, $b = I^{1/q'}$, we obtain

$$I \leq C_0 \|f\|_{p,v}^q + \frac{C_0^q}{q} \|f\|_{p,v}^q + \frac{I}{q'},$$

which implies

$$I \lesssim \|f\|_{p,v}^q,$$

i.e.,

$$\|Kf\|_{q,u} \lesssim \|f\|_{p,v},$$

and this is the required estimate (2.34). Moreover, it follows from (2.44) and (2.46) that $C \lesssim \max(A_0, A_1)$ and hence together with (2.42) we have (2.37).

Now, we can easily prove the corresponding result for the conjugate operator $\widetilde{K}$, defined by

$$(\widetilde{K}g)(x) := \int_x^\infty k(t,x)g(t)dt, \quad x > 0. \tag{2.47}$$

Theorem 2.12. *Let* $1 < p \le q < \infty$. *Let* $\widetilde{K}$ *be the conjugate Hardy-type operator defined by formula* (2.47). *Let* K_s *and* $\widetilde{K}_s$ *be given by formula* (2.30). *Then the inequality*

$$\|\widetilde{K}g\|_{q,u} \le C\|g\|_{p,v} \tag{2.48}$$

holds for all $g \ge 0$ *if and only if*

$$\widetilde{A}_0 := \sup_{t>0}(K_q u)^{1/q}(t)(\widetilde{K}_0 v^{1-p'})^{1/p'}(t) < \infty \tag{2.49}$$

and

$$\widetilde{A}_1 := \sup_{t>0}(K_0 u)^{1/q}(t)(\widetilde{K}_{p'} v^{1-p'})^{1/p'}(t) < \infty. \tag{2.50}$$

The constant C *is the same as in* (2.34).

Proof. (i) The proof of the *necessity* part is the same as in the proof of Theorem 2.10, and it is therefore omitted.

(ii) *Sufficiency.* By duality, we have

$$\|\widetilde{K}g\|_{q,u} = \sup_{\|f\|_{q',u^{1-q'}}=1}\left|\int_0^\infty (\widetilde{K}g)(x)f(x)dx\right|$$

$$= \sup_{\|f\|_{q',u^{1-q'}}=1}\left|\int_0^\infty g(x)(Kf)(x)dx\right|.$$

Hölder's inequality yields

$$\left|\int_0^\infty g(x)(Kf)(x)dx\right| \le \|g\|_{p,v}\|Kf\|_{p',v^{1-p'}}.$$

Finally, since $\widetilde{A}_0$, $\widetilde{A}_1$ correspond to A_0, A_1 with u, v and p, q replaced by $v^{1-p'}$, $u^{1-q'}$ and q', p', respectively, Theorem 2.10 implies that

$$\|Kf\|_{p',v^{1-p'}} \le C\|f\|_{q',u^{1-q'}}$$

holds if and only if conditions (2.49) and (2.50) are satisfied.

Combining all these estimates, we obtain (2.48). □

Example 2.13. We state the results for the Riemann-Liouville operator and for the Weyl fractional integral operator, mentioned in Remark 2.8 (iii) (see p. 85).

Let $1 < p \leq q < \infty$.

(i) If

$$(Kf)(x) = \int_0^x (x - t)^\alpha f(t)dt, \quad \alpha \geq 0, \ x > 0,$$

then the inequality

$$\|Kf\|_{q,u} \leq C\|f\|_{p,v}$$

is satisfied for all $f \geq 0$ if and only if

$$\sup_{t>0} \left(\int_t^\infty (s - t)^{\alpha q} u(s)ds \right)^{1/q} \left(\int_0^t v^{1-p'}(s)ds \right)^{1/p'} < \infty$$

and

$$\sup_{t>0} \left(\int_t^\infty u(s)ds \right)^{1/q} \left(\int_0^t (t - s)^{\alpha p'} v^{1-p'}(s)ds \right)^{1/p'} < \infty.$$

(ii) If

$$(\widetilde{K}g)(x) = \int_x^\infty (t - x)^\alpha g(t)dt, \quad \alpha \geq 0, \ x > 0,$$

then the inequality

$$\|\widetilde{K}g\|_{q,u} \leq C\|g\|_{p,v}$$

is satisfied for all $g \geq 0$ if and only if

$$\sup_{t>0} \left(\int_0^t (t - s)^{\alpha q} u(s)ds \right)^{1/q} \left(\int_t^\infty v^{1-p'}(s)ds \right)^{1/p'} < \infty$$

and

$$\sup_{t>0} \left(\int_0^t u(s)ds \right)^{1/q} \left(\int_t^\infty (s - t)^{\alpha p'} v^{1-p'}(s)ds \right)^{1/p'} < \infty.$$

2.5. General Hardy-type Operators. The Case $p > q$

Notation and Remarks 2.14. (i) Let us now consider the case

$$0 < q < p < \infty, \quad p > 1.$$

Define r by

$$\frac{1}{r} := \frac{1}{q} - \frac{1}{p}$$

and introduce numbers

$$B_0 := \left\{ \int_0^\infty \left[(\widetilde{K}_q u)^{1/q}(t)(K_0 v^{1-p'})^{1/q'}(t) \right]^r v^{1-p'}(t) dt \right\}^{1/r},$$

$$B_1 := \left\{ \int_0^\infty \left[(\widetilde{K}_0 u)^{1/p}(t)(K_{p'} v^{1-p'})^{1/p'}(t) \right]^r u(t) dt \right\}^{1/r}. \tag{2.51}$$

If K is the Hardy operator H, i.e., if we take $k(x,t) \equiv 1$ for $t < x$, $k(x,t) \equiv 0$ for $t > x$, then B_1 is a multiple of B_0 and B_0 coincides with the number A from (0.21).

(ii) Recall that for the Hardy operator and its conjugate, a characterization was given for the full range $0 < q < p < \infty$, $p > 1$. In what follows, we first restrict ourselves to the range

$$1 < q < p < \infty.$$

Theorem 2.15. *Suppose* $1 < q < p < \infty$, $\frac{1}{r} = \frac{1}{q} - \frac{1}{p}$. *Let K be a general Hardy-type operator (see Definition 2.5). Then the inequality* (2.34) *holds for all $f \geq 0$ if and only if*

$$\max(B_0, B_1) < \infty \tag{2.52}$$

where B_0, B_1 are defined in (2.51).

Moreover, the best constant C in (2.34) *satisfies*

$$C \approx \max(B_0, B_1). \tag{2.53}$$

Proof. (*i*) *Necessity of* (2.52). Choosing u and v appropriately, we may assume that $B_0 < \infty$, without increasing the constant C

in (2.34) (e.g., replace u and $v^{1-p'}$ by u_ε and $v_\varepsilon^{1-p'}$ supported in $(\varepsilon, 1/\varepsilon)$, $\varepsilon > 0$). Once the result is proved in this case, the additional conditions on u and v can be removed by approximating the general weights by these *appropriate* weights.

Now define $\bar{f}$ by

$$\bar{f}(t) = (\widetilde{K}_q u)^{r/(pq)}(t)(K_0 v^{1-p'})^{r/(pq')}(t)v^{1-p'}(t).$$

Then $B_0^{r/p} = \|\bar{f}\|_{p,v}$ and substituting $\bar{f}$ into (2.34) which is satisfied by assumption, we have

$$CB_0^{r/p} = C\|\bar{f}\|_{p,v} \geq \|K\bar{f}\|_{q,u}$$

$$= \left(\int_0^\infty \left(\int_0^x k(x,t)\bar{f}(t)dt \right)^q u(x)dx \right)^{1/q}$$

$$= \left(\int_0^\infty u(x) \left(\int_0^x k(x,t)\bar{f}(t)dt \right) \left(\int_0^x k(x,s)\bar{f}(s)ds \right)^{q-1} dx \right)^{1/q}$$

$$= \left(\int_0^\infty \bar{f}(t) \left(\int_t^\infty k(x,t)u(x) \left(\int_0^x k(x,s)\bar{f}(s)ds \right)^{q-1} dx \right) dt \right)^{1/q}$$

$$\geq \left(\int_0^\infty \bar{f}(t) \left(\int_t^\infty k(x,t)u(x) \left(\int_0^t k(x,s)\bar{f}(s)ds \right)^{q-1} dx \right) dt \right)^{1/q},$$

where we have used Fubini's theorem and the fact that $q > 1$. Since $s < t < x$, we have $k(x,s) \approx k(x,t) + k(t,s)$ and in particular $k(x,s) \geq C_2 k(x,t)$ for some $C_2 > 0$. Hence

$$CB_0^{r/p} \geq C_2^{1/q'} \left(\int_0^\infty \bar{f}(t) \left(\int_t^\infty k(x,t)u(x) \right. \right.$$

$$\left. \left. \times \left(\int_0^t k(x,t)\bar{f}(s)ds \right)^{q-1} dx \right) dt \right)^{1/q}$$

$$= C_2^{1/q'} \left(\int_0^\infty \bar{f}(t)(\widetilde{K}_q u)(t) \left(\int_0^t \bar{f}(s)ds \right)^{q-1} dt \right)^{1/q}$$

$$= C_2^{1/q'} \left(\int_0^\infty \bar{f}(t)(\widetilde{K}_q u)(t) \right.$$

$$\left. \times \left(\int_0^t (\widetilde{K}_q u)^{r/(pq)}(s)(K_0 v^{1-p'})^{r/(pq')}(s)v^{1-p'}(s)ds \right)^{q-1} dt \right)^{1/q}$$

$$\geq C_2^{1/q'} \left(\int_0^\infty (\widetilde{K}_q u)^{1+r(q-1)/(pq)}(t)\bar{f}(t) \right.$$

$$\left. \times \left(\int_0^t (K_0 v^{1-p'})^{r/(pq')}(s)v^{1-p'}(s)ds \right)^{q-1} dt \right)^{1/q}$$

since $\widetilde{K}_q u$ is a decreasing function. Furthermore,

$$\int_0^t (K_0 v^{1-p'})^{r/(pq')}(s)v^{1-p'}(s)ds$$

$$= \int_0^t v^{1-p'}(s) \left(\int_0^s v^{1-p'}(z)dz \right)^{r/(pq')} ds$$

$$= \frac{1}{r/(pq') + 1} \left(\int_0^t v^{1-p'}(s)ds \right)^{r/(pq')+1}$$

$$= \frac{1}{r/(p'q)} (K_0 v^{1-p'})^{r/(p'q)}(t)$$

so that

$$CB_0^{r/p} \geq C_2^{1/q'} \left(\frac{p'q}{r} \right)^{1/q'}$$

$$\times \left(\int_0^\infty (\widetilde{K}_q u)^{1+r/(pq')}(t)\bar{f}(t)(K_0 v^{1-p'})^{r/(p'q')}(t)dt \right)^{1/q}$$

$$= \left(C_2 \frac{p'q}{r} \right)^{1/q'} \left(\int_0^\infty (\widetilde{K}_q u)^{1+r/(pq')+r/(pq)}(t) \right.$$

$$\left. \times (K_0 v^{1-p'})^{r/(p'q')+r/(pq')}(t)v^{1-p'}(t)dt \right)^{1/q}$$

$$= \left(C_2 \frac{p'q}{r}\right)^{1/q'} \left(\int_0^\infty (\widetilde{K}_q u)^{r/q}(t)(K_0 v^{1-p'})^{r/q'}(t) v^{1-p'}(t) dt\right)^{1/q}$$

$$= \left(C_2 \frac{p'q}{r}\right)^{1/q'} \left(\int_0^\infty \bar{f}^p(t) v(t) dt\right)^{1/q} = \left(C_2 \frac{p'q}{r}\right)^{1/q'} B_0^{r/q}.$$

Therefore

$$C \geq \left(C_2 \frac{p'q}{r}\right)^{1/q'} B_0^{r/q - r/p} = \left(C_2 \frac{p'q}{r}\right)^{1/q'} B_0,$$

i.e., $B_0 \lesssim C < \infty$ if (2.34) holds.

To prove an analogous assertion for B_1, recall that (2.34) is equivalent to the inequality $\|\widetilde{K}g\|_{p',v^{1-p'}} \leq C\|g\|_{q',u^{1-q'}}$, where $\widetilde{K}$ is the conjugate to K, with the same constant C as in (2.34). Now define

$$\bar{g}(t) = (\widetilde{K}_0 u)^{r/(pq')}(t)(K_{p'}v^{1-p'})^{r/(p'q')}(t) u(t).$$

Then $B_1^{r/q'} = \|\bar{g}\|_{q',u^{1-q'}}$ and we proceed as before, using of course the dual inequality. The result $B_1 \lesssim C$ proves the necessity part.

(ii) *Sufficiency* of (2.52). Note that by Lemma 2.9,

$$I := \int_0^\infty u(x)(Kf)^q(x) dx$$

$$\leq C_1 \left(\int_0^\infty f(x)(K_0 f)^{q-1}(x)(\widetilde{K}_q u)(x) dx \right.$$

$$\left. + \int_0^\infty f(x)(Kf)^{q-1}(x)(\widetilde{K}u)(x) dx\right)$$

$$=: C_1(J_0 + J_1).$$

To estimate J_0, we apply Hölder's inequality for the product of three factors with exponents p, $p/(q-1)$ and r/q. We obtain

$$J_0 = \int_0^\infty [f(x)v^{1/p}(x)] \, [v^{(1-p')/p}(x)(K_0 v^{1-p'})^{-1}(x)(K_0 f)(x)]^{q-1}$$

$$\times [(\widetilde{K}_q u)(x)(K_0 v^{1-p'})^{q-1}(x)v^{(1-p')q/r}(x)] dx$$

$$\leq \|f\|_{p,v} \left(\int_0^\infty v^{1-p'}(x)(K_0 v^{1-p'})^{-p}(x)(K_0 f)^p(x) dx\right)^{(q-1)/p} B_0^q.$$

Now, we can use the Hardy inequality

$$\left(\int_0^\infty v^{1-p'}(x)(K_0 v^{1-p'})^{-p}(x)\left(\int_0^x f(t)dt\right)^p dx\right)^{1/p}$$
$$\leq C_p \left(\int_0^\infty f^p(x)v(x)dx\right)^{1/p},$$

which holds since the necessary and sufficient condition for its validity is satisfied:

$$\sup_{x>0}\left(\int_x^\infty v^{1-p'}(t)(K_0 v^{1-p'})^{-p}(t)dt\right)^{1/p}\left(\int_0^x v^{1-p'}(t)dt\right)^{1/p'}$$

$$= \sup_{x>0}\left(\int_x^\infty v^{1-p'}(t)\left(\int_0^t v^{1-p'}(s)ds\right)^{-p}dt\right)^{1/p}\left(\int_0^x v^{1-p'}(t)dt\right)^{1/p'}$$

$$= \sup_{x>0}\left(\frac{1}{p-1}\right)^{1/p}\left(\int_0^x v^{1-p'}(t)dt\right)^{-1+\frac{1}{p}}\left(\int_0^x v^{1-p'}(t)dt\right)^{1/p'}$$

$$= \left(\frac{1}{p-1}\right)^{1/p} < \infty.$$

Therefore

$$J_0 \leq C_p^{q-1}\|f\|_{p,v}^{1+q-1}B_0^q = C_p^{q-1}\|f\|_{p,v}^q B_0^q. \tag{2.54}$$

To estimate J_1, we first use Hölder's inequality which yields

$$J_1 = \int_0^\infty f(x)(Kf)^{q-1}(x)(\widetilde{K}u)(x)dx$$

$$= \int_0^\infty [f(x)v^{1/p}(x)][(Kf)^{q-1}(x)(\widetilde{K}u)(x)v^{-1/p}(x)]dx$$

$$\leq \|f\|_{p,v}J_2^{1/p'},$$

where

$$J_2 := \int_0^\infty (Kf)^{p'(q-1)}(x)(\widetilde{K}u)^{p'}(x)v^{1-p'}(x)dx$$

$$= \int_0^\infty (Kf)^{p'(q-1)}(x)d\left(-\int_x^\infty (\widetilde{K}u)^{p'}v^{1-p'}(t)dt \right)$$

$$= \int_0^\infty \left(\int_x^\infty (\widetilde{K}u)^{p'}(t)v^{1-p'}(t)dt \right) d(Kf)^{p'(q-1)}(x).$$

Let us denote the inner integral by $J_2(x)$. Then, by Minkowski's integral inequality, we have

$$(J_2(x))^{1/p'} := \left(\int_x^\infty v^{1-p'}(t) \left(\int_t^\infty k(s,t)u(s)ds \right)^{p'} dt \right)^{1/p'}$$

$$\leq \left(\int_x^\infty u(s) \left(\int_0^s k^{p'}(s,t)v^{1-p'}(t)dt \right)^{1/p'} ds =: J_3(x).$$

Therefore

$$J_1 \leq \|f\|_{p,v} \left(\int_0^\infty J_2(x)d(Kf)^{p'(q-1)}(x) \right)^{1/p'}$$

$$\leq \|f\|_{p,v} \left(\int_0^x J_3^{p'}(x)d(Kf)^{p'(q-1)}(x) \right)^{1/p'}. \qquad (2.55)$$

It is $p' < r$, and hence we can fix an s, $p' < s < r$, and use Hölder's inequality:

$$J_3(x) = \int_x^\infty u(t)(K_{p'}v^{1-p'})^{1/p'}(t)dt$$

$$= \int_x^\infty [(\widetilde{K}_0u)^{1/p}(t)(K_{p'}v^{1-p'})^{1/p'}(t)u^{1/s}(t)][(\widetilde{K}_0u)^{-1/p}(t)u^{1/s'}(t)]dt$$

$$\leq \left(\int_x^\infty [(\widetilde{K}_0 u)^{1/p}(t)(K_{p'} v^{1-p'})^{1/p'}(t)]^s u(t) dt \right)^{1/s}$$

$$\times \left(\int_x^\infty (\widetilde{K}_0 u)^{-s'/p}(t) u(t) dt \right)^{1/s'}$$

$$= \left(\frac{p}{p - s'} \right)^{1/s'} (\widetilde{K}_0 u)^{1/p' - 1/s}(x)$$

$$\times \left(\int_x^\infty [(\widetilde{K}_0 u)^{1/p}(t)(K_{p'} v^{1-p'})^{1/p'}(t)]^s u(t) dt \right)^{1/s}$$

since

$$\int_x^\infty (\widetilde{K}_0 u)^{-s'/p}(t) u(t) dt = - \int_x^\infty (\widetilde{K}_0 u)^{-s'/p}(t) d(\widetilde{K}_0 u)(t)$$

$$= \frac{p}{p - s'} (K_0 u)^{-s'/p+1}(x)$$

and $-1/p + 1/s' = 1/p' - 1/s$. Denote

$$J_4(x) := \int_x^\infty [(\widetilde{K}_0 u)^{1/p}(t)(K_{p'} v^{1-p'})^{1/p'}(t)]^s u(t) dt.$$

Then it follows from (2.55) that

$$J_1 \leq \left(\frac{p}{p - s'} \right)^{1/s'} \|f\|_{p,v}$$

$$\times \left(\int_0^\infty (\widetilde{K}_0 u)^{1-p'/s}(x) J_4^{p'/s}(x) d(Kf)^{p'(q-1)}(x) \right)^{1/p'}.$$

Applying again the Hölder inequality, now with exponents s/p', $s/(s - p')$, we get

$$\int_0^\infty (\widetilde{K}_0 u)^{1-p'/s}(x) J_4^{p'/s}(x) d(Kf)^{p'(q-1)}(x)$$

$$= p'(q - 1) \int_0^\infty (\widetilde{K}_0 u)^{1-p'/s}(x) J_4^{p'/s}(x)$$

$$\times (Kf)^{p'(q-1)-1}(x) d(Kf)(x)$$

$$= p'(q-1) \int_0^\infty \left[J_4^{p'/s}(x)(Kf)^{p'(q-1)(1/p+1/s)-1}(x) \right.$$

$$\times \left(\frac{d(Kf)(x)}{dx} \right)^{p'/s} \right] \left[(\widetilde{K}_0 u)^{1-p'/s}(x)(Kf)^{(q-1)(1-p'/s)} \right.$$

$$\times \left. \left(\frac{d(Kf)(x)}{dx} \right)^{1-p'/s} \right] dx$$

$$\leq p'(q-1) \left(\int_0^\infty J_4(x)(Kf)^{(q-1)s(\frac{1}{p}+\frac{1}{s})-s/p'}(x)d(Kf)(x) \right)^{p'/s}$$

$$\times \left(\int_0^\infty (\widetilde{K}_0 u)(x)(Kf)^{q-1}(x)d(Kf)(x) \right)^{1-p'/s}$$

$$= \frac{p'}{q'} \left(\int_0^\infty u(x)(Kf)^q(x)dx \right)^{1-p'/s}$$

$$\times \left(\int_0^\infty J_4(x)(Kf)^{(q-1)/(s/p+1)-s/p'}(x)d(Kf)(x) \right)^{p'/s}.$$

Now, integrating by parts and using Hölder's inequality with exponents r/s and $r/(r-s)$, we obtain

$$\int_0^\infty J_4(x)(Kf)^{(q-1)/(s/p-1)-s/p'}(x)d(Kf)(x)$$

$$= \int_0^\infty \left(\int_x^\infty (\widetilde{K}u_0)^{s/p}(t)(K_{p'}v^{1-p'})^{s/p'}(t)u(t)dt \right)$$

$$\times (Kf)^{(q-1)(s/p+1)-s/p'}(x)d(Kf)(x)$$

$$= \frac{(Kf)^{qs(1/s-1/r)}(x)}{qs(1/s-1/r)} \left(\int_x^\infty (\widetilde{K}_0 u)^{s/p}(t)(K_{p'}v^{1-p'})^{s/p'}(t)u(s)ds \right) \Big|_0^\infty$$

$$+ \frac{1}{qs(1/s-1/r)} \int_0^\infty (Kf)^{qs(1/s-1/r)}(x)(\widetilde{K}_0 u)^{s/p}(x)$$

$$\times (K_{p'}v^{1-p'})^{s/p'}(x)u(x)dx$$

$$= \frac{1}{qs(1/s-1/r)} \int_0^\infty [(\widetilde{K}_0 u)^{s/p}(x)(K_{p'}v^{1-p'})^{s/p'}(x)u^{s/r}(x)]$$

$$\times [(Kf)^{q(1-s/r)}(x)u^{1-s/r}(x)]dx$$

$$\leq \frac{1}{qs(1/s - 1/r)} \left(\int_0^\infty (\widetilde{K}_0 u)^{r/p}(x) \right.$$

$$\left. \times (K_{p'} v^{1-p'})^{r/p'}(x) u(x) dx \right)^{s/r} \left(\int_0^\infty u(x)(Kf)^q(x) dx \right)^{1-s/r}$$

$$= \frac{1}{qs(1/s - 1/r)} B_1^s \|Kf\|_{q,u}^{q(1-s/r)}.$$

Therefore

$$J_1 \leq C(p,q,s)\|f\|_{p,v} B_1 \|Kf\|_{q,u}^{q-1} = C(p,q,s)\|f\|_{p,v} B_1 I^{1/q'}.$$

However, using (2.54) and Young's inequality as at the end of the proof of the Theorem 2.10, we obtain

$$I \leq C_1(J_0 + J_1) \leq C_p^{q-1} C_1 \|f\|_{p,v}^q B_0^q + C(p,q,s)C_1\|f\|_{p,v} B_1 I^{1/q'}$$

$$\leq C_p^{q-1} C_1 \|f\|_{p,v}^q B_0^q + \frac{(C(p,q,s)C_1\|f\|_{p,v} B_1)^q}{q} + \frac{I}{q'},$$

i.e.,

$$I \|f\|_{p,v}^q (B_0^q + B_1^q),$$

and (2.34) follows with

$$C(B_0^q + B_1^q)^{1/q}. \qquad \square$$

Remark 2.16. (i) The formulation (and proof) of the corresponding theorem for the conjugate operator $\widetilde{K}$ is similar and left to the reader.

(ii) As mentioned in Remark 2.14 (ii), Theorem 2.15 (and its counterpart for the operator $\widetilde{K}$) was proved for $q < p$ with $q > 1$. The next theorem deals with the case

$$0 < q < 1 < p < \infty$$

and holds under weaker conditions on the kernel $k = k(x,t)$ of the Hardy-type operator K, defined by

$$(Kf)(x) = \int_0^x k(x,t)f(t)dt, \quad x > 0, \ f \geq 0$$

we only suppose that $k(x, t) \geq 0$ is *decreasing* in t, and no other conditions are imposed. On the other hand, the sufficient conditions are different from the necessary ones.

Theorem 2.17. *Suppose $0 < q < 1 < p < \infty$ and denote $\frac{1}{r} = \frac{1}{q} - \frac{1}{p}$. Let K be the operator from Remark 2.16 (ii). If*

$$B_0 := \left(\int_0^\infty [(\widetilde{K}_q u)^{1/q}(t)(K_0 v^{1-p'})^{1/q'}(t)]^r v^{1-p'}(t) dt \right)^{1/r} < \infty,$$

then the inequality

$$\|Kf\|_{q,u} \leq C\|f\|_{p,v} \tag{2.56}$$

holds for all $f \geq 0$ with $C \leq B_0$.

Conversely, if (2.56) holds for all $f \geq 0$, then $B_2 \leq C$ where

$$B_2 := \left(\int_0^\infty (\widetilde{K}_q u)^{p'/q}(t) v^{1-p'}(t) dt \right)^{1/p'}.$$

Remark 2.18. (i) Observe that in Theorem 2.17 we require *only one* condition for the sufficiency part, namely the finiteness of the same B_0 as in Theorem 2.15, and also only one condition for the necessity is required.

(ii) For the proof of sufficiency, we need the concept of the so-called *level function* introduced by I. Halperin [H1]. The proof of the following lemma can be found, e.g., in the Appendix to [OK]. Another proof was given independently in the Ph.D. thesis of G. Sinnamon [Si1], cf. also [Si2].

Lemma 2.19. *Let (a, b) be an interval and w a weight function on (a, b). Suppose that $0 < w(x) < \infty$ for a.e. $x \in (a, b)$ and $\int_a^b w(x) dx < \infty$.*

Then for each measurable function $f \geq 0$ there exists a non-negative function f° (called "level function of f") such that

(i) $\int_a^x f(t) dt \leq \int_a^x f^\circ(t) dt$ *for $x \in (a, b)$,*

(ii) $\frac{f^\circ(x)}{w(x)}$ *is decreasing on (a, b),*

(iii) $\int_a^b \left(\frac{f^\circ(x)}{w(x)} \right)^p w(x) dx \leq \int_a^b \left(\frac{f(x)}{w(x)} \right)^p w(x) dx$ *for $p \geq 1$.*

Proof of Theorem 2.17.

(i) Suppose that (2.56) is satisfied for all functions $f \geq 0$. Then by the reverse Minkowski integral inequality we find

$$C\|f\|_{p,v} \geq \left(\int_0^\infty u(x) \left(\int_0^x k(x,t)f(t)dt \right)^q dx \right)^{1/q}$$

$$\geq \int_0^\infty f(t) \left(\int_t^\infty k^q(x,t)u(x)dx \right)^{1/q} dt.$$

Applying the duality of $L^p(v)$ it follows that

$$C \geq \sup_{\|f\|_{p,v}=1} \int_0^\infty f(t) \left(\int_t^\infty k^q(k,t)u(x)dx \right)^{1/q} dt$$

$$= \sup_{\|f\|_{p,v}=1} \int_0^\infty f(t)(\widetilde{K}_q u)^{1/q}(t)dt = \|(\widetilde{K}_q u)^{1/q}\|_{p',v^{1-p'}}$$

$$= \left(\int_0^\infty (\widetilde{K}_q u)^{p'/q}(t)v^{1-p'}(t)dt \right)^{1/p'} = B_2.$$

(ii) To prove that $B_0 < \infty$ implies (2.56) we assume without loss of generality that f is compactly supported in $(0,\infty)$ and that $v^{1-p'} \in L^1$. Property (i) of Lemma 2.19 shows that

$$(Kf)(x) = \int_0^x k(x,t)d\left(\int_0^t f(s)ds \right)$$

$$= k(x,x)\int_0^x f(s)ds + \int_0^x \left(\int_0^t f(s)ds \right) d_t(-k(x,t))$$

$$\leq k(x,x)\int_0^x f^o(s)ds + \int_0^x \left(\int_0^t f^o(s)ds \right) d_t(-k(x,t))$$

$$= (Kf^o)(x),$$

where f^o is the level function of f. Therefore

$$\int_0^\infty (Kf)^q(x)u(x)dx \leq \int_0^\infty (Kf^o)^q(x)u(x)dx$$

$$= \int_0^\infty u(x)(Kf^o)^{q-1}(x) \int_0^x k(x,t)f^o(t)dtdx$$

$$= \int_0^\infty f^\circ(t) \int_t^\infty k(x,t)u(x) \left[\left(\int_0^t + \int_t^x \right) k(x,s)f^\circ(s)ds \right]^{q-1} dxdt$$

$$\leq \int_0^\infty f^\circ(t) \int_t^\infty k(x,t)u(x) \left[\int_0^t k(x,s)f^\circ(s)ds \right]^{q-1} dxdt$$

$$\leq \int_0^\infty f^\circ(t) \int_t^\infty k^q(x,t)u(x) \left(\int_0^t f^\circ(s)ds \right)^{q-1} dxdt$$

since $0 < q < 1$ and $k(x,s) \geq k(x,t)$ for $0 < s \leq t$. By property (ii) of Lemma 2.19 with $w = v^{1-p'}$ we have

$$\int_0^t f^\circ(s)ds = \int_0^t \left(\frac{f^\circ(s)}{v^{1-p'}(s)} \right) v^{1-p'}(s)ds$$

$$\geq f^\circ(t)v^{p'-1}(t) \int_0^t v^{1-p'}(s)ds$$

so that

$$\int_0^\infty u(x)(Kf)^q(x)dx \leq \int_0^\infty (f^\circ(t))^q v^{(p'-1)(q-1)-q/p}(t)v^{q/p}(t)$$

$$\times \left(\int_0^t v^{1-p'}(s)ds \right)^{q-1} (\widetilde{K}_q u)(t)dt.$$

Since $[(p'-1)(q-1) - q/p](r/q) = [(p'-1)/q' - 1/p]r = 1 - p'$, Hölder's inequality with exponents p/q and $(p/q)' = r/q$ yields

$$\int_0^\infty u(x)(Kf)^q(x)dx$$

$$\leq \left(\int_0^\infty (f^\circ)^p(x)v(x)dx \right)^{q/p}$$

$$\times \left(\int_0^\infty v^{1-p'}(x) \left(\int_0^x v^{1-p'}(t)dt \right)^{(q-1)r/q} (\widetilde{K}_q u)^{r/q}(x)dx \right)^{q/r}$$

$$= \left(\int_0^\infty (f^\circ)^p(x)v(x)dx \right)^{q/p}$$

$$\times \left(\int_0^\infty v^{1-p'}(x)(K_0 v^{1-p'})^{r/q'}(x)(\widetilde{K}_q u)^{r/q}(x)dx \right)^{q/r}$$
$$= \|f^\circ\|_{p,v}^q B_0^q.$$

The result now follows from the fact that, due to property (iii) of Lemma 2.19,

$$\|f^\circ\|_{p,v}^p = \int_0^\infty (f^\circ)^p(x)v(x)dx = \int_0^\infty \left(\frac{f^\circ(x)}{v^{1-p'}(x)} \right)^p v^{(1-p')p+1}(x)dx$$

$$= \int_0^\infty \left(\frac{f^\circ(x)}{v^{1-p'}(x)} \right)^p v^{1-p'}(x)dx$$

$$\leq \int_0^\infty \left(\frac{f(x)}{v^{1-p'}(x)} \right)^p v^{1-p'}(x)dx$$

$$= \int_0^\infty f^p(x)v(x)dx = \|f\|_{p,v}^p. \qquad \square$$

The corresponding result for the adjoint operator $\widetilde{K}$ was in previous edition of this book only "Conjecture 2.20" but this conjecture has now been solved in [JJG1] by P. Jain *et al.*

Theorem 2.20. *Suppose $0 < q < 1 < p < \infty$ and denote $\frac{1}{r} = \frac{1}{q} - \frac{1}{p}$. Let $\widetilde{K}$ be defined by*

$$(\widetilde{K}g)(x) = \int_x^\infty k(t,x)g(t)dt, \quad x > 0, \ g \geq 0,$$

where $k(t,x)$ is increasing in t. If

$$\widetilde{B}_0 \equiv \left(\int_0^\infty (\widetilde{K}_0 v^{1-p'})^{r/q'}(t)(K_q u)^{r/q}(t)v^{1-p'}(t)dt \right)^{1/r} < \infty,$$

then

$$\|\widetilde{K}g\|_{q,u} \leq C\|g\|_{p,v} \qquad (2.57)$$

for all $g \geq 0$ with $C \leq B_0$.

Conversely, if (2.57) holds for all $g \geq 0$, then $\widetilde{B}_2 \leq C < \infty$, where

$$\widetilde{B}_2 \equiv \left(\int_0^\infty (K_q u)^{p'/q}(t) v^{1-p'}(t) dt \right)^{1/p'}.$$

2.6. Some Modifications and Extensions

The general interval (a, b)

Let us mention the results corresponding to the results of Secs. 2.3–2.5 for the case of a general interval (a, b), i.e., for the operators

$$(Kf)(x) = \int_a^x k(x, t) f(t) dt,$$

$$(\widetilde{K}g)(x) = \int_x^b k(t, x) g(t) dt, \tag{2.58}$$

with $k(x, t)$ defined for $a < t < x < b$ and satisfying

$$k(x, t) \geq 0 \quad \text{for } x > t > a,$$

k is increasing in x and decreasing in t,

$$k(x, t) \approx k(x, z) + k(z, t) \quad \text{for } a < t < z < x < b. \tag{2.59}$$

Furthermore, analogously to (2.30) define operators K_s and $\widetilde{K}_s$ for $s \geq 0$ as

$$(K_s h)(x) = \int_a^x k^s(x, t) h(t) dt, \quad (\widetilde{K}_s h)(x) = \int_x^b k^s(t, x) h(t) dt. \tag{2.60}$$

Then we have the following assertion which corresponds to Theorem 2.10:

Theorem 2.21. *Let $1 < p \leq q < \infty$. Let K be the operator from (2.58) satisfying conditions (2.59) and let K_s and $\widetilde{K}_s$ be given by formula (2.60). Then the inequality*

$$\left(\int_a^b (Kf)^q(x) u(x) dx \right)^{1/q} \leq C \left(\int_a^b f^p(x) v(x) dx \right)^{1/p}, \tag{2.61}$$

i.e., the inequality

$$\|Kf\|_{q,u} \leq C\|f\|_{p,v}$$

holds for all $f \geq 0$ if and only if

$$A_0 := \sup_{a<t<b} (\widetilde{K}_q u)^{1/q}(t)(K_0 v^{1-p'})^{1/p'}(t) < \infty,$$

$$A_1 := \sup_{a<t<b} (\widetilde{K}_0 u)^{1/q}(t)(K_{p'} v^{1-p'})^{1/p'}(t) < \infty. \qquad (2.62)$$

We omit the proof since we can reduce Theorem 2.21 to Theorem 2.10, i.e., to the case of the interval $(0, \infty)$, by the same steps as it was done in Sec. 1.5. For this reason, the proof of Theorem 2.21 as well as the formulation and proof of the analogues of Theorem 2.12 and Theorem 2.15 are left to the reader.

Strong and weak type inequalities

Inequality (2.1), i.e., the inequality

$$\left(\int_a^b |(Tf)(x)|^q u(x)dx \right)^{1/q} \leq C \left(\int_a^b |f(x)|^p v(x)dx \right)^{1/p}$$

with a rather general (integral) operator T is sometimes called a *strong type inequality* or a *strong (p, q) inequality*.

Many authors deal also with the so-called *weak type inequalities* or *weak (p, q) inequalities* of the form

$$u(\{x \in (a, b) \ (Tf)(x) > \lambda\}) \leq C\lambda^{-q} \left(\int_a^b |f(x)|^p v(x)dx \right)^{q/p},$$

$$(2.63)$$

where $\lambda > 0$ is arbitrary, C is independent of f and λ and $u(E) = \int_E u(x)dx$ for any measurable set $E \subset (a, b)$.

We will not deal here with weak type inequalities, we just refer to some early results by K. Andersen and B. Muckenhoupt [AM1], V.M. Kokilashvili [Ko2], I.Z. Genebashvili *et al.* [GGK1] and F.J. Martín-Reyes [Mar1].

A more general case

The first two papers just mentioned deal in fact with strong and weak type inequalities for the more special operator T defined as

$$(Tf)(x) = \sup_{t \geq 0} \left| \int_X k(x, y, t) f(y) d\mu \right| \qquad (2.64)$$

with (X, d, μ) a general measure space with a quasi-metric d, a measure μ and k a positive measurable kernel on $X \times X \times [0, \infty)$.

For $X = (a, b)$, $d\mu = u(y)dy$ and k depending only on x and y, we have the operator investigated in this chapter.

Remark 2.22. Some new information concerning Hardy-type inequalities for Hardy operators with kernels can be found in Sec. 7.5. In particular, it is pointed out that in several cases inequalities called Hilbert-type or Hardy-Hilbert-type inequalities follow from some general results when the kernels are homogeneous of degree -1 (of degree $-n$ for the n-dimensional case). We also refer to the Ph.D. theses of L. Arendarenko [Ar1] and E. Ushakova [U1] (see also references therein) and to the papers [KPS1], [PSU1], [SU2], [U2], [LPSW1] and [LPS2]. Concerning the discrete case we refer to the Ph.D. thesis by A. Temirkhanova [Te1].

2.7. Comments and Remarks

2.7.1. The proof of Theorem 2.3 follows the ideas of P.A. Zharov [Z1] (where only the case $p, q > 1$ was considered, but the idea can be used also for $0 < q \leq 1$, $p > 1$).

2.7.2. The general Hardy-type operator in the form from Definition 2.5 was introduced and investigated by S. Bloom and R. Kerman [BK1] and independently by R. Oinarov [O1]. The result involving the Riemann-Liouville operator (see Example 2.7 (iii) or Remark 2.8 (iii)) was proved by F.J. Martín-Reyes and E. Sawyer [MS1] and independently by V. Stepanov [St1]; in fact, in the first mentioned paper, a somewhat more general kernel was considered, namely the convolution kernel given in Example 2.7 (i).

The proofs of the results given here follow closely those of V. Stepanov [St2], [St5].

2.7.3. The case $0 < q < 1 < p < \infty$ in Theorem 2.17 requires the concept of the *level function* and its properties described in Lemma 2.19. This lemma was proved by I. Halperin [H1] and used first in connection with the Hardy inequality in the Ph.D. thesis by G. Sinnamon [Si1] (see also [Si2], [Si3]).

2.7.4. The result of Theorem 2.10 extends further to the index range $0 < q < p$, $p > 1$. To see this we have to apply the modular inequality of Q. Lai [L3, Theorem 1], with $P(x) = x^p$, $p > 1$, and $Q(x) = x^q$, $0 < q < p$. Then this result, together with Schur's lemma, shows that inequality (2.34) holds if and only if for all covering sequences $\{x_k\}_{k\in z}$ (see Q. Lai [L3]),

$$\sum_{k\in z}\left[\int_{x_k}^{x_{k+1}}\left|\int_{x_{k-1}}^{x_k}k^{p'}(x_k,t)v^{1-p'}(t)dt\right|^{q/p'}u(x)dx\right]^{r/q} \leq C$$

and

$$\sum_{k\in z}\left[\int_{x_k}^{x_{k+1}}\left|\int_{x_{k-1}}^{x_k}v^{1-p'}(t)dt\right|^{q/p'}k^q(x,x_k)u(x)dx\right]^{r/q} \leq C$$

are satisfied, where $\frac{1}{r} = \frac{1}{q} - \frac{1}{p}$.

2.7.5. As in the case of the Hardy inequalities, the question of sharp constants is largely open. However, certain estimates for the constants can be given in terms of the operator norm. In the case of the Riemann-Liouville operator, such estimates (which for specific weights are sharp) have been given by V.M. Manakov [Ma1].

2.7.6. The weight characterization for the Riemann-Liouville operator, which we rewrite here in the form

$$(I_\alpha f)(x) = \int_0^x (x-t)^{\alpha-1}f(t)dt, \quad \alpha > 0,$$

has been given in this chapter only in the case $\alpha \geq 1$, and similarly for the conjugate Weyl fractional integral operator. The case $0 < \alpha < 1$

does not follow here since then the kernel $k(x, y) = (x - y)^{\alpha-1}$ is *not* an Oinarov kernel.

Properties of the modified operator

$$(T_\alpha f)(x) = \frac{v(x)}{x^\alpha} \int_0^x (x - t)^{\alpha-1} f(t) dt, \ \alpha > 0$$

as a mapping between *nonweighted* L^p and L^q spaces with $0 < p, q < \infty$ and $p > \max(1/\alpha, 1)$ are considered by D.V. Prokhorov [Pr1].

2.7.7. Integral operators, in particular fractional integrals, in more general homogeneous spaces (see Sec. 2.6) are investigated by many authors. Let us mention at least an important Georgian school, represented, e.g., by the paper of V.M. Kokilashvili and A. Meshkhi [KM1], the book [KoMP] and the references therein.

3

The Hardy-Steklov Operator

3.1. Introduction

In this chapter, we will deal mainly with functions defined on the (standard) interval

$$(0, \infty).$$

Let us start with an example.

Example 3.1. The classical Hardy operator H, defined for functions $f = f(t) \geq 0, t \in (0, \infty)$, as

$$(Hf)(x) := \int_0^x f(t)dt, \qquad 0 < x < \infty, \tag{3.1}$$

is obviously related to the triangular domain $\Delta = \{(x, t), \ 0 < t < x < \infty\}$ (see Fig. 3.1). This can be modified by considering the operator T, defined by

$$(Tf)(x) := \int_0^{b(x)} f(t)dt, \quad 0 < x < \infty, \tag{3.2}$$

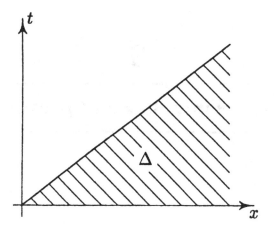

Fig. 3.1

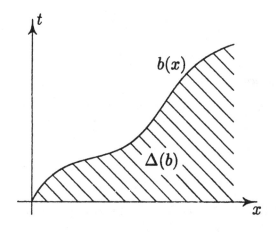

Fig. 3.2

where $b(x)$ is a strictly increasing differentiable function on $[0, \infty]$ satisfying $b(0) = 0, b(\infty) = \infty$. The operator T is related to a "perturbed" triangular domain $\Delta(b) = \{(x, t) : 0 < x < \infty, \ 0 < t < b(x)\}$ (see Fig. 3.2), but the investigation of the weighted norm inequality

$$\| Tf \|_{q,u} \leq C \| f \|_{p,v}, \tag{3.3}$$

i.e., the inequality

$$\left(\int_0^\infty |(Tf)(x)|^q u(x) dx \right)^{1/q} \leq C \left(\int_0^\infty |f(x)|^p v(x) dx \right)^{1/p}, \quad (3.4)$$

can be easily reduced to the investigation of the Hardy inequality for the pair of weight functions U, v where

$$U(x) = u(b^{-1}(x))(b^{-1})'(x) \quad (3.5)$$

with b^{-1} the inverse function to b.

Indeed, we have

$$\int_0^\infty (Tf)^q(x) u(x) dx = \int_0^\infty \left(\int_0^{b(x)} f(t) dt \right)^q u(x) dx$$

$$= \int_0^\infty \left(\int_0^y f(t) dt \right)^q u(b^{-1}(y))(b^{-1})'(y) dy$$

$$= \int_0^\infty (Hf)^q(y) U(y) dy$$

by the obvious substitution $y = b(x)$, i.e., $x = b^{-1}(y)$.

Consequently, the necessary and sufficient condition for the validity of the Hardy inequality

$$\| Hf \|_{q,U} \leq C \| f \|_{p,v}$$

in the range of parameters $1 < p \leq q < \infty$, i.e., the condition

$$\sup_{y>0} \left(\int_y^\infty U(t) dt \right)^{1/q} \left(\int_0^y v^{1-p'}(t) dt \right)^{1/p'} < \infty$$

can be rewritten — in the case of inequality (3.4) — as

$$\sup_{x>0} \left(\int_x^\infty u(s) ds \right)^{1/q} \left(\int_0^{b(x)} v^{1-p'}(s) ds \right)^{1/p'} < \infty. \quad (3.6)$$

It is easy to see that the conjugate operator $\widetilde{T}$ of T has the form

$$(\widetilde{T}f)(x) := \int_{a(x)}^{\infty} f(t)dt \quad \text{with} \quad a(x) = b^{-1}(x)$$

(the inverse of b). The weight characterization of this operator follows the same lines as the characterization of the conjugate Hardy operator $\widetilde{H}$.

This motivates the introduction of a new operator T, defined by

$$(Tf)(x) := \int_{a(x)}^{b(x)} f(t)dt, \tag{3.7}$$

where the functions a and b are defined below.

Definition 3.2. Let $a = a(x)$, $b = b(x)$ be strictly increasing differentiable functions on $[0, \infty]$ satisfying

$$\begin{aligned} a(0) &= b(0) = 0, \\ a(x) &< b(x) \quad \text{for} \quad 0 < x < \infty, \\ a(\infty) &= b(\infty). \end{aligned} \tag{3.8}$$

The operator T defined for $f = f(t) \geq 0, 0 < t < \infty$, by formula (3.7), will be called the *Hardy-Steklov operator*.

Remark 3.3. (i) Similarly as in the foregoing chapters, it is our aim to characterize weight functions u, v for which the mapping T from (3.7) is bounded,

$$T : L^p(v) \to L^q(u),$$

i.e., for which inequality (3.4) holds for p, q such that

$$0 < q < \infty, \quad 1 < p < \infty.$$

Let us mention that for the case $1 < p \leq q < \infty$ it will be shown (see Theorem 3.7) that the inequality

$$\left(\int_0^{\infty} \left(\int_{a(x)}^{b(x)} f(t)dt \right)^q u(x)dx \right)^{1/q} \leq C \left(\int_0^{\infty} f^p(x)v(x)dx \right)^{1/p} \tag{3.9}$$

holds for all $f \geq 0$ if and only if

$$A := \sup \left(\int_t^x u(s)ds \right)^{1/q} \left(\int_{a(x)}^{b(t)} v^{1-p'}(s)ds \right)^{1/p'} < \infty, \qquad (3.10)$$

where the supremum is taken over all x and t such that

$$0 < t < x < \infty \quad \text{and} \quad a(x) < b(t). \qquad (3.11)$$

(ii) To see that the class of admissible weights u, v defined by (3.10) is *strictly larger* than that of (3.6) take

$$a(x) = \frac{1}{2}x, \ b(x) = x, \ u(x) = x^\alpha, \ v(x) = x^\beta. \qquad (3.12)$$

Thus we compare two operators:

$$(T_1 f)(x) = \int_0^x f(t)dt \quad \text{and} \quad (T_2 f)(x) = \int_{x/2}^x f(t)dt.$$

T_1 is the Hardy operator, and the inequality $\| T_1 f \|_{q,u} \leq C \| f \|_{p,v}$ is satisfied if and only if

$$\frac{\beta}{p} = \frac{\alpha+1}{q} + \frac{1}{p'} \qquad (3.13)$$

and additionally $\alpha < -1$ (i.e., $\beta < p-1$; see Example 0.5). To verify the validity of the inequality $\| T_2 f \|_{q,u} \leq C \| f \|_{p,v}$, we must show that condition (3.10) is satisfied. For our choice of u and v we use the estimate

$$A = \sup_{t<x<2t} \left(\int_t^x s^\alpha ds \right)^{1/q} \left(\int_{x/2}^t s^{\beta(1-p')}ds \right)^{1/p'}$$

$$\leq \sup_{x>0} \left(\int_{x/2}^x s^\alpha ds \right)^{1/q} \left(\int_{x/2}^x s^{\beta(1-p')}ds \right)^{1/p'}$$

$$= C_0 \sup_{x>0} x^{\frac{\alpha+1}{q} + \frac{\beta(1-p')+1}{p'}};$$

the last expression (and consequently also A) is finite if (3.13) holds. Hence, for the Hardy-Steklov operator T_2 we have *no additional condition* on α (and β) and thus, a bigger class of admissible weights.

(iii) In Theorem 3.7 we will deal with the case $1 < p \leq q < \infty$. Let us mention that for the case $p = q = 1$, inequality (3.9) is trivial with

$$v(x) = \int_{b^{-1}(x)}^{a^{-1}(x)} u(t)dt.$$

Indeed, as Fubini's theorem shows,

$$\int_0^\infty \left(\int_{a(x)}^{b(x)} f(t)dt \right) u(x)dx = \int_0^\infty f(t) \left(\int_{b^{-1}(t)}^{a^{-1}(t)} u(x)dx \right) dt,$$

i.e., (3.9) becomes equality with $C = 1$.

On the other hand, the Hardy inequality with $p = q = 1$ and $u(x) = \frac{1}{x}$ fails for *every* choice of v, as the following example illustrates for the case $v(x) \equiv 1$.

Example 3.4. The foregoing considerations indicate that there is a substantial distinction between the Hardy and the Hardy-Steklov operators. To give another example, let us consider again $a(x)$ and $b(x)$ as in (3.12), but $u(x) = x^{-q}$ and $v(x) \equiv 1$. The weighted norm inequality for the operator T from (3.2) is then the Hardy inequality for the averaging operator:

$$\left(\int_0^\infty \left(\frac{1}{x} \int_0^x f(t)dt \right)^q dx \right)^{1/q} \leq C \left(\int_0^\infty f^p(x)dx \right)^{1/p}$$

which is *not satisfied* if we take $p = q = 1$, since the corresponding necessary and sufficient condition,

$$\sup_{x>0} \left(\int_x^\infty u(t)dt \right) < \infty$$

(see, e.g., [OK, Lemma 5.4]) does not hold. On the other hand, the corresponding inequality (3.9) for $p = q = 1$, i.e., the inequality

$$\int_0^\infty \left(\frac{1}{x} \int_{x/2}^x f(t)dt \right) dx \leq C \int_0^\infty f(x)dx,$$

is satisfied with $C = \log 2$ and the equality sign, as can be shown by Fubini's theorem.

An application to financial markets

The Hardy-Steklov operator gives rise to the *moving averaging operator*

$$(S_a^b f)(x) = \frac{1}{b(x) - a(x)} \int_{a(x)}^{b(x)} f(t)dt, \quad f \geq 0. \tag{3.14}$$

This operator in its various forms is of considerable importance to the *technical analysts* in the study of equity markets. These *technical analysts* try to predict the future of the stock price or the future of an equity market solely on the base of the past performance of the stock price or market valuation, respectively. For example, some analysts consider an equity (stock) f whose price at time t is $f(t)$, a recommended "buy" if $(S_{t-200}^t f)(t) \leq f(t)$ while the reversal of this inequality is a "sell".

Similarly, if $f(t)$ represents, say, the Dow Jones Industrial Average in New York at time t and if $(S_{t-200}^t f)(t) \leq f(t)$, then it was observed that the return of the market in the year following t gained 12% while a reversal of this inequality showed that the market lost 7% in the year following t.

The introduction of weights and the pointwise estimate replaced by mean estimates with good control of the constants may therefore yield additional quantitative information in the study of financial markets.

The Steklov operator

For functions f defined on $(-\infty, \infty)$ and for $\gamma > 0$, the *Steklov operator* S_γ is defined as

$$(S_\gamma f)(x) = \int_{x-\gamma}^{x+\gamma} f(t)dt. \tag{3.15}$$

As a consequence of our results concerning the Hardy-Steklov shift operator, we will deduce later a characterization of weights u, v

for which

$$\left(\int_{-\infty}^{\infty} \left(\int_{x-\gamma}^{x+\gamma} f(t)dt \right)^q u(x)dx \right)^{1/q} \le C \left(\int_{-\infty}^{\infty} f^p(x)v(x)dx \right)^{1/p}$$

is satisfied.

An application to fractional order Hardy inequalities

Although these inequalities will be dealt with in detail in Chap. 5, let us here mention one particular result.

The integral

$$\int_0^{\infty} |f(x)|^p x^{-\lambda p} dx$$

can be estimated for $f \in C^1(0, \infty)$ with $f(0) = f(\infty) = 0$ and for $0 < \lambda < 1$, $\lambda \ne 1/p$, $1 < p < \infty$, either by

$$C_1 \int_0^{\infty} |f'(x)|^p x^{(1-\lambda)p} dx \tag{3.16}$$

(in this case, it is the Hardy inequality (1.25) for $(a, b) = (0, \infty)$, $p = q$, $v(x) = x^{(1-\lambda)p}$ and $u(x) = x^{-\lambda p}$), or by

$$C_2 \int_0^{\infty} \int_0^{\infty} \frac{|f(x) - f(y)|^p}{|x - y|^{1+\lambda p}} dx dy \tag{3.17}$$

(see, e.g., G.N. Jakovlev [J1], [J2] or P. Grisvard [Gr1]). The question arises which one of these two estimates is "better", i.e., which one of the expressions (3.16) and (3.17) is larger. The answer will be given in Sec. 3.5 using the Hardy-Steklov operator. In particular, it will be shown that

$$\int_0^{\infty} \int_0^{\infty} \frac{|f(x) - f(y)|^p}{|x - y|^{1+\lambda p}} dx dy \le C \int_0^{\infty} |f'(x)|^p x^{(1-\lambda)p} dx.$$

This result was derived earlier (see, e.g., A. Kufner [Ku4], [Ku5]), and in Chap. 5 we will give a direct proof (see Theorem 5.3). Here it is mentioned simply as an interesting application of the Hardy-Steklov operator.

3.2. Some Auxiliary Results

Definition 3.5. The functions a and b introduced in Definition 3.2 are strictly increasing and differentiable, and consequently, the inverse functions a^{-1} and b^{-1} exist and are strictly increasing and differentiable, too. We define a sequence $\{m_k\}_{k\in\mathbb{Z}}$ as follows:

For fixed $m > 0$ define $m_0 = m$ and

$$
\begin{aligned}
m_{k+1} &= a^{-1}(b(m_k)) && \text{if} \quad k \geq 0, \\
m_k &= b^{-1}(a(m_{k+1})) && \text{if} \quad k < 0.
\end{aligned}
\tag{3.18}
$$

Clearly

$$
a(m_{k+1}) = b(m_k) \quad \text{for all} \quad k \in \mathbb{Z}
\tag{3.19}
$$

(see Fig. 3.3).

The following technical lemma will be required in the sequel:

Lemma 3.6. *Fix $m > 0$ and define $\{m_k\}_{k\in\mathbb{Z}}$ by (3.18). Then $m_k < m_{k+1}$ for $k \in \mathbb{Z}$ and*

$$
\lim_{k\to\infty} m_k = \infty, \qquad \lim_{k\to-\infty} m_k = 0.
$$

Proof. Since $a(m_k) < b(m_k) = a(m_{k+1})$, it follows that $m_k < m_{k+1}$ for $k \in \mathbb{Z}$. Hence $\{m_k\}_{k\in\mathbb{Z}}$ is increasing, and this implies the existence

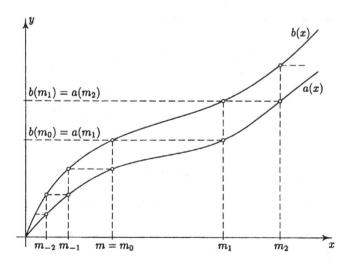

Fig. 3.3

of $M^- \in [0, \infty)$ and $M^+ \in (0, \infty]$ such that

$$\lim_{k \to -\infty} m_k = M^-, \quad \lim_{k \to \infty} m_k = M^+.$$

Since a and b are continuous, we have

$$b(M^-) = \lim_{k \to -\infty} b(m_k) = \lim_{k \to -\infty} a(m_{k+1}) = a(M^-),$$
$$b(M^+) = \lim_{k \to \infty} b(m_k) = \lim_{k \to \infty} a(m_{k+1}) = a(M^+).$$

But, since $a(x) < b(x)$ for $x \in (0, \infty)$, we conclude that $M^- = 0$ and $M^+ = \infty$. $\qquad \square$

Some notation

(i) Define functions u_a and u_b by

$$u_a(y) = u(a^{-1}(y))(a^{-1})'(y),$$
$$u_b(y) = u(b^{-1}(y))(b^{-1})'(y). \tag{3.20}$$

Then

$$u_a(y)dy = u(x)dx \quad \text{and} \quad u_b(y)dy = u(x)dx$$

if $y = a(x)$ and $y = b(x)$, respectively.

(ii) As mentioned, we will deal with the inequality

$$\left(\int_0^\infty \left(\int_{a(x)}^{b(x)} f(t)dt \right)^q u(x)dx \right)^{1/q}$$
$$\leq C \left(\int_0^\infty f^p(x)v(x)dx \right)^{1/p}, \quad f \geq 0. \tag{3.21}$$

If we denote

$$w = v^{1-p'}$$

and replace f in (3.21) by fw, we obtain an *equivalent* inequality

$$\left(\int_0^\infty \left(\int_{a(x)}^{b(x)} f(t)w(t)dt \right)^q u(x)dx \right)^{1/q} \leq C \left(\int_0^\infty f^p(x)w(x)dx \right)^{1/p}.$$
$$\tag{3.22}$$

Inequalities (3.21) and (3.22) will be used interchangeably in the sequel.

3.3. The Case $p \leq q$

The first main result of this chapter reads as follows.

Theorem 3.7. *If* $1 < p \leq q < \infty$, *then* (3.21) *holds for all functions* $f \geq 0$ *if and only if*

$$A < \infty, \tag{3.23}$$

where

$$A := \sup \left(\int_t^x u(s)ds \right)^{1/q} \left(\int_{a(x)}^{b(t)} v^{1-p'}(s)ds \right)^{1/p'} \tag{3.24}$$

and the supremum is taken over all x *and* t *such that*

$$0 < t \leq x < \infty \quad and \quad a(x) < b(t) \tag{3.25}$$

(*see Fig. 3.4*).

Moreover, if C *is the least constant for which* (3.21) *holds then*

$$C \approx A. \tag{3.26}$$

Proof. (i) *Necessity* of (3.23). Suppose that (3.21) holds, fix t and x according to (3.25) and let $\{w_n\}_{k=1}^{\infty}$ be a sequence of L^1-weights

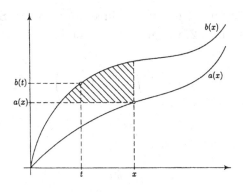

Fig. 3.4

such that $w_n(x) \le w(x)$ and $w_n \uparrow w$ as $n \to \infty$. For a fixed $n \in \mathbb{N}$, let

$$f(s) = \chi_{(a(x),b(t))}(s)w_n(s)/w(s).$$

If $t < s \le x$, then

$$\left(\int_{a(x)}^{b(t)} w_n(s)ds \right) \left(\int_t^x u(s)ds \right)^{1/q}$$

$$= \left(\int_t^x \left(\int_{a(x)}^{b(t)} f(t)w(t)dt \right)^q u(s)ds \right)^{1/q}$$

$$\le \left(\int_0^\infty \left(\int_{a(s)}^{b(s)} f(t)w(t)dt \right)^q u(s)ds \right)^{1/q}$$

$$\le C \left(\int_0^\infty f^p(s)w(s)ds \right)^{1/p}$$

$$= C \left(\int_{a(x)}^{b(t)} w_n^p(s)w^{1-p}(s)ds \right)^{1/p}$$

$$\le C \left(\int_{a(x)}^{b(t)} w_n(s)ds \right)^{1/p}, \tag{3.27}$$

where we have used the equivalent inequality (3.22). Since $w_n \in L^1$, the last integral in (3.27) is finite and therefore

$$\left(\int_t^x u(s)ds \right)^{1/q} \left(\int_{a(x)}^{b(t)} w_n(s)ds \right)^{1-1/p} \le C,$$

where C is independent of n. Since $w_n \uparrow w$ as $n \to \infty$, the Monotone Convergence Theorem implies

$$\left(\int_t^x u(s)ds \right)^{1/q} \left(\int_{a(x)}^{b(t)} w(s)ds \right)^{1/p'} \le C$$

for all x and t satisfying (3.25). Taking the supremum over all such x and t we have (3.24) with $w = v^{1-p'}$ and

$$A \leq C < \infty. \tag{3.28}$$

(ii) *Sufficiency* of (3.23). Fix t and write $y = a(x)$, $w = v^{1-p'}$ in (3.24). Then

$$\sup_{a(t) \leq y \leq b(t)} \left(\int_{a(t)}^{y} u_a(s)ds \right)^{1/q} \left(\int_{y}^{b(t)} w(s)ds \right)^{1/p'} \leq A < \infty.$$

But this is nothing else than the necessary and sufficient condition for the validity of the Hardy inequality in the interval $(a(t), b(t))$ and with weights u_a and v:

$$\widetilde{A}(t) = \widetilde{A}(a(t), b(t); u_a, v) \leq A < \infty$$

(see (0.24) and (1.34)), and thus we have

$$\left(\int_{a(t)}^{b(t)} \left(\int_{y}^{b(t)} f(t)dt \right)^{q} u_a(s)ds \right)^{1/q} \leq C \left(\int_{a(t)}^{b(t)} f^{p}(s)v(s)ds \right)^{1/p},$$

which can be rewritten, similarly as in p. 130, into the form

$$\left(\int_{a(t)}^{b(t)} \left(\int_{y}^{b(t)} f(t)w(t)dt \right)^{q} u_a(s)ds \right)^{1/q} \leq C \left(\int_{a(t)}^{b(t)} f^{p}(s)w(s)ds \right)^{1/p} \tag{3.29}$$

with $w = v^{1-p'}$ and with $C \leq k(p,q)\widetilde{A}(t) \leq k(p,q)A$ (here we used (0.26)).

Similarly, if x is fixed and $y = b(t)$ in (3.24), then

$$\sup_{a(x) \leq y \leq b(x)} \left(\int_{y}^{b(x)} u_b(s)ds \right)^{1/q} \left(\int_{a(x)}^{y} w(s)ds \right)^{1/p'} \leq A < \infty$$

and similarly as above this means that

$$A(x) = A(a(x), b(x); u_b, v) \leq A < \infty,$$

which implies

$$\left(\int_{a(x)}^{b(x)} \left(\int_{a(x)}^{y} f(t)w(t)dt \right)^q u_b(s)ds \right)^{1/q} \leq C \left(\int_{a(x)}^{b(x)} f^p(s)w(s)ds \right)^{1/p} \tag{3.30}$$

with $C \leq k(p,q)A(x) \leq k(p,q)A$.

Now fix $m > 0$ and define $\{m_k\}_{k \in \mathbb{Z}}$ as in (3.18). Denote

$$E_k = (m_k, m_{k+1}).$$

Then $a(x) < a(m_{k+1}) = b(m_k) \leq b(x)$ for $x \in E_k$, and if we denote

$$a_k = a(m_k), \ b_k = b(m_k),$$

we obtain via Minkowski's inequality

$$\left(\int_0^\infty \left(\int_{a(x)}^{b(x)} f(t)w(t)dt \right)^q u(x)dx \right)^{1/q}$$

$$= \left(\int_0^\infty \left(\sum_{k \in \mathbb{Z}} \chi_{E_k}(x) \int_{a(x)}^{b_k} f(t)w(t)dt \right. \right.$$

$$+ \left. \left. \sum_{k \in \mathbb{Z}} \chi_{E_k}(x) \int_{a_{k+1}}^{b(x)} f(t)w(t)dt \right)^q u(x)dx \right)^{1/q}$$

$$\leq \left(\int_0^\infty \left(\sum_{k \in \mathbb{Z}} \chi_{E_k}(x) \int_{a(x)}^{b_k} f(t)w(t)dt \right)^q u(x)dx \right)^{1/q}$$

$$+ \left(\int_0^\infty \left(\sum_{k \in \mathbb{Z}} \chi_{E_k}(x) \int_{a_{k+1}}^{b(x)} f(t)w(t)dt \right)^q u(x)dx \right)^{1/q} =: I_1 + I_2.$$

For each $x \in (0, \infty)$, only one term in the sum $\sum_{k \in \mathbb{Z}}$ under the integral sign can be zero, and since $(0, \infty) = \bigcup_{k \in \mathbb{Z}} \overline{E}_k$

$= \bigcup[m_k, m_{k+1}]$, the change of variable $y = a(x)$ and (3.29) yield

$$I_1 = \left(\sum_{k \in \mathbb{Z}} \int_{m_k}^{m_{k+1}} \left(\int_{a(x)}^{b_k} f(t)w(t)dt \right)^q u(x)dx \right)^{1/q}$$

$$= \left(\sum_{k \in \mathbb{Z}} \int_{a_k}^{b_k} \left(\int_{y}^{b_k} f(t)w(t)dt \right)^q u_a(y)dy \right)^{1/q}$$

$$\leq \left(\sum_{k \in \mathbb{Z}} C^q \left(\int_{a_k}^{b_k} f^p(s)w(s)ds \right)^{q/p} \right)^{1/q}$$

$$\leq C \left(\int_0^\infty f^p(s)w(s)ds \right)^{1/p},$$

where the last inequality follows from the fact that $q/p \geq 1$.

The estimate of I_2 is similar. The change of variables $y = b(x)$ and (3.30) yield

$$I_2 = \left(\sum_{k \in \mathbb{Z}} \int_{m_k}^{m_{k+1}} \left(\int_{a_{k+1}}^{b(x)} f(t)w(t)dt \right)^q u(x)dx \right)^{1/q}$$

$$= \left(\sum_{k \in \mathbb{Z}} \int_{a_{k+1}}^{b_{k+1}} \left(\int_{a_{k+1}}^{y} f(t)w(t)dt \right)^q u_b(y)dy \right)^{1/q}$$

$$\leq \left(\sum_{k \in \mathbb{Z}} C^q \left(\int_{a_{k+1}}^{b_{k+1}} f^p(s)w(s)ds \right)^{q/p} \right)^{1/q}$$

$$\leq C \left(\int_0^\infty f^p(s)w(s)ds \right)^{1/p}.$$

Therefore (3.22) — and consequently (3.21) — follows with the constant $2C \leq 2k(p,q)A$.

The last inequality and (3.28) imply (3.26). $\qquad \square$

Remark 3.8. (i) We can easily see that the operator $\widetilde{T}$ conjugate to

$$(Tf)(x) = \int_{a(x)}^{b(x)} f(t)dt$$

has the form

$$(\widetilde{T}f)(x) = \int_{b^{-1}(x)}^{a^{-1}(x)} f(t)dt.$$

(Use the duality formula $\langle f, Tg \rangle = \langle \widetilde{T}f, g \rangle$ and Fubini's theorem.) Since the operator $\widetilde{T}$ has a form similar to the operator T and maps $L^{q'}(u^{1-q'})$ continuously into $L^{p'}(v^{1-p'})$ provided the mapping $T : L^p(v) \to L^q(u)$ is bounded, we immediately obtain from Theorem 3.7 a characterization of admissible weights rewriting condition (3.23) into

$$\widetilde{A} < \infty,$$

where

$$\widetilde{A} := \sup \left(\int_t^x v^{1-p'}(s)ds \right)^{1/p'} \left(\int_{b^{-1}(x)}^{a^{-1}(t)} u^{(1-q')(1-q)}(s)ds \right)^{1/q}$$

$$= \sup \left(\int_t^x v^{1-p'}(s)ds \right)^{1/p'} \left(\int_{b^{-1}(x)}^{a^{-1}(t)} u(s)ds \right)^{1/q},$$

the supremum being taken over all x and t such that $0 < t \le x < \infty$ and $b^{-1}(x) < a^{-1}(t)$.

(ii) If we suppose that

$$1 = p \le q < \infty,$$

then (3.21) holds for $f \ge 0$ if

$$A := \operatorname*{ess\,sup}_{t>0} \frac{1}{v(t)} \left(\int_{b^{-1}(t)}^{a^{-1}(t)} u(s)ds \right)^{1/q} < \infty.$$

Indeed, using Minkowski's integral inequality, we obtain

$$
\| Tf \|_{q,u} = \left(\int_0^\infty \left(\int_{a(x)}^{b(x)} f(t)dt \right)^q u(x)dx \right)^{1/q}
$$

$$
\leq \int_0^\infty f(t) \left(\int_{b^{-1}(t)}^{a^{-1}(t)} u(x)dx \right)^{1/q} dt
$$

$$
= \int_0^\infty f(t)v(t) \left[\frac{1}{v(t)} \left(\int_{b^{-1}(t)}^{a^{-1}(t)} u(s)ds \right)^{1/q} \right] dt
$$

$$
\leq A \int_0^\infty f(t)v(t)dt = A\|f\|_{1,v}.
$$

3.4. The Case $p > q$

In order to provide a weight characterization for the Hardy-Steklov operator in the index range

$$
0 < q < p, \ 1 < p < \infty,
$$

we need the following known result (see [HS1]).

Proposition 3.9. *Suppose* $0 < q < p$, $1 < p < \infty$, $m > 0$, *and denote* $1/r = 1/q - 1/p$, $a_k = a(m_k)$ *and* $b_k = b(m_k)$ *with* a, b *from Definition 3.2 and* m_k *from the sequence* $\{m_k\}_{k \in \mathbb{Z}}$ *constructed in Definition 3.5. Let* u_a, u_b *be given by* (3.20). *If* $C = C(m_k)$ *and* $\widetilde{C} = \widetilde{C}(m_k)$ *are the best constants in*

$$
\left(\int_{a_k}^{b_k} \left(\int_{a_k}^{y} f(s)w(s)ds \right)^q u_b(y)dy \right)^{1/q} \leq C \left(\int_{a_k}^{b_k} f^p(y)w(y)dy \right)^{1/p}
$$
(3.31)

and

$$
\left(\int_{a_k}^{b_k} \left(\int_{y}^{b_k} f(s)w(s)ds \right)^q u_a(y)dy \right)^{1/q} \leq \widetilde{C} \left(\int_{a_k}^{b_k} f^p(y)w(y)dy \right)^{1/p},
$$
(3.32)

respectively, then

$$C(m_k) \approx D(m_k) \quad and \quad \widetilde{C}(m_k) \approx \widetilde{D}(m_k), \qquad (3.33)$$

where

$$D(m_k) = \left(\int_{a_k}^{b_k} \left(\int_{a_k}^{y} w(s)ds \right)^{r/p'} \left(\int_{y}^{b_k} u_b(s)ds \right)^{r/p} u_b(y)dy \right)^{1/r},$$

$$\widetilde{D}(m_k) = \left(\int_{a_k}^{b_k} \left(\int_{y}^{b_k} w(s)ds \right)^{r/p'} \left(\int_{a_k}^{y} u_a(s)ds \right)^{r/p} u_a(y)dy \right)^{1/r}.$$

$$(3.34)$$

Remark and Notation 3.10. (i) Observe that $D(m_k)$ and $\widetilde{D}(m_k)$ in Proposition 3.9 are the numbers $A = A(a_k, b_k; u_b, w)$ and $\widetilde{A} = \widetilde{A}(a_k, b_k; u_a, w)$ from (0.20) and (0.25) with $w = v^{1-p'}$ (more precisely, the corresponding numbers A^* and $\widetilde{A}^*$ from Remark 1.5 (iv)) whose finiteness is necessary and sufficient for the Hardy inequality and the conjugate Hardy inequality, i.e., the inequalities (3.31) and (3.32), to hold. The equivalence relations (3.33) follow from the corresponding estimates — see e.g. (0.27).

(ii) Before formulating and proving the second main result of this chapter, let us introduce a sequence $\{M_k\}_{k \in \mathbb{Z}}$, which is constructed as the sequence $\{m_k\}_{k \in \mathbb{Z}}$ in Definition 3.5 but with $M_0 = b^{-1}(1)$, and the so-called *normalizing function*

$$\sigma(t) := \sum_{k \in \mathbb{Z}} \chi_{(M_k, M_{k+1})}(t) \frac{d}{dt} (b^{-1} \circ a)^k(t), \qquad (3.35)$$

where $(b^{-1} \circ a)^k$ for $k > 0$ denotes the k times repeated composition, while for $k < 0$, $(b^{-1} \circ a)^k$ is to be interpreted as $\frac{1}{(b^{-1} \circ a)^{-k}}$.

Theorem 3.11. *Suppose* $0 < q < p$, $1 < p < \infty$, $\frac{1}{r} = \frac{1}{q} - \frac{1}{p}$. *Then inequality (3.21) (or equivalently inequality (3.22)) holds for* *all functions $f \geq 0$ if and only if*

$$\widetilde{A} = \max(A_1, A_2) < \infty, \tag{3.36}$$

where

$$A_1 := \left(\int_0^\infty \left(\int_{b^{-1}(a(t))}^t \left(\int_{a(t)}^{b(x)} w(s)ds \right)^{r/p'} \right. \right.$$

$$\left. \left. \times \left(\int_x^t u(s)ds \right)^{r/p} u(x)dx \right) \sigma(t)dt \right)^{1/r} \tag{3.37}$$

and

$$A_2 := \left(\int_0^\infty \left(\int_t^{a^{-1}(b(t))} \left(\int_{a(x)}^{b(t)} w(s)ds \right)^{r/p'} \right. \right.$$

$$\left. \left. \times \left(\int_t^x u(s)ds \right)^{r/p} u(x)dx \right) \sigma(t)dt \right)^{1/r} \tag{3.38}$$

with $w = v^{1-p'}$ and $\sigma(t)$ defined by (3.35).

Proof. (i) *Necessity* of condition (3.36). Suppose that (3.22) is satisfied. Let $\{u_n\}_{n=1}^\infty$, $\{w_n\}_{n=1}^\infty$ be sequences of L^1-weights such that $u_n(x) < u(x)$, $w_n(x) < w(x)$ and $u_n \uparrow u$, $w_n \uparrow w$ as $n \to \infty$. If $u_{b,n}(y) = u_n(b^{-1}(y))(b^{-1})'(y)$ and $u_b(y) = u(b^{-1}(y))(b^{-1})'(y)$, then $u_{b,n}(y) < u_b(y)$.

Fix $m > 0$ and define $\{m_k\}_{k \in \mathbb{Z}}$ as in Definition 3.8. For $D(m_k)$ and $\widetilde{D}(m_k)$ from (3.34) define

$$\mathcal{D}(m) = \left(\sum_{k \in \mathbb{Z}} D^r(m_k) \right)^{1/r}, \quad \widetilde{\mathcal{D}}(m) = \left(\sum_{k \in \mathbb{Z}} \widetilde{D}^r(m_k) \right)^{1/r}$$

and suppose

$$\max\left\{\sup_{m>0}\mathcal{D}(m),\ \sup_{m>0}\widetilde{\mathcal{D}}(m)\right\}<\infty. \tag{3.39}$$

Then

$$\left(\int_0^\infty\left(\int_{a(t)}^{b(t)}\left(\int_{a(t)}^y w(s)ds\right)^{r/p'}\right.\right.$$

$$\left.\left.\times\left(\int_y^{b(t)}u_b(s)ds\right)^{r/p}u_b(y)dy\right)\sigma(t)dt\right)^{1/r}$$

$$=\left(\sum_{k\in\mathbb{Z}}\int_{M_k}^{M_{k+1}}\left(\int_{a(t)}^{b(t)}\left(\int_{a(t)}^y w(s)ds\right)^{r/p'}\right.\right.$$

$$\left.\left.\times\left(\int_y^{b(t)}u_b(s)ds\right)^{r/p}u_b(y)dy\right)\frac{d}{dt}(b^{-1}\circ a)^k(t)dt\right)^{1/r}.$$

If $m=(b^{-1}\circ a)^k(t)$ then $t=(a^{-1}\circ b)^k(m)=m_k$. Hence if $t=M_{k+1}$ then $M_{k+1}=m_k$, which implies that $m=a^{-1}(1)$. If $t=M_k$, then $(a^{-1}\circ b)^k(m)=M_k=(a^{-1}\circ b)^k M_0$, and since $M_0=b^{-1}(1)$, it follows that $m=b^{-1}(1)$. Therefore the last sum is equal to

$$\left(\sum_{k\in\mathbb{Z}}\int_{b^{-1}(1)}^{a^{-1}(1)}\left(\int_{a_k}^{b_k}\left(\int_{a_k}^y w(s)ds\right)^{r/p'}\right.\right.$$

$$\left.\left.\times\left(\int_y^{b_k}u_b(s)ds\right)^{r/p}u_b(y)dy\right)dm\right)^{1/r}$$

$$=\left(\sum_{k\in\mathbb{Z}}\int_{b^{-1}(1)}^{a^{-1}(1)}D^r(m_k)dm\right)^{1/r}$$

$$=\left(\int_{b^{-1}(1)}^{a^{-1}(1)}\sum_{k\in\mathbb{Z}}D^r(m_k)dm\right)^{1/r}$$

$$\leq (a^{-1}(1) - b^{-1}(1))^{1/r} \sup_{m>0} \left(\sum_{k\in\mathbb{Z}} D^r(m_k) \right)^{1/r}$$

$$= (a^{-1}(1) - b^{-1}(1))^{1/r} \sup_{m>0} \mathcal{D}(m) < \infty$$

due to (3.39). But now the change of variables $y = b(x)$ shows that $A_1 < \infty$.

In the same way we can show that if $\sup_{m>0} \widetilde{\mathcal{D}}(m) < \infty$, then $A_2 < \infty$.

To complete the proof of necessity we must show that (3.39) holds. Write $a_k = a(m_k)$, $b_k = b(m_k)$ and $\chi_k = \chi_{(a_k,b_k)}$. Since $r/(pq') + 1 = r/(qp')$, we have

$$\left(\int_{a_k}^y w_n(s)ds \right)^{r/p'}$$

$$= \left(\left(\int_{a_k}^y w_n(s)ds \right)^{r/(qp')} \right)^q$$

$$= \left(\int_{a_k}^y \frac{d}{dt} \left(\int_{a_k}^t w_n(s)ds \right)^{r/(pq')+1} dt \right)^q$$

$$= (r/(pq') + 1)^q \left(\int_{a_k}^y \left(\int_{a_k}^t w_n(s)ds \right)^{r/(pq')} w_n(t)dt \right)^q$$

$$= \left(\frac{r}{p'q} \right)^q \left(\int_{a_k}^y \left(\int_{a_k}^t w_n(s)ds \right)^{r/(pq')} w_n(t)dt \right)^q,$$

which together with the fact that $t < y$, yields

$$\left(\sum_{k\in\mathbb{Z}} \int_{a_k}^{b_k} \left(\int_{a_k}^y w_n(s)ds \right)^{r/p'} \left(\int_y^{b_k} u_{b,n}(s)ds \right)^{r/p} u_{b,n}(y)dy \right)^{1/q}$$

$$= \frac{r}{p'q} \left(\sum_{k\in\mathbb{Z}} \int_{a_k}^{b_k} \left(\int_{a_k}^y \left(\int_{a_k}^t w_n(s)ds \right)^{r/(pq')} w_n(t)dt \right)^q \right.$$

$$\times \left(\int_y^{b_k} u_{b,n}(s)ds \right)^{r/p} u_{b,n}(y)dy \right)^{1/q}$$

$$\leq \frac{r}{p'q} \left(\sum_{k\in\mathbb{Z}} \int_{a_k}^{b_k} \left(\int_{a_k}^y \left(\int_{a_k}^t w_n(s)ds \right)^{r/(pq')} \right. \right.$$

$$\times \left(\int_t^{b_k} u_{b,n}(s)ds \right)^{r/(pq)} w_n(t)dt \right)^q u_{b,n}(y)dy \right)^{1/q}. \quad (3.40)$$

Here for each k, $a_k < y < b_k$ and $a_k < t < y$ so that $a_k < t < b_k$, and consequently, the last expression is equal to

$$\frac{r}{p'q} \left(\int_0^\infty \sum_{k\in\mathbb{Z}} \chi_k(y) \left(\int_{a_k}^y \left(\int_{a_k}^t w_n(s)ds \right)^{r/(pq')} \right. \right.$$

$$\times \left(\int_t^{b_k} u_{b,n}(s)ds \right)^{r/(pq)} \chi_k(t)w_n(t)dt \right)^q u_{b,n}(y)dy \right)^{1/q}$$

$$= \frac{r}{p'q} \left(\int_0^\infty \left(\sum_{k\in\mathbb{Z}} \chi_k(y) \int_{a_k}^y \left(\int_{a_k}^t w_n(s)ds \right)^{r/(pq')} \right. \right.$$

$$\times \left(\int_t^{b_k} u_{b,n}(s)ds \right)^{r/(pq)} \chi_k(t)w_n(t)dt \right)^q u_{b,n}(y)dy \right)^{1/q}$$

since for each y only one term of the sum can be non-zero. Now, if $y < b_k = b(m_k)$, then $b^{-1}(y) < m_k$ and $a(b^{-1}(y)) < a(m_k) = a_k$. Increasing the interval of integration (with respect to t) from (a_k, y) to $(a(b^{-1}(y)), y)$ and replacing χ_k by 1 we conclude that the last expression is not larger than

$$\frac{r}{p'q} \left(\int_0^\infty \left(\sum_{k\in\mathbb{Z}} \int_{a(b^{-1}(y))}^y \left(\int_{a_k}^t w_n(s)ds \right)^{r/(pq')} \right. \right.$$

$$\times \left(\int_t^{b_k} u_{b,n}(s)ds \right)^{r/(pq)} \chi_k(t)w_n(t)dt \right)^q u_{b,n}(y)dy \right)^{1/q}$$

$$= \frac{r}{p'q} \left(\int_0^\infty \left(\int_{a(b^{-1}(y))}^y \left(\sum_{k \in \mathbb{Z}} \left(\int_{a_k}^t w_n(s)ds \right)^{r/q'} \right. \right. \right.$$

$$\left. \left. \left. \times \left(\int_t^{b_k} u_{b,n}(s)ds \right)^{r/q} \chi_k(t) \right)^{1/p} w_n(t)dt \right)^q u_{b,n}(y)dy \right)^{1/q}$$

(3.41)

since only one term of the sum can be non-zero.

Now take

$$f(t) = \left(\sum_{k \in \mathbb{Z}} \left(\int_{a_k}^t w_n(s)ds \right)^{r/q'} \right.$$

$$\left. \times \left(\int_t^{b_k} u_{b,n}(s)ds \right)^{r/q} \chi_k(t) \right)^{1/p} \frac{w_n(t)}{w(t)}.$$

Then the change of variables $y = b(x)$ shows that (3.41) is equal to

$$\frac{r}{p'q} \left(\int_0^\infty \left(\int_{a(x)}^{b(x)} f(t)w(t)dt \right)^q u_n(x)dx \right)^{1/q}$$

$$\leq \frac{r}{p'q} \left(\int_0^\infty \left(\int_{a(x)}^{b(x)} f(t)w(t)dt \right)^q u(x)dx \right)^{1/q}$$

$$\leq C \frac{r}{p'q} \left(\int_0^\infty f^p(x)w(x)dx \right)^{1/p}$$

(3.42)

by (3.22). But, since $w_n < w$, the definition of f shows that

$$\left(\int_0^\infty f^p(x)w(x)dx \right)^{1/p} = \left(\int_0^\infty \sum_{k \in \mathbb{Z}} \left(\int_{a_k}^t w_n(s)ds \right)^{r/q'} \right.$$

$$\left. \times \left(\int_t^{b_k} u_{b,n}(s)ds \right)^{r/q} \chi_k(t)w_n^p(t)w^{1-p}(t)dt \right)^{1/p}$$

$$\leq \left(\int_0^\infty \sum_{k\in\mathbb{Z}} \left(\int_{a_k}^t w_n(s)ds \right)^{r/q'} \right.$$

$$\left. \times \left(\int_t^{b_k} u_{b,n}(s)ds \right)^{r/q} \chi_n(t)w_n(t)dt \right)^{1/p}$$

$$= \left(\sum_{k\in\mathbb{Z}} \int_{a_k}^{b_k} \left(\int_{a_k}^t w_n(s)ds \right)^{r/q'} \left(\int_t^{b_k} u_{b,n}(s)ds \right)^{r/q} w_n(t)dt \right)^{1/p}$$

$$= \left(\frac{p'}{q} \right)^{1/p} \left(\sum_{k\in\mathbb{Z}} \int_{a_k}^{b_k} \left(\int_{a_k}^t w_n(s)ds \right)^{r/p'} \right.$$

$$\left. \times \left(\int_t^{b_k} u_{b,n}(s)ds \right)^{r/p} u_{b,n}(t)dt \right)^{1/p},$$

where the last equality follows by integration by parts (see also Remark 1.5 (iv)). Since $u_{b,n}$ and w_n are in L^1, the sum is finite, and comparing the last estimate via (3.42) with the first term in (3.40), we obtain by dividing that

$$\left(\sum_{k\in\mathbb{Z}} \int_{a_k}^{b_x} \left(\int_{a_k}^t w_n(s)ds \right)^{r/p'} \left(\int_t^{b_k} u_{b,n}(s)ds \right)^{r/p} u_{b,n}(t)dt \right)^{1/q-1/p}$$

$$\leq \frac{Cr}{p'q} \left(\frac{p'}{q} \right)^{1/p}.$$

Now, since $w_n \uparrow w$ and $u_{b,n} \uparrow u_b$, the Monotone Convergence Theorem applies and the last estimate holds with w_n, $u_{b,n}$ replaced by w, u_b, respectively, which reads

$$\left(\sum_{k\in\mathbb{Z}} D^r(m_k) \right)^{1/r} \leq \frac{Cr}{p'q} \left(\frac{p'}{q} \right)^{1/p}.$$

Consequently,

$$\sup_{m>0} \mathcal{D}(m) < \infty.$$

The same arguments with minor modifications show that also $\sup_{m>0}\widetilde{D}(m) < \infty$. This proves (3.39) and hence the necessity of condition (3.36).

(ii) *Sufficiency* of (3.36). Suppose that A_1 and A_2 from (3.37) and (3.38) are finite. Then also

$$\left(\int_0^\infty \left(\int_{a(t)}^{b(t)} \left(\int_{a(t)}^y w(s)ds \right)^{r/p'} \right. \right.$$

$$\times \left. \left. \left(\int_y^{b(t)} u_b(s)ds \right)^{r/p} u_b(y)dy \right) \sigma(t)dt \right)^{1/r} < \infty$$

and

$$\left(\int_0^\infty \left(\int_{a(t)}^{b(t)} \left(\int_y^{b(t)} w(s)ds \right)^{r/p'} \right. \right.$$

$$\times \left. \left. \left(\int_{a(t)}^y u_a(s)ds \right)^{r/p} u_a(y)dy \right) \sigma(t)dt \right)^{1/r} < \infty.$$

But this means — as was just shown — that

$$\left(\int_{b^{-1}(1)}^{a^{-1}(1)} \mathcal{D}^r(m)dm \right)^{1/r} \quad \text{and} \quad \left(\int_{b^{-1}(1)}^{a^{-1}(1)} \widetilde{\mathcal{D}}^r(m)dm \right)^{1/r}$$

are finite. Hence $\mathcal{D}(m)$ and $\widetilde{\mathcal{D}}(m)$ are finite a.e. in $(b^{-1}(1), a^{-1}(1))$, and therefore there is an $m \in (b^{-1}(1), a^{-1}(1))$ where *both* $\mathcal{D}(m)$ and $\widetilde{\mathcal{D}}(m)$ are finite. Starting with this m construct a sequence $\{m_k\}_{k\in\mathbb{Z}}$ according to Definition 3.5. Hence, if $m_k < x < m_{k+1}$, then $a(x) < a(m_{k+1}) = b(m_k) < b(x)$ and writing again $a_k = a(m_k)$, $b_k = b(m_k)$, $k \in \mathbb{Z}$, it follows that

$$
\int_0^\infty \left(\int_{a(x)}^{b(x)} f(t)w(t)dt \right)^q u(x)dx
$$

$$
= \sum_{k \in \mathbb{Z}} \int_{m_k}^{m_{k+1}} \left(\int_{a(x)}^{b_k} f(t)w(t)dt + \int_{a_{k+1}}^{b(x)} f(t)w(t)dt \right)^q u(x)dx
$$

$$
\lesssim \left(\sum_{k \in \mathbb{Z}} \int_{m_k}^{m_{k+1}} \left(\int_{a(x)}^{b_k} f(t)w(t)dt \right)^q u(x)dx \right.
$$

$$
\left. + \sum_{k \in \mathbb{Z}} \int_{m_k}^{m_{k+1}} \left(\int_{a_{k+1}}^{b(x)} f(t)w(t)dt \right)^q u(x)dx \right)
$$

$$
=: (S_1 + S_2). \tag{3.43}
$$

In S_1, the change of variable $y = a(x)$, Proposition 3.9 and Hölder's inequality with indices r/q and p/q yield

$$
S_1 = \sum_{k \in \mathbb{Z}} \int_{a_k}^{b_k} \left(\int_y^{b_k} f(t)w(t)dt \right)^q u_a(y)dy
$$

$$
\leq \sum_{k \in \mathbb{Z}} \widetilde{C}^q(m_k) \left(\int_{a_k}^{b_k} f^p(t)w(t)dt \right)^{q/p}
$$

$$
\leq \left(\sum_{k \in \mathbb{Z}} \widetilde{C}^r(m_k) \right)^{q/r} \left(\sum_{k \in \mathbb{Z}} \int_{a_k}^{b_k} f^p(t)w(t)dt \right)^{q/p}
$$

$$
\lesssim \left(\sum_{k \in \mathbb{Z}} \widetilde{D}^r(m_k) \right)^{q/r} \left(\int_0^\infty f^p(t)w(t)dt \right)^{q/p}
$$

$$
= \widetilde{\mathcal{D}}^q(m) \left(\int_0^\infty f^p(t)w(t)dt \right)^{q/p}.
$$

Similarly, in S_2, the change of variable $y = b(x)$, Proposition 3.9 and Hölder's inequality with indices r/q and p/q yield

$$S_2 = \sum_{k \in \mathbb{Z}} \int_{a_{k+1}}^{b_{k+1}} \left(\int_{a_{k+1}}^{y} f(t)w(t)dt \right)^q u_b(y)dy$$

$$\leq \sum_{k \in \mathbb{Z}} C^q(m_{k+1}) \left(\int_{a_{k+1}}^{y} f^p(t)w(t)dt \right)^{q/p}$$

$$\leq \left(\sum_{k \in \mathbb{Z}} C^r(m_{k+1}) \right)^{q/r} \left(\sum_{k \in \mathbb{Z}} \int_{a_{k+1}}^{b_{k+1}} f^p(t)w(t)dt \right)^{q/p}$$

$$\lesssim \left(\sum_{k \in \mathbb{Z}} D^r(m_{k+1}) \right)^{q/r} \left(\int_0^\infty f^p(t)w(t)dt \right)^{q/p}$$

$$= \mathcal{D}^q(m) \left(\int_0^\infty f^p(t)w(t)dt \right)^{q/p}.$$

Using the last two estimates in (3.43), we obtain inequality (3.22) with a constant

$$C \lesssim (\widetilde{\mathcal{D}}^q(m) + \mathcal{D}^q(m))^{1/q}.$$

$\square$

Remark 3.12. Notice that if $0 < q < 1 = p$ and u, v satisfy

$$A^\# := \operatorname*{ess\,sup}_{t>0} \frac{1}{v(t)} \int_{b^{-1}(t)}^{a^{-1}(t)} u(s)ds < \infty$$

and if $u \in L^1$, then

$$\left(\int_0^\infty \left(\int_{a(x)}^{b(x)} f(t)dt \right)^q u(x)dx \right)^{1/q}$$

$$\leq A^\# \|u\|_1^{-1/q'} \int_0^\infty f(x)v(x)dx.$$

Indeed, if we rewrite the left-hand side in the last inequality as

$$\left(\int_0^\infty \left(\int_{a(x)}^{b(x)} f(t)v(t)\frac{1}{v(t)}dt \right)^q u^q(x)u^{1-q}(x)dx \right)^{1/q}$$

and use Hölder's inequality with indices $\rho = \frac{1}{q}$ and $\rho' = 1/(1-q)$, we can estimate it by

$$\left(\left(\int_0^\infty \left(\int_{a(x)}^{b(x)} f(t)v(t)\frac{1}{v(t)}dt \right)^{q\rho} u^{q\rho}(x)dx \right)^{1/\rho} \right.$$

$$\left. \times \left(\int_0^\infty u^{(1-q)\rho'}(x)dx \right)^{1/\rho'} \right)^{1/q}$$

$$= \left(\int_0^\infty \int_{a(x)}^{b(x)} f(t)v(t)\frac{1}{v(t)}u(x)dtdx \right) \left(\int_0^\infty u(x)dx \right)^{(1-q)/q}$$

$$= \left(\int_0^\infty \frac{1}{v(t)} \left(\int_{b^{-1}(t)}^{a^{-1}(t)} u(x)dx \right) f(t)v(t)dt \right) \|u\|_1^{-1/q'},$$

where we use Fubini's theorem. Now, the inequality follows due to the definition of $A^\#$.

3.5. Some Applications

First, we make a special choice of the functions $a(x)$ and $b(x)$ and formulate this example as a corollary.

Corollary 3.13. *Let a and b be real numbers, $0 < a < b$. Then inequality*

$$\left(\int_0^\infty \left(\int_{ax}^{bx} f(t)dt \right)^q u(x)dx \right)^{1/q} \leq C \left(\int_0^\infty f^p(x)v(x)dx \right)^{1/p}$$

$$(3.44)$$

holds for all functions $f \geq 0$ if and only if

(i) *for* $1 < p \le q < \infty$,

$$A := \sup_{t \le x \le bt/a} \left(\int_t^x u(s)ds \right)^{1/q} \left(\int_{ax}^{bt} v^{1-p'}(s)ds \right)^{1/p'} < \infty,$$

(ii) *for* $0 < q < p, 1 < p < \infty$,

$$\max(A_1, A_2) < \infty,$$

where

$$A_1 := \left(\int_0^\infty \frac{1}{t} \int_{at/b}^t \left(\int_{at}^{bx} v^{1-p'}(s)ds \right)^{r/p'} \right.$$
$$\left. \times \left(\int_x^t u(s)ds \right)^{r/p} u(x)dxdt \right)^{1/r}$$

and

$$A_2 := \left(\int_0^\infty \frac{1}{t} \int_t^{bt/a} \left(\int_{ax}^{bt} v^{1-p'}(s)ds \right)^{r/p'} \right.$$
$$\left. \times \left(\int_t^x u(s)ds \right)^{r/p} u(x)dxdt \right)^{1/r}$$

with $\frac{1}{r} = \frac{1}{q} - \frac{1}{p}$.

Moreover, the best constant C in (3.44) satisfies $C \approx A$ for $1 < p \le q < \infty$ and $C \approx \max(A_1, A_2)$ in the remaining index-range.

Proof. (i) If $1 < p \le q < \infty$, the result follows at once from Theorem 3.7 with $a(x) = ax$ and $b(x) = bx$.

(ii) If $0 < q < p, p > 1$, we have to verify with the same choice of $a(x)$ and $b(x)$ the conditions of Theorem 3.11. We have $b^{-1}(x) = x/b$ so that $M_0 = b^{-1}(1) = 1/b$ and $(b^{-1} \circ a)(t) = b^{-1}(a(t)) = at/b$ and

$(b^{-1} \circ a)^k(t) = (a/b)^k t$, $k \in \mathbb{Z}$. Therefore,

$$\frac{d}{dt}(b^{-1} \circ a)^k(t) = \left(\frac{a}{b}\right)^k$$

and $M_k = (b/a)^k \frac{1}{b}$, $k \in \mathbb{Z}$. Hence if $t \in (M_k, M_{k+1})$, then we have

$$\frac{1}{b} = \left(\frac{a}{b}\right)^k M_k \le t\sigma(t) \le \left(\frac{a}{b}\right)^k M_{k+1} = \frac{1}{a}$$

with $\sigma(t)$ the normalizing function from (3.35). Consequently, $\sigma(t) \approx 1/t$, and substituting this in (3.37) and (3.38) with $w = v^{1-p'}$, the corollary follows. $\qquad\square$

Now we apply the foregoing corollary to obtain a weighted characterization involving the Steklov operator S_γ from (3.15).

Corollary 3.14. *Let U and V be weight functions on $\mathbb{R}$. Then the inequality*

$$\left(\int_{-\infty}^{\infty} \left(\int_{x-\gamma}^{x+\gamma} F(t)dt\right)^q U(x)dx\right)^{1/q} \le C \left(\int_{-\infty}^{\infty} F^p(x)V(x)dx\right)^{1/p}$$

with $\gamma > 0$ fixed holds for every $F \ge 0$ with a constant $C = C(\gamma) > 0$ independent of F if and only if
(i) for $1 < p \le q < \infty$,

$$\sup_{t \le x \le 2\gamma + t} \left(\int_t^x U(s)ds\right)^{1/q} \left(\int_{x-\gamma}^{t+\gamma} V^{1-p'}(s)ds\right)^{1/p'} < \infty, \quad (3.45)$$

(ii) for $0 < q < p, 1 < p < \infty$,

$$\max(A_1, A_2) < \infty,$$

where

$$A_1 := \left(\int_{-\infty}^{\infty} \left(\int_{t-2\gamma}^{t} \left(\int_{t-\gamma}^{x+\gamma} V^{1-p'}(s)ds\right)^{r/p'}\right.\right.$$
$$\left.\left. \times \left(\int_x^t U(s)ds\right)^{r/p} U(x)dx\right) dt\right)^{1/r},$$

$$A_2 := \left(\int_{-\infty}^{\infty} \left(\int_{t}^{t+2\gamma} \left(\int_{x-\gamma}^{t+\gamma} V^{1-p'}(s)ds \right)^{r/p'} \right. \right.$$
$$\left. \left. \times \left(\int_{t}^{x} U(s)ds \right)^{r/p} U(x)dx \right) dt \right)^{1/r}$$

with $\frac{1}{r} = \frac{1}{q} - \frac{1}{p}$.

Proof. We apply Corollary 3.13 with $a = e^{-\gamma}$, $b = e^{\gamma}$, $\gamma > 0$, $v(x) = x^{p-1}V(\log x)$, $u(x) = \frac{1}{x}U(\log x)$ and $f(x) = \frac{1}{x}F(\log x)$. Since F is non-negative on R if and only if f is non-negative on $(0, \infty)$, the result follows after several changes of variables.   □

A Cauchy problem

Consider the initial value problem

$$w_{tt} - w_{xx} = 0, \ t > 0, \ x \in \mathbb{R},$$
$$w(x, 0) = 0, \ \frac{dw}{dt}(x, 0) = f(x), \ x \in \mathbb{R}.$$

It is well known that the solution of this problem is given by d'Alembert's formula

$$w(x, t) = \frac{1}{2} \int_{x-t}^{x+t} f(s)ds.$$

But this is $\frac{1}{2}(S_t f)(x)$, where S_t is the Steklov operator, and by Corollary 3.14 it follows that the estimate

$$\|w\|_{q,U} \le C(t)\|f\|_{p,V}, \quad 1 < p < \infty, \ 0 < q < \infty,$$

with t arbitrary but fixed, holds if and only if the weight conditions (i) or (ii) from this corollary are satisfied.

For example, if we take $p = q = 2$ and $U(x) = V(x) \equiv 1$, then condition (3.45) is satisfied since

$$\sup_{y \le x \le 2t+y} (y + t - x + t)^{1/2}(x - y)^{1/2} = t < \infty,$$

and it follows that

$$\|w\|_2 \le Ct\|f\|_2,$$

i.e.,

$$\left(\int_{-\infty}^{\infty} w^2(x,t)dx\right)^{1/2} \leq Ct \left(\int_{-\infty}^{\infty} f^2(x)dx\right)^{1/2}$$

for every $t > 0$ with C independent of t.

A two-dimensional weight

The next theorem which may be of independent interest involves a special Hardy-Steklov operator with a two-dimensional weight. More precisely, we give conditions on the weights under which the inequality

$$\left(\int_0^1 \int_0^\infty w(x,t)\left(\int_{xt}^x g(s)ds\right)^q dxdt\right)^{1/q} \leq C\left(\int_0^\infty g^p(x)v(x)dx\right)^{1/p} \tag{3.46}$$

is satisfied.

Theorem 3.15. *Let $1 < p \leq q < \infty$ and let $v(x)$ and $w(x,t)$ with $t \in (0,1)$ be weight functions on $(0,\infty)$. If*

$$A := \left(\int_0^1 \sup_{0<\alpha<\beta<\alpha/t} \left(\int_\beta^{\alpha/t} w(x,t)dx\right)\right.$$

$$\left. \times \left(\int_\alpha^\beta v^{1-p'}(x)dx\right)^{q/p'} dt\right)^{1/q} < \infty \tag{3.47}$$

then inequality (3.46) holds for all functions $g \geq 0$.
  Conversely, if (3.46) holds for all $g \geq 0$ then

$$\sup_{0<\alpha<\beta<\infty} \left(\int_0^{\alpha/\beta} \left(\int_\beta^{\alpha/t} w(x,t)dx\right) dt\right)^{1/q}$$

$$\times \left(\int_\alpha^\beta v^{1-p'}(x)dx\right)^{1/p'} < \infty. \tag{3.48}$$

Proof. Fix $t \in (0,1)$ and apply Corollary 3.13 with $a = t$, $b = 1$. Then it follows that

$$\int_0^1 \left(\int_0^\infty w(x,t) \left(\int_{xt}^x g(s)ds \right)^q dx \right) dt$$

$$\leq \int_0^1 C^q(t) \left(\int_0^\infty g^p(x)v(x)dx \right)^{q/p} dt,$$

i.e.,

$$\left(\int_0^1 \left(\int_0^\infty w(x,t) \left(\int_{xt}^x g(s)ds \right)^q dx \right) dt \right)^{1/q}$$

$$\leq \left(\int_0^1 C^q(t)dt \right)^{1/q} \left(\int_0^\infty g^p(x)v(x)dx \right)^{1/p},$$

where

$$C(t) \leq k(p,q) \sup_{y \leq x \leq y/t} \left(\int_y^x w(s,t)ds \right)^{1/q} \left(\int_{tx}^y v^{1-p'}(s)ds \right)^{1/p'}$$

$$= k(p,q) \left(\sup_{0<\alpha<\beta<\alpha/t} \left(\int_\beta^{\alpha/t} w(s,t)ds \right) \left(\int_\alpha^\beta v^{1-p'}(s)ds \right)^{q/p'} \right)^{1/q}.$$

Here we have made the change $y = \beta$, $x = \alpha/t$; the last expression is finite for a.e. $t \in (0,1)$ due to (3.47). Consequently, we have obtained (3.46) with

$$C = k(p,q)A.$$

Conversely, if (3.46) is satisfied for all $g \geq 0$, then for fixed $\alpha, \beta, 0 < \alpha < \beta$, let

$$g(x) = \chi_{(\alpha,\beta)}(x)v^{1-p'}(x)dx.$$

For this function g, (3.46) yields

$$C \left(\int_\alpha^\beta v^{1-p'}(x)dx \right)^{1/p} = C \left(\int_0^\infty g^p(x)v(x) \right)^{1/p}$$

$$\geq \left(\int_0^1 \left(\int_0^\infty w(x,t) \left(\int_{tx}^x g(s)ds \right)^q dx \right) dt \right)^{1/q}$$

$$\geq \left(\int_0^{\alpha/\beta} \left(\int_\beta^{\alpha/t} w(x,t) \left(\int_{tx}^x v^{1-p'}(s)\chi_{(\alpha,\beta)}(s)ds \right)^q dx \right) dt \right)^{1/q}$$

$$= \left(\int_0^{\alpha/\beta} \left(\int_\beta^{\alpha/t} w(x,t) \left(\int_\alpha^\beta v^{1-p'}(s)ds \right)^q dx \right) dt \right)^{1/q}$$

$$= \left(\int_\alpha^\beta v^{1-p'}(s)ds \right) \left(\int_0^{\alpha/\beta} \left(\int_\beta^{\alpha/t} w(x,t)dx \right) dt \right)^{1/q}$$

and (3.48) follows by dividing by $\left(\int_\alpha^\beta v^{1-p'}(s)ds \right)^{1/p}$.   □

Remark 3.16. Let us point out that (3.48) is not sufficient for (3.46) and that (3.47) is not necessary for (3.46). For details, see H.P. Heinig and G. Sinnamon [HS1].

The following result is an easy consequence of Theorem 3.15. It answers the question raised on p. 128 and complements some of the results which will be dealt with in Chap. 5.

Corollary 3.17. *Let* $1 < p \leq q < \infty$ *and* $f \in C^1(0,\infty)$. *Then the inequality*

$$\left(\int_0^\infty \int_0^\infty \frac{|f(x) - f(y)|^q}{u(|x-y|)} dxdy \right)^{1/q} \leq C \left(\int_0^\infty |f'(x)|^p v(x)dx \right)^{1/p}$$

$$(3.49)$$

holds provided

$$A := \left(\int_0^1 \sup_{0 < \alpha < \beta < \alpha/t} \left(\int_\beta^{\alpha/t} \frac{x}{u(x(1-t))} dx \right) \right.$$

$$\left. \times \left(\int_\alpha^\beta v^{1-p'}(x) dx \right)^{q/p'} dt \right)^{1/q} < \infty. \qquad (3.50)$$

Proof. Fubini's theorem and an obvious change of variables yield

$$\left(\int_0^\infty \int_0^\infty \frac{|f(x) - f(y)|^q}{u(|x-y|)} dx dy \right)^{1/q}$$

$$= \left(\int_0^\infty \int_0^x \frac{|f(x) - f(y)|^q}{u(|x-y|)} dy dx \right.$$

$$\left. + \int_0^\infty \int_x^\infty \frac{|f(x) - f(y)|^q}{u(|x-y|)} dy dx \right)^{1/q}$$

$$= \left(2 \int_0^\infty \int_0^x \frac{|f(x) - f(y)|^q}{u(|x-y|)} dy dx \right)^{1/q}$$

$$= 2^{1/q} \left(\int_0^1 \int_0^\infty \frac{|f(x) - f(xt)|^q}{u(x(1-t))} x dx dt \right)^{1/q}$$

$$\leq 2^{1/q} \left(\int_0^1 \int_0^\infty \frac{x}{u(x(1-t))} \left(\int_{xt}^x |f'(s)| ds \right)^q dx dt \right)^{1/q}.$$

Apply Theorem 3.15 with $w(x,t) = x/u(x(1-t))$ for $g(x) = |f'(x)|$, and the result follows. $\square$

Example 3.18. Take $u(x) = x^{1+\lambda p}$, $v(x) = x^{(1-\lambda)p}$ with $\lambda \in (0,1)$ and $p = q$. Then the number A from (3.50) has the form

$$\left(\int_0^1 \frac{1}{(1-t)^{1+\lambda p}} \sup_{0 < \alpha < \beta < \alpha/t} \left(\int_\beta^{\alpha/t} x^{-\lambda p} dx \right) \right.$$

$$\left. \times \left(\int_\alpha^\beta x^{(\lambda-1)p'} dx \right)^{p-1} dt \right)^{1/q}$$

and it can be shown that this number is finite. Consequently, the corresponding inequality (3.49) is the inequality mentioned on p. 128 and shows that (3.17) is dominated by (3.16).

3.6. Some Generalizations and Extensions

An N-dimensional Hardy-Steklov operator

Similarly as on pp. 58 and 59, we can consider the following N-dimensional analogue of the Hardy-Steklov operator:

$$(\mathscr{H}_N f)(x) = \int_{a(|x|)<|y|<b(|x|)} f(y)dy, \quad x, y \in \mathbb{R}^N, \tag{3.51}$$

where a, b are the functions from Definition 3.2. The investigation of the inequality

$$\left(\int_{\mathbb{R}^N} (\mathscr{H}_N f)^q(x) u(x) dx \right)^{1/q} \le C \left(\int_{\mathbb{R}^N} f^p(x) v(x) dx \right)^{1/p} \tag{3.52}$$

for $f \ge 0$ with $u(x), v(x)$ weight function on $\mathbb{R}^N$ can be reduced to the one-dimensional case by using polar coordinates,

$$x = t\tau, \ t \in (0, \infty), \ \tau \in \Sigma_N,$$

with Σ_N the unit sphere in $\mathbb{R}^N$, and the notation

$$U(t) = \int_{\Sigma_N} u(t\tau) t^{N-1} d\tau,$$

$$V(t) = \left(\int_{\Sigma_N} v^{1-p'}(t\tau) t^{N-1} d\tau \right)^{1-p}. \tag{3.53}$$

Theorem 3.19. *Let $0 < q < \infty$, $1 < p < \infty$. Then inequality (3.52) holds for all $f \ge 0$ if and only if*

$$\left(\int_0^\infty \left(\int_{a(t)}^{b(t)} F(s) ds \right)^q U(t) dt \right)^{1/q} \le C \left(\int_0^\infty F^p(t) V(t) dt \right)^{1/p} \tag{3.54}$$

is satisfied for all $F \ge 0$. The constants C in (3.54) and (3.52) are the same.

Proof.   (i) *Sufficiency* of (3.54). Fix $f = f(x)$, $x \in \mathbb{R}^N$, and define

$$F(t) = \int_{\Sigma_N} f(t\tau)t^{N-1}d\tau.$$

Then, by Hölder's inequality and by (3.53),

$$F(t) \le \left(\int_{\Sigma_N} f^p(t\tau)v(t\tau)t^{N-1}d\tau\right)^{1/p} \left(\int_{\Sigma_N} v^{1-p'}(t\tau)t^{N-1}d\tau\right)^{1/p'}$$

$$= \left(\int_{\Sigma_N} f^p(t\tau)v(t\tau)t^{N-1}d\tau\right)^{1/p} \cdot V^{-1/p}(t).$$

Now the change to polar coordinates $x = t\tau$, $y = s\sigma$ and inequality (3.54) yield, together with (3.53),

$$\left(\int_{\mathbb{R}^N} \left(\int_{a(|x|)<|y|<b(|x|)} f(y)dy\right)^q u(x)dx\right)^{1/q}$$

$$= \left(\int_0^\infty \int_{\Sigma_N} \left(\int_{a(t)}^{b(t)} \int_{\Sigma_N} f(s\sigma)s^{N-1}d\sigma ds\right)^q u(t\tau)t^{N-1}d\tau dt\right)^{1/q}$$

$$= \left(\int_0^\infty \left(\int_{a(t)}^{b(t)} F(s)ds\right)^q U(t)dt\right)^{1/q}$$

$$\le C \left(\int_0^\infty F^p(t)V(t)dt\right)^{1/p}$$

$$\le C \left(\int_0^\infty \left(\int_{\Sigma_N} f^p(t\tau)v(t\tau)t^{N-1}d\tau\right)dt\right)^{1/p}$$

$$= C \left(\int_{\mathbb{R}^N} f^p(x)v(x)dx\right)^{1/p}.$$

(ii) *Necessity* of (3.54). Suppose that (3.52) is satisfied for $f \geq 0$, and define F by

$$f(t\tau) = F(t)v^{1-p'}(t\tau)\left(\int_{\Sigma_N} v^{1-p'}(t\sigma)t^{N-1}d\sigma\right)^{-1},$$

$$t > 0, \ \tau \in \Sigma_N.$$

Then

$$\int_{\Sigma_N} f(t\tau)t^{N-1}d\tau = F(t)$$

and hence

$$\left(\int_0^\infty \left(\int_{a(t)}^{b(t)} F(s)ds\right)^q U(t)dt\right)^{1/q}$$

$$= \left(\int_0^\infty U(t)\left(\int_{a(t)}^{b(t)} \int_{\Sigma_N} f(s\tau)s^{N-1}d\tau ds\right)^q dt\right)^{1/q}$$

$$= \left(\int_0^\infty \left(\int_{\Sigma_n} u(t\sigma)t^{N-1}d\sigma\right)\right.$$

$$\times \left.\left(\int_{a(t)}^{b(t)} \int_{\Sigma_N} f(s\tau)s^{N-1}d\tau ds\right)^q dt\right)^{1/q}$$

$$= \left(\int_{\mathbb{R}^N} u(x)\left(\int_{a(|x|)<|y|<b(|x|)} f(y)dy\right)^q dx\right)^{1/q}$$

$$\leq C\left(\int_{\mathbb{R}^N} f^p(x)v(x)dx\right)^{1/p}$$

$$= C\left(\int_0^\infty \int_{\Sigma_N} f^p(t\tau)t^{N-1}v(t\tau)d\tau dt\right)^{1/p}$$

$$= C\left(\int_0^\infty \int_{\Sigma_N} t^{N-1}v(t\tau)\left[F^p(t)v^{p(1-p')}(t\tau)\right.\right.$$

$$\times \left.\left.\left(\int_{\Sigma_N} v^{1-p'}(t\sigma)t^{N-1}d\sigma\right)^{-p}\right]d\tau dt\right)^{1/p}$$

$$= C \left(\int_0^\infty F^p(t) \left(\int_{\Sigma_N} t^{N-1} v^{1-p'}(t\tau) d\tau \right) \right.$$

$$\left. \times \left(\int_{\Sigma_N} t^{N-1} v^{1-p'}(t\sigma) d\sigma \right)^{-p} dt \right)^{1/p}$$

$$= C \left(\int_0^\infty F^p(t) V(t) dt \right)^{1/p},$$

where we have used the notation from (3.53). □

Theorem 3.19 allows to characterize weights u and v for which (3.52) is satisfied. Let us formulate the result emphasizing again that the functions a, b are defined on $(0, \infty)$ and satisfy the properties described in Definition 3.2. For $x \in \mathbb{R}^N$ and $x = t\tau$ with $t > 0$ and $\tau \in \Sigma_N$, we have $|x| = t$.

Corollary 3.20. *Inequality (3.52) is satisfied for all $f = f(x) \geq 0$, $x \in \mathbb{R}^N$, if and only if*

(i) *for $1 < p \leq q < \infty$ and $x, y, z \in \mathbb{R}^N$,*

$$\sup \left(\int_{|y|<|z|<|x|} u(z) dz \right)^{1/q} \left(\int_{a(|x|)<|z|<b(|y|)} v^{1-p'}(z) dz \right)^{1/p'} < \infty,$$

(3.55)

where the supremum is taken over all $x, y \in \mathbb{R}^N$ with $|y| < |x|$ and $a(|x|) < b(|y|)$;

(ii) *for $0 < q < p, 1 < p < \infty, \frac{1}{r} = \frac{1}{q} - \frac{1}{p}$,*

$$\left(\int_{\mathbb{R}^N} S(x) \left[\int_{b^{-1}(a(|x|))<|y|<|x|} \left(\int_{|y|<|y_1|<|x|} u(y_1) dy_1 \right)^{r/p} u(y) \right. \right.$$

$$\left. \left. \times \left(\int_{a(|x|)<|y_2|<b(|y|)} v^{1-p'}(y_2) dy_2 \right)^{r/p'} dy \right] dx \right)^{1/r} < \infty$$

(3.56)

and

$$\left(\int_{\mathbb{R}^N} S(x) \left[\int_{|x|<|y|<a^{-1}(b|x|)} u(y) \left(\int_{|x|<|y_1|<|y|} u(y_1)dy_1 \right)^{r/p} \right. \right.$$

$$\left. \left. \times \left(\int_{a(|y|)<|y_2|<b(|x|)} v^{1-p'}(y_2)dy_2 \right)^{r/p'} dy \right] dx \right)^{1/r} < \infty$$

$$(3.57)$$

are satisfied. Here S is a radial function defined by

$$S(x) = S(|x|) = |x|^{1-N}\sigma(|x|)/|\Sigma_N|$$

with σ the normalizing function defined by (3.35) and $|\Sigma_N|$ the surface area of the unit sphere in $\mathbb{R}^N$.

Proof. By Theorem 3.19 it suffices to show that conditions (3.55), (3.56) and (3.57) are equivalent to the conditions of Theorems 3.7 and 3.11 with u, v replaced by U, V defined by (3.53). But this follows at once via changes to polar coordinates. For example, if we change to polar coordinates in (3.56), taking $x = t\tau$, $y = s\rho$, $y_1 = s_1\rho_1$, $y_2 = s_2\rho_2$ with $\tau, \rho, \rho_1, \rho_2 \in \Sigma_N$ and $t, s, s_1, s_2 \in (0, \infty)$, then we obtain

$$\left(\int_0^\infty \int_{\Sigma_N} t^{N-1} S(t) d\tau \left[\int_{b^{-1}(a(t))}^t \int_{\Sigma_N} s^{N-1} u(s\rho) d\rho \right. \right.$$

$$\left. \times \left(\int_s^t \int_{\Sigma_N} s_1^{N-1} u(s_1\rho_1) d\rho_1 ds_1 \right)^{r/p} \right.$$

$$\left. \left. \times \left(\int_{a(t)}^{b(s)} \int_{\Sigma_N} s_2^{N-1} v^{1-p'}(s_2\rho_2) d\rho_2 ds_2 \right)^{r/p'} ds \right] dt \right)^{1/r}$$

$$= \left(\int_0^\infty \sigma(t) \left[\int_{b^{-1}(a(t))}^t U(s) \left(\int_s^t U(s_1) ds_1 \right)^{r/p} \right. \right.$$

$$\left. \left. \times \left(\int_{a(t)}^{b(t)} V^{1-p'}(s_2) ds_2 \right)^{r/p'} ds \right] dt \right)^{1/r}.$$

But this is the expression A_1 from (3.37) as required. The arguments for (3.55) and (3.57) are similar.   □

Generalized Hardy-type operators

The operator K is defined by

$$(Kf)(x) = \int_0^x k(x,t)f(t)dt,$$

investigated in Chap. 2, can be in a natural way extended to operators $K_{a,b}$ of the form

$$(K_{a,b}f)(x) = \int_{a(x)}^{b(x)} k(x,t)f(t)dt, \tag{3.58}$$

where the non-negative kernel $k = k(x,t)$ satisfies the following conditions:

(i) $k(x,t)$ is increasing in x and decreasing in t,
(ii) $k(x,z) \le C_1(k(x,b(y)) + k(y,z))$ for all $y \le x$ and $a(x) \le z \le b(y)$ with a constant $C_1 \ge 1$.

Weighted inequalities for this operator are studied among others in A.S. Gogatishvili and J. Lang [GL1]. Let us mention one particular result which for $k(x,t) \equiv 1$ reduces to Theorem 3.7.

Theorem 3.21. *Let* $1 < p \le q < \infty$. *Then the inequality*

$$\left(\int_0^\infty \left(\int_{a(x)}^{b(x)} k(x,t)f(t)dt \right)^q u(x)dx \right)^{1/q}$$

$$\le C \left(\int_0^\infty f^p(x)v(x)dx \right)^{1/p}$$

holds for all functions $f \ge 0$ *if and only if*

$$\sup_{x \le y, a(y) \le b(x)} \left(\int_x^y u(s)k^q(s,b(x))ds \right)^{1/q} \left(\int_{a(y)}^{b(x)} v^{1-p'}(s)ds \right)^{1/p'} < \infty$$

and

$$\sup_{x \le y, a(y) \le b(x)} \left(\int_x^y u(s)ds \right)^{1/q} \left(\int_{a(y)}^{b(x)} k^{p'}(x,s)v^{1-p'}(s)ds \right)^{1/p'} < \infty.$$

Monotone functions

In Chap. 6, we will deal with Hardy and Hardy-type inequalities for monotone functions. The methods used there are also applicable for Hardy-Steklov operators. For example, if T is such an operator defined on *decreasing* functions f, then the characterization of weights u, v for which

$$T : \ L^p(v) \to L^q(u), \quad 1 < p, q < \infty$$

(boundedly) requires (in the case $\int_0^\infty u(x)dx = \infty$) conditions on u, v for which the inequality

$$\left(\int_0^\infty \left(\int_0^x (\widetilde{T}g)(t)dt \right)^{p'} \left(\int_0^x u(t)dt \right)^{-p} u(x)dx \right)^{1/p'}$$

$$\le C \left(\int_0^\infty g^{q'}(x)v^{1-p'}(x)dx \right)^{1/q'} \tag{3.59}$$

is satisfied for $g \ge 0$, where $\widetilde{T}$ is the conjugate Hardy-Steklov operator (see Chap. 6 for details).

In order to solve (3.59) we need the knowledge of the mapping properties for the operator $K_{a,b}$ given in (3.58) with some kernel $k(x,t)$. However, there are *four* weight conditions which characterize the weights for which $K_{a,b} : \ L^p(v) \to L^q(u)$ (boundedly) if $1 < q < p < \infty$ and *two* conditions for the index range $1 < p \le q < \infty$ (see T. Chen and G. Sinnamon [ChS1]), so the result is unwieldy. It may therefore be appropriate to state here a simple (sufficient) condition for which the Hardy-Steklov operator defined on decreasing functions is bounded. For details, see the paper just mentioned.

Theorem 3.22. *If u, v are weight functions on $(0, \infty)$ such that $\int_0^\infty u(x)dx = \infty$, then the inequality*

$$\left(\int_0^\infty \left(\int_{a(x)}^{b(x)} f(t)dt \right)^q u(x)dx \right)^{1/q} \leq C \left(\int_0^\infty f^p(x)v(x)dx \right)^{1/p}$$

is satisfied for all decreasing functions $f \geq 0$ whenever

$$\sup_{x>0} \left(\int_0^{a(x)} u(s)ds \right)^{1-p} \left(\int_0^x (b(s) - a(s))^q v(s)ds \right)^{1/q} < \infty$$

holds in the case $1 < p \leq q < \infty$, and

$$\int_0^\infty \left(\int_0^{a(x)} u(s)ds \right)^{-r/q} \left(\int_0^x (b(s) - a(s))^q v(s)ds \right)^{r/q}$$

$$\times u(a(x))a'(x)dx < \infty$$

holds in the case $1 < q < p < \infty$ with $\frac{1}{r} = \frac{1}{q} - \frac{1}{p}$.

3.7. Comments and Remarks

3.7.1. Most of the results mentioned above, namely, the main Theorems 3.7 and 3.11, are taken from H.P. Heinig and G. Sinnamon [HS1]. If T is the Hardy-Steklov operator (3.7) or the moving averaging operator (3.14), then for the special case $b(x) = x$, $a(x) = x/\lambda$, $\lambda > 1$, the characterization of weights u, v for which $T : L^p(v) \to L^q(u)$, $1 < p \leq q < \infty$, is bounded, was given by E. Sawyer (see H.P. Heinig and G. Sinnamon [HS1]).

3.7.2. An application of the moving averaging operator to financial markets can be found in Journal of Finance 1992, which focused largely on the Dow Jones industrial average from 1897 to 1986.

3.7.3. The weight characterization of the Steklov operator (see Corollary 3.14) was given also by E.N. Batuev and V.D. Stepanov [BS1] in the index range $1 < p, q < \infty$. In addition, they discussed there the Cauchy Problem, see p. 151.

3.7.4. Theorem 3.15 can again be found in H.P. Heinig and G. Sinnamon [HS1] together with examples which show that (3.47) is not necessary for (3.46) and (3.48) is not sufficient for (3.46).

3.7.5. In Corollary 3.17, the weight $u(|x - y|)$ may be replaced by any weight u satisfying $u(x, y) = u(y, x)$.

3.7.6. Theorem 3.19 is a modification of the corresponding results for the Hardy operator in higher dimensions by G. Sinnamon [Si6], cf. also [Si5].

3.7.7. As mentioned on p. 161, A.S. Gogatishvili and J. Lang [GL1] studied weighted inequalities for the operator $K_{a,b}$ from (3.58). In fact, they proved weight characterizations for which $K_{a,b}$ is bounded between certain weighted Banach function spaces. As a consequence, a characterization of weights u, v follows for which $K_{a,b} : L^p(v) \to L^q(u)$ boundedly if $1 < p \le q < \infty$, while a characterization for the range $1 < p, q < \infty$ together with the discrete analogue was given by T. Chen and G. Sinnamon [ChS1]. For related results see also E. Lomakina and V. Stepanov [LS1].

3.7.8. The implication derived in Remark 3.12 may be regarded as a complement of Theorem 3.11 for $p = 1$. To derive necessary and sufficient conditions in this case seems to be an open question.

3.7.9. Some new information can be found in the recent book by D. V. Prokhorov *et al.* [PrSU1]. Some more information concerning the mapping properties of Hardy-Steklov operators can be found in the paper [NU1] by M. Nasyrova and E. Ushakova. See also the papers [SU1], [SU2], [SU3] and [U3] by V. Stepanov and E. Ushakova, and the paper [JG1] by P. Jain and B. Gupta and the paper [BJT1] by V. Burenkov *et al.*

4

Higher Order Hardy Inequalities

4.1. Preliminaries

In this chapter, we consider the k-th order Hardy inequality

$$\left(\int_a^b |g(x)|^q u(x) dx \right)^{1/q} \leq C \left(\int_a^b |g^{(k)}(x)|^p v(x) dx \right)^{1/p} \tag{4.1}$$

for functions g from certain subclasses of AC^{k-1} with $k > 1$. More precisely, since inequality (4.1) is meaningless if g is a polynomial of order less than k, we will suppose that g satisfies the following *boundary conditions*:

$$g^{(i)}(a) = 0 \quad \text{for } i \in M_0,$$
$$g^{(j)}(b) = 0 \quad \text{for } j \in M_1, \tag{4.2}$$

where M_0, M_1 are certain subsets of the set

$$N_k = \{0, 1, \ldots, k - 1\}.$$

The corresponding set of functions g for which inequality (4.1) will be investigated is denoted by

$$AC^{k-1}(M_0, M_1).$$

Example 4.1. Not all choices of the index sets M_0, M_1 are admissible. If we take $k = 2$, $M_0 = M_1 = \{1\}$, then $AC^1(M_0, M_1)$ consists of functions $g \in AC^1$ such that $g'(a) = g'(b) = 0$. The function $g(x) = \text{const} \ (\neq 0)$ obviously satisfies conditions (4.2) but the right hand side in (4.1) *vanishes*.

The Pólya condition

Let us introduce the *incidence matrix* E of the pair (M_0, M_1) as the $2 \times k$ matrix with elements $e_{\alpha i}$, $\alpha = 0, 1$, $i = 0, 1, \ldots, k-1$ defined as

$$e_{\alpha i} = \begin{cases} 1 & \text{if } i \in M_\alpha, \\ 0 & \text{if } i \notin M_\alpha. \end{cases}$$

The pair (M_0, M_1) is said to satisfy the *Pólya condition* if

$$\sum_{i=0}^{r}(e_{0i} + e_{1i}) \geq r + 1, \quad r = 0, 1, \ldots, k-1. \tag{4.3}$$

(This condition states that there are at least i 1's in the first i columns of E for $i = 1, 2, \ldots, k$.)

Remark 4.2. (i) As was shown in P. Drábek and A. Kufner [DK1], the Hardy inequality (4.1) is meaningful for $u \in AC^{k-1}(M_0, M_1)$ if and only if (M_0, M_1) satisfies the Pólya condition.

(ii) There is another more important reason for introducing the Pólya condition. Namely, for (M_0, M_1) satisfying (4.3), the boundary value problem (BVP)

$$\begin{aligned} g^{(k)} &= f \quad \text{in } (a, b); \\ g^{(i)}(a) &= 0 \quad \text{for } i \in M_0, \\ g^{(j)}(b) &= 0 \quad \text{for } j \in M_1 \end{aligned} \tag{4.4}$$

with f a locally integrable function is *uniquely solvable*. Therefore, there is a Green function $G(x, t)$ for this BVP such that the solution of (4.4) is given by

$$g(x) = \int_a^b G(x,t)f(t)dt \ (= (Tf)(x)). \qquad (4.5)$$

Consequently, instead of the Hardy inequality (4.1) for f from $AC^{k-1}(M_0, M_1)$ we investigate the equivalent *weighted norm inequality*

$$\left(\int_a^b |(Tf)(x)|^q u(x)dx \right)^{1/q} \le C \left(\int_a^b |f(x)|^p v(x)dx \right)^{1/p}. \qquad (4.6)$$

Let us denote by $|M_i|$ the number of elements of the index set M_i ($i = 0, 1$). We will consider the following special case:

$$|M_0| + |M_1| = k, \qquad (4.7)$$

i.e., exactly k boundary conditions appear in (4.2).

4.2. Some Special Cases I

The case $M_0 = \mathbb{N}_k$, $M_1 = \emptyset$ or $M_0 = \emptyset$, $M_1 = \mathbb{N}_k$

If *all* boundary conditions appear at *one* end-point, then the Pólya condition is obviously satisfied and the Green function $G(x,t)$ of the corresponding BVP (4.4) can be easily determined: we have

$$G(x,t) = \frac{(x-t)^{k-1}}{(k-1)!} \chi_{(a,x)}(t)$$

if $M_1 = \emptyset$ and

$$G(x,t) = \frac{(t-x)^{k-1}}{(k-1)!} \chi_{(x,b)}(t)$$

if $M_0 = \emptyset$. Consequently, inequality (4.6) has in the first case the form

$$\frac{1}{(k-1)!} \left(\int_a^b \left| \int_a^x (x-t)^{k-1} f(t)dt \right|^q u(x)dx \right)^{1/q}$$

$$\le C \left(\int_a^b |f(x)|^p v(x)dx \right)^{1/p},$$

and analogously for the other case:

$$\frac{1}{(k-1)!} \left(\int_a^b \left| \int_x^b (t-x)^{k-1} f(t) dt \right|^q u(x) dx \right)^{1/q}$$

$$\leq C \left(\int_a^b |f(x)|^p v(x) dx \right)^{1/p}.$$

But these inequalities are special cases of certain weighted norm inequalities investigated in Chap. 2 (see Theorem 2.21), and so we can immediately formulate our first result:

Theorem 4.3. *The inequality* (4.1) *holds for all functions* $g \in AC^{k-1}$ *satisfying* $g(a) = g'(a) = \cdots = g^{(k-1)}(a) = 0$ *if and only if*
(i) *either* $1 < p \leq q < \infty$ *and*

$$\sup_{a<x<b} \left(\int_x^b (t-x)^{(k-1)q} u(t) dt \right)^{1/q} \left(\int_a^x v^{1-p'}(t) dt \right)^{1/p'} < \infty,$$

$$(4.8)$$

$$\sup_{a<x<b} \left(\int_x^b u(t) dt \right)^{1/q} \left(\int_a^x (x-t)^{(k-1)p'} v^{1-p'}(t) dt \right)^{1/p'} < \infty;$$

(ii) *or* $1 < q < p < \infty$ *and*

$$\left(\int_a^b \left(\int_x^b (t-x)^{(k-1)q} u(t) dt \right)^{r/q} \right.$$

$$\times \left(\int_a^x v^{1-p'}(t) dt \right)^{r/q'} \left. v^{1-p'}(x) dx \right)^{1/r} < \infty,$$

$$(4.9)$$

$$\left(\int_a^b \left(\int_a^x (x-t)^{(k-1)p'} v^{1-p'}(t) dt \right)^{r/p'} \right.$$

$$\times \left(\int_x^b u(t) dt \right)^{r/p} \left. u(x) dx \right)^{1/r} < \infty$$

with $\frac{1}{r} = \frac{1}{q} - \frac{1}{p}$.

Remark 4.4. Part (i) of Theorem 4.3 is precisely Theorem 2.21 with the particular kernel $k(x,t) = (x-t)^{k-1}$, while part (ii) corresponds to Theorem 2.15 with the same kernel and the interval (a,b). Conditions (4.8) are then conditions (2.62) while (4.9) corresponds to (2.52). Using the notation from Introduction and from Chap. 1, and we can rewrite conditions (4.8) as

$$A(a,b; (t-x)^{(k-1)q}u(t), v(t)) < \infty,$$
$$A(a,b; u(t), (x-t)^{(1-k)p}v(t)) < \infty$$

(see (0.20)) and conditions (4.9) as

$$A(a,b; (t-x)^{(k-1)q}u(t), \; v(t)) < \infty,$$
$$A^*(a,b; u(t), (x-t)^{(1-k)p}v(t)) < \infty$$

(see (0.21) and Remark 1.5 (iv)). Of course, in a similar way we can describe necessary and sufficient conditions for the case of functions $g \in AC^{k-1}$ satisfying $g(b) = g'(b) = \ldots = g^{(k-1)}(b) = 0$: we only have to replace A, A^* by $\widetilde{A}, \widetilde{A}^*$.

Remark 4.5. The k-th order Hardy inequality (4.1) can be also investigated using the (first order) Hardy inequality (1.25) successively for the functions $g, g', \ldots, g^{(k-1)}$. For example, taking $k = 2$ and investigating the inequality

$$\left(\int_a^b |g(x)|^q u(x)dx \right)^{1/q} \le C \left(\int_a^b |g''(x)|^p v(x)dx \right)^{1/p} \tag{4.10}$$

for functions $g \in AC^1$ satisfying the conditions

$$g(a) = 0, \quad g'(b) = 0$$

(i.e., taking $M_0 = \{0\}$, $M_1 = \{1\}$ so that the Pólya condition is satisfied) we can try to solve this problem considering *two* inequalities: the inequality

$$\left(\int_a^b |g(x)|^q u(x)dx \right)^{1/q} \le C_1 \left(\int_a^b |g'(x)|^r w(x)dx \right)^{1/r} \tag{4.11}$$

for functions g satisfying $g(a) = 0$, and the inequality

$$\left(\int_a^b |g'(x)|^r w(x)dx \right)^{1/r} \le C_2 \left(\int_a^b |g''(x)|^p v(x)dx \right)^{1/p} \qquad (4.12)$$

for functions g' satisfying $g'(b) = 0$, with some "intermediate" weight function w and an auxiliary parameter r. But besides the presence of these rather undetermined and from the point of view of the *initial* inequality (4.10) redundant auxiliary data w, r, this approach has another disadvantage: it cannot be used for every choice of boundary conditions. E.g., the conditions under which inequalities (4.11) and (4.12) should hold *simultaneously* may have an empty intersection. See also the following example.

Example 4.6. $[(a, b) = (-\infty, \infty)]$. Consider inequality (4.10) with $p = q$, $a = -\infty$, $b = +\infty$ and $u(x) = e^{\alpha x}$, $v(x) = e^{\beta x}$, where $\alpha, \beta \in \mathbb{R}$. If $g \in AC^1$ satisfies

$$g(-\infty) = 0, \quad g'(+\infty) = 0, \qquad (4.13)$$

then inequality (4.11) holds — with $r = p$, $w(x) = e^{\gamma x}$ and for g such that $g(-\infty) = 0$ — if and only if

$$\gamma = \alpha < 0, \qquad (4.14)$$

while inequality (4.12) holds — with the same r, w and for g such that $g'(+\infty) = 0$ — if and only if

$$\gamma = \beta > 0 \qquad (4.15)$$

(see [OK, Example 6.12]). The contradiction between conditions (4.14), (4.15) indicates that this approach — i.e., the successive application of first order Hardy inequalities — does not work for the interval $(-\infty, \infty)$ if we consider boundary conditions of the type (4.13), i.e., with M_0 and M_1 *both nonempty*. (Notice that the approach just described *works* if we consider g such that $g(-\infty) = g'(-\infty) = 0$, i.e. $M_0 = \{0, 1\}$, $M_1 = \emptyset$: we have to take $\alpha = \gamma = \beta < 0$.)

Consequently, this suggests that for $(a, b) = (-\infty, \infty)$, the choice of M_0, M_1 such that either $M_0 = \emptyset$ or $M_1 = \emptyset$ is the only reasonable one provided we want to use succesively first order Hardy inequalities.

The case $(a,b) = (0, \infty)$

Similarly as the infinite interval $(-\infty, \infty)$, also the semibounded interval $(0, \infty)$ plays an exceptional role. Namely, similarly as in Example 4.6, we can show that for some choices of the index sets M_0, M_1, the successive application of first order Hardy inequalities does not give satisfactory results. This approach works for the interval $(0, \infty)$ if we choose $M_0 = \mathbb{N}_k$, $M_1 = \emptyset$ (see Theorem 4.3) or $M_0 = \emptyset$, $M_1 = \mathbb{N}_k$ (see Remark 4.4). Also the choice

$$M_0 = \{0, 1, \ldots, m - 1\}, \quad M_1 = \{m, m + 1, \ldots, k - 1\} \qquad (4.16)$$

with $0 < m < k$ is reasonable as will be shown in Theorem 4.8 below.

We will illustrate it in Example 4.10. Moreover, the exceptional role of the interval $(0, \infty)$ will be dealt with in Sec. 4.9.

Of course, the formulation of analogous results for intervals of the type (a, ∞), $a \neq 0$, and $(-\infty, b)$, $b \in \mathbb{R}$, is only a matter of routine.

Example 4.7. Let us consider the second order Hardy inequality

$$\int_0^\infty |g(x)|^p x^{\alpha - 2p} dx \leq C \int_0^\infty |g''(x)|^p x^\alpha dx \qquad (4.17)$$

via two first order Hardy inequalities

$$\int_0^\infty |g(x)|^p x^{\alpha - 2p} dx \leq C_1 \int_0^\infty |g'(x)|^p x^{\alpha - p} dx \qquad (4.18)$$

and

$$\int_0^\infty |g'(x)|^p x^{\alpha - p} dx \leq C_2 \int_0^\infty |g''(x)|^p x^\alpha dx. \qquad (4.19)$$

Inequality (4.17) holds

$$\begin{aligned}
&\text{for } \alpha < p - 1 &&\text{if} \quad g(0) = 0, \ g'(0) = 0, \\
&\text{for } \alpha \in (p - 1, 2p - 1) &&\text{if} \quad g(0) = 0, \ g'(\infty) = 0, \\
&\text{for } \alpha > 2p - 1 &&\text{if} \quad g(\infty) = 0, \ g'(\infty) = 0
\end{aligned}$$

(i.e., if $M_0 = \{0, 1\}$, $M_1 = \emptyset$ or $M_0 = \{0\}$, $M_1 = \{1\}$ or $M_0 = \emptyset$, $M_1 = \{0, 1\}$, respectively; the second case corresponds to (4.16)). On the other hand, if we consider functions g satisfying

$$g(\infty) = 0, \quad g'(0) = 0$$

(i.e., choosing $M_0 = \{1\}$, $M_1 = \{0\}$), we cannot use this approach since inequality (4.18) holds for g satisfying $g(\infty) = 0$ if and only if $\alpha \in (2p - 1, \infty)$ while (4.19) holds for g satisfying $g'(0) = 0$ if and only if $\alpha \in (-\infty, p - 1)$, so that there is *no common value* of α.

Nevertheless, inequality (4.17) holds even for the choice $g(\infty) = g'(0) = 0$ if the parameter α satisfies

$$\alpha > 2p - 1,$$

see Example 4.48.

Theorem 4.8. *Let $0 < m < k$, $a = 0$, $b = \infty$. Then inequality (4.1) holds with $0 < p, q < \infty$, $p > 1$ for all functions $g \in AC^{k-1}(M_0, M_1)$ with M_0, M_1 defined by (4.16) if and only if*

$$A(0, \infty; t^{(m-1)q} u(t), \ t^{(k-m)p} v(t)) < \infty, \tag{4.20}$$

$$\widetilde{A}(0, \infty; t^{mq} u(t), \ t^{(k-m+1)p} v(t)) < \infty. \tag{4.21}$$

Proof. The function $g = Gf$, where

$$(Gf)(x) = \frac{1}{(m-1)!(m-k-1)!} \int_0^x (x - t)^{m-1}$$

$$\times \left(\int_t^\infty (s - t)^{m-k-1} f(s) ds \right) dt \tag{4.22}$$

satisfies the conditions

$$g(0) = g'(0) = \cdots = g^{(m-1)}(0) = 0,$$
$$g^{(m)}(\infty) = g^{(m+1)}(\infty) = \cdots = g^{(k-1)}(\infty) = 0,$$

i.e., conditions (4.2) with M_0, M_1 from (4.16). Moreover, we have $g^{(k)} = f$, and inequality (4.1) can be rewritten as the weighted norm inequality

$$\left(\int_0^\infty |(Gf)(x)|^q u(x) dx \right)^{1/q} \le C \left(\int_0^\infty |f(x)|^p v(x) dx \right)^{1/p}. \quad (4.23)$$

Obviously, it is possible to consider inequality (4.23) only for $f \ge 0$.

Now, by Fubini's theorem we have

$$(Gf)(x) = \frac{1}{(m-1)!(m-k-1)!} \left(\int_0^x K_1(x,t) f(t) dt \right.$$
$$\left. + \int_x^\infty K_2(x,t) f(t) dt \right),$$

where

$$K_1(x,t) = \int_0^t (x-s)^{m-1} (t-s)^{m-k-1} ds, \quad 0 < t < x < \infty,$$
$$K_2(x,t) = \int_0^x (x-s)^{m-1} (t-s)^{m-k-1} ds, \quad 0 < x < t < \infty,$$

and it can be easily shown that

$$K_1(x,t) \approx x^{m-1} t^{m-k},$$
$$K_2(x,t) \approx x^m t^{m-k-1}.$$

Consequently, inequality (4.23) is equivalent to inequality

$$\left(\int_0^\infty \left(x^{m-1} \int_0^x t^{m-k} f(t) dt + x^m \int_x^\infty t^{m-k-1} f(t) dt \right)^q u(x) dx \right)^{1/q}$$
$$\le C \left(\int_0^\infty f^p(x) v(x) dx \right)^{1/p}. \quad (4.24)$$

On the left-hand side of (4.24) we have the operator T given in (2.7) with

$$\varphi_1(x) = x^{m-1}, \ \psi_1(x) = t^{m-k}, \ \varphi_2(x) = x^m, \ \psi_2(t) = t^{m-k-1}.$$

Hence inequality (4.24) holds by Theorem 2.3 if and only if (4.20) and (4.21) are satisfied.   □

4.3. The General Case

Unlike the unbounded intervals $(-\infty, \infty)$ and $(0, \infty)$, we have greater variety of choices of the index sets M_0, M_1 for a *bounded* interval (a, b). Without loss of generality, we assume from now on that the interval is

$$(0, 1).$$

Thus, the k-th order Hardy inequality (4.1) has now the form (4.30) below.

Let us again emphasize that we suppose that (M_0, M_1) satisfies the Pólya condition and that $|M_0| + |M_1| = k$, $M_0 \neq \emptyset$, $M_1 \neq \emptyset$.

Definition 4.9. For a pair (M_0, M_1) satisfying the Pólya condition and such that $|M_0| > 0$, $|M_1| > 0$, we define non-negative integers a, b, c, d as follows: Let a be the number of consecutive 1's at the beginning of the top row of the incidence matrix E; b the number of consecutive 1's at the beginning of the bottom row of E; c the number of consecutive 0's at the end of the top row of E; and d the number of consecutive 0's at the end of the bottom row of E.

Furthermore, we define non-negative integers A, B, C, D as follows:

$$A = a - 1, \quad C = c - 1 \quad \text{if} \quad a + c = k,$$
$$A = a, \qquad C = c \qquad \text{if} \quad a + c < k,$$
$$B = b - 1, \quad D = d - 1 \quad \text{if} \quad b + d = k,$$
$$B = b, \qquad D = d \qquad \text{if} \quad b + d < k.$$

Example 4.10. Take $k = 6$, $M_0 = \{0, 1, 2, 4\}$, $M_1 = \{1, 3\}$ [i.e., $g \in AC^5(M_0, M_1)$ satisfies $g(0) = g'(0) = g''(0) = g^{(4)}(0) = 0$,

$g'(1) = g'''(1) = 0]$. Then the incidence matrix is

$$E = \begin{pmatrix} 1 & 1 & 1 & 0 & 1 & 0 \\ 0 & 1 & 0 & 1 & 0 & 0 \end{pmatrix}.$$

The Pólya condition is obviously satisfied and we have $a = 3$, $b = 0$, $c = 1$, $d = 2$.

Obviously, for any pair (M_0, M_1), we have

$$a + c \leq k, \quad b + d \leq k \tag{4.25}$$

and

$$a + c \neq k - 1.$$

We have mentioned that the boundary value problem (4.4) is uniquely solvable. For the corresponding Green function we have

$$G(x, t) = \frac{1}{(k-1)!}(x - t)^{k-1}\chi_{(0,x)}(t) + \sum_{i=0}^{k-1}\sum_{j=0}^{k-1} w_{ij}\frac{x^i}{i!}\frac{t^j}{j!}, \tag{4.26}$$

where the constants w_j depend on (M_0, M_1). If we denote

$$G(x, t) = \begin{cases} K_1(x, t) & \text{for } 0 < t < x < 1, \\ K_2(x, t) & \text{for } 0 < x < t < 1, \end{cases} \tag{4.27}$$

then the investigation of the Hardy inequality (4.1), i.e., of the equivalent weighted norm inequality

$$\left(\int_0^1 \left|\int_0^x K_1(x, t)f(t)dt + \int_x^1 K_2(x, t)f(t)dt\right|^q u(x)dx\right)^{1/q}$$

$$\leq C\left(\int_0^1 |f(x)|^p v(x)dx\right)^{1/p} \tag{4.28}$$

is based on the following result:

Theorem 4.11. *Let (M_0, M_1) satisfy $|M_0| + |M_1| = k$, $|M_0| > 0$, $|M_1| > 0$. Then the kernels K_1, K_2 from (4.27) satisfy*

$$|K_1(x, t)| \approx x^A(1 - x)^b t^c(1 - t)^D \quad \text{for} \quad 0 < t < x < 1,$$

$$|K_2(x, t)| \approx x^a(1 - x)^B t^C(1 - t)^d \quad \text{for} \quad 0 < x < t < 1. \tag{4.29}$$

Before proving Theorem 4.11, we will show how the approximation of the Green function G gives necessary and sufficient conditions for the validity of inequality (4.1) with $g \in AC^{k-1}(M_0, M_1)$. This characterization is simple if both $a + c < k$ and $b + d < k$ since then a, b, c, d coincide with A, B, C, D and $|G(x, t)|$ can be approximated by the special polynomial

$$x^a(1 - x)^b t^c(1 - t)^d$$

on the *whole* square $(0, 1) \times (0, 1)$.

Theorem 4.12. *Let $0 < q < \infty$, $1 < p < \infty$. Let (M_0, M_1) satisfy the Pólya condition, $|M_0| + |M_1| = k$, $|M_0| > 0$, $|M_1| > 0$ and $a + c < k$, $b + d < k$. Then the Hardy inequality*

$$\left(\int_0^1 |g(x)|^q u(x) dx \right)^{1/q} \leq C \left(\int_0^1 |g^{(k)}(x)|^p v(x) dx \right)^{1/p} \quad (4.30)$$

holds for all $g \in AC^{k-1}(M_0, M_1)$ if and only if

$$A := \left(\int_0^1 x^{aq}(1 - x)^{bq} u(x) dx \right)^{1/q}$$

$$\times \left(\int_0^1 t^{cp'}(1 - t)^{dp'} v^{1-p'}(t) dt \right)^{1/p'} < \infty. \quad (4.31)$$

Proof. Writing $g(x) = \int_0^1 G(x, t) f(t) dt$, we have $g^{(k)}(x) = f(x)$ and the Hardy inequality (4.30) becomes

$$\left(\int_0^1 \left| \int_0^1 G(x, t) f(t) dt \right|^q u(x) dx \right)^{1/q} \leq C \left(\int_0^1 |f(t)|^p v(t) dt \right)^{1/p'}.$$

It follows from (4.29) that $G(x, t)$ does not change sign in $(0, 1) \times (0, 1)$, and so we may restrict our attention to non-negative functions f. Since in our case

$$|G(x, t)| \approx x^a(1 - x)^b t^c(1 - t)^d,$$

the above inequality is equivalent to

$$R(f) := \left(\int_0^1 x^{aq}(1-x)^{bq}u(x)dx \right)^{1/q} \left(\int_0^1 t^c(1-t)^d f(t)dt \right)$$

$$\leq C' \left(\int_0^1 f^p(t)v(t)dt \right)^{1/p}.$$

(i) If we put here $f(t) = v^{1-p'}(t)t^{c(p'-1)}(1-t)^{d(p'-1)}$, we obtain
(4.31) — more precisely, $A \leq C'$ — since then

$$\int_0^1 t^c(1-t)^d f(t)dt = \int_0^1 t^{cp'}(1-t)^{dp'}v^{1-p'}(t)dt$$

and also

$$\int_0^1 f^p(t)v(t)dt = \int_0^1 t^{cp'}(1-t)^{dp'}v^{1-p'}(t)dt.$$

Thus condition (4.31) is necessary.

(ii) By the Hölder inequality,

$$\int_0^1 t^c(1-t)^d f(t)dt = \int_0^1 f(t)v^{1/p}(t)t^c(1-t)^d v^{-1/p}(t)dt$$

$$\leq \left(\int_0^1 f^p(t)v(t)dt \right)^{1/p}$$

$$\times \left(\int_0^1 t^{cp'}(1-t)^{dp'}v^{1-p'}(t)dt \right)^{1/p'},$$

and for $f \geq 0$ we have

$$\|Gf\|_{q,u} \lesssim R(f) \leq A\|f\|_{p,v}.$$

Thus condition (4.31) is sufficient. □

In the foregoing case, it was not necessary to distinguish between the cases $p \leq q$ and $p > q$. In the next theorem, which describes the general case, we have to consider these two possibilities separately.

Theorem 4.13. *Let (M_0, M_1) satisfy the Pólya condition, $|M_0| + |M_1| = k$, $|M_0| > 0$, $|M_1| > 0$. Let a, b, c, d, A, B, C, D be as in*

Definition 4.9. Then the Hardy inequality (4.30) *holds for all* $g \in AC^{k-1}(M_0, M_1)$ *if and only if*

(i) *either* $1 < p \leq q < \infty$ *and*

$$\sup_{0<x<1} \left(\int_x^1 t^{Aq}(1-t)^{bq}u(t)dt \right)^{1/q}$$

$$\times \left(\int_0^x t^{cp'}(1-t)^{Dp'}v^{1-p'}(t)dt \right)^{1/p'} < \infty,$$

$$\sup_{0<x<1} \left(\int_0^x t^{aq}(1-t)^{Bq}u(t)dt \right)^{1/q}$$

$$\times \left(\int_x^1 t^{Cp'}(1-t)^{dp'}v^{1-p'}(t)dt \right)^{1/p'} < \infty, \tag{4.32}$$

(ii) *or* $0 < q < p$, $1 < p < \infty$ *and*

$$\left(\int_0^1 \left(\int_x^1 t^{Aq}(1-t)^{bq}u(t)dt \right)^{r/q} \right.$$

$$\times \left(\int_0^x t^{cp'}(1-t)^{Dp'}v^{1-p'}(t)dt \right)^{r/q'}$$

$$\left. \times x^{Aq}(1-x)^{bq}u(x)dx \right)^{1/r} < \infty,$$

$$\left(\int_0^1 \left(\int_0^x t^{aq}(1-t)^{Bq}u(t)dt \right)^{r/q} \right. \tag{4.33}$$

$$\times \left(\int_x^1 t^{Cp'}(1-t)^{dp'}v^{1-p'}(t)dt \right)^{r/q'}$$

$$\left. \times x^{aq}(1-x)^{Bq}u(x)dx \right)^{1/r} < \infty$$

with $\frac{1}{r} = \frac{1}{q} - \frac{1}{p}$.

Proof. As mentioned above, the Hardy inequality is equivalent to inequality (4.28) which again — due to the estimates (4.29) and to the fact that the kernels K_1, K_2 do not change sign — is equivalent to

$$\left(\int_0^1 \left(x^A(1-x)^b \int_0^x t^c(1-t)^D f(t)dt \right. \right.$$

$$\left. \left. + x^a(1-x)^B \int_x^1 t^C(1-t)^d f(t)dt \right)^q u(x)dx \right)^{1/q}$$

$$\leq C' \left(\int_0^1 f^p(t)v(t)dt \right)^{1/p}$$

for $f \geq 0$. The result now follows from Theorem 2.3 analogously as in the proof of Theorem 4.8. □

Remark 4.14. Necessary and sufficient conditions (4.32) and/or (4.33) can again be rewritten in terms of the expressions $A, \widetilde{A}, A^*, \widetilde{A}^*$ (see (0.20), (0.25) and Remark 1.5 (iv)). More precisely, (4.32) means that

$$A(0, 1; t^{Aq}(1-t)^{bq}u(t), t^{-cp}(1-t)^{-Dp}v(t)) < \infty,$$

$$\widetilde{A}(0, 1; t^{aq}(1-t)^{Bq}u(t), t^{-Cp}(1-t)^{-dp}v(t)) < \infty$$

while (4.33) is

$$A^*(0, 1; t^{Aq}(1-t)^{bq}u(t), t^{-cp}(1-t)^{-Dp}v(t)) < \infty,$$

$$\widetilde{A}^*(0, 1; t^{aq}(1-t)^{Bq}u(t), t^{-Cp}(1-t)^{-dp}v(t)) < \infty.$$

What now remains is the proof of Theorem 4.11. This will be done in several auxiliary assertions. Let us start with the Green function $G(x, t)$ given by formula (4.26) with fixed coefficients w_{ij}.

Lemma 4.15. *Let $t \in (0, 1)$ be fixed and denote $h(x) = G(x, t)$. Then for $0 \leq i \leq k - 1$,*

(i) $h(x) \neq 0$ *for $x \in (0, 1)$,*
(ii) $h^{(i)}(0) = 0$ *if and only if $i \in M_0$,*
(iii) $h^{(i)}(1) = 0$ *if and only if $i \in M_1$.*

Proof. From formula (4.26) it is clear that h has $k - 2$ continuous derivatives and that $h^{(k-2)}$ is not differentiable. In particular, none of $h, h', \ldots, h^{(k-2)}$ is identically zero. Thus if we denote by S_i the (closed) support of $h^{(i)}$, then S_i is not empty. Since h is a polynomial on $[0, t]$ and on $[t, 1]$, S_i is one of the intervals $[0, 1], [0, t]$ or $[t, 1]$.

Let z_i be the number of zeros of $h^{(i)}$ in S_i. Take L_i to be 1 if $h^{(i)}$ has a zero at the left endpoint of S_i and 0 otherwise. Take R_i to be 1 if $h^{(i)}$ has a zero at the right endpoint of S_i and 0 otherwise.

Obviously,

$$z_0 \geq L_0 + R_0. \tag{4.34}$$

If $0 < i \leq k - 2$, then $h^{(i-1)}$ is constant off S_i so the z_{i-1} zeros of $h^{(i-1)}$ that lie in S_{i-1} are, in fact, in S_i. By the mean value theorem, $h^{(i)}$ has at least $z_{i-1} - 1$ zeros in the interior of S_i. Thus

$$z_i \geq z_{i-1} - 1 + L_i + R_i, \quad 0 < i \leq k - 2. \tag{4.35}$$

If $L_i = 0$, then either $h^{(i)}(0) \neq 0$ or $S_i = [t, 1]$ and $h^{(i)}(t) \neq 0$. The latter is impossible since $h^{(i)}$ is continuous. Since h is a section of Green's function of the boundary value problem (4.4), h must satisfy the boundary conditions, so the former implies $i \notin M_0$. Thus

$$L_i \geq \chi_{M_0}(i), \quad 0 \leq i \leq k - 2. \tag{4.36}$$

Similarly

$$R_i \geq \chi_{M_1}(i), \quad 0 \leq i \leq k - 2. \tag{4.37}$$

Now the graph of $h^{(k-2)}$ is a line segment on $[0, t]$ and another line segment on $[t, 1]$ so it has at most two zeros in its support, that is $z_{k-2} \leq 2$. If $k - 1 \in M_0$, then $h^{(k-1)}(0) = 0$ so $h^{(k-2)}$ is constant on $[0, t]$. Whether the constant is zero or non-zero, there is no contribution to z_{k-2} from $[0, t]$ so in this case $z_{k-2} \leq 1$. Similarly, if $k - 1 \in M_1$, then $z_{k-2} \leq 1$. It would violate the Pólya condition for $k - 1$ to be in both M_0 and M_1, so we may conclude that

$$2 \geq z_{k-2} + \chi_{M_0}(k - 1) + \chi_{M_1}(k - 1). \tag{4.38}$$

Adding the inequalities (4.34)–(4.38) and using the fact that

$$\sum_{i=0}^{k-1}(\chi_{M_0}(i) + \chi_{M_1}(i)) = |M_0| + |M_1| = k,$$

we get the trivial inequality $2 \leq 2$. It follows that all the inequalities (4.34)–(4.38) are in fact equalities.

Suppose that $S_0 = [0, t]$. Then $S_1 = \cdots = S_{k-2} = [0, t]$ and, by continuity, $R_0 = R_1 = \cdots = R_{k-2} = 1$. Equality in (4.37) implies that $\{0, 1, \ldots, k-2\} \subset M_1$. Also, since $S_{k-2} = [0, t]$, the only zero of $h^{(k-2)}$ in S_{k-2} is at $x = t$. Thus $z_{k-2} = 1$ and $g^{(k-1)}(0) \neq 0$, so $k - 1 \notin M_0$. Equality in (4.38) implies that $k - 1 \in M_1$ and we have $M_1 = \{0, 1, \ldots, k-1\}$, i.e. $M_0 = \emptyset$, which was explicitly excluded. Thus $S_0 \neq [0, t]$. Similarly we can show that $S_0 \neq [t, 1]$, so $S_0 = [0, 1]$.

Now equality in (4.34) implies that h has no zeros in the interior of S_0, which is statement (i) of our lemma.

If $i \in M_0$, then standard properties of Green's function imply that $h^{(i)}(0) = 0$. On the other hand, if $i \notin M_0$, then equality in (4.36) implies that $L_i = 0$ and as argued above, $h^{(i)} \neq 0$. This proves statement (ii), and (iii) follows analogously. $\square$

Corollary 4.16. $x^{-a}G(x, t)$ *is a polynomial for* $x \leq t$ *and* $(1-x)^{-b}G(x, t)$ *is a polynomial for* $x \geq t$.

Proof. Let t and h be as in Lemma 4.15. For $x \leq t$, formula (4.26) becomes

$$h(x) = \sum_{i=0}^{k-1}\sum_{j=0}^{k-1} w_{ij}\frac{x^i}{i!}\frac{t^j}{j!}.$$

From Lemma 4.15 (ii) and from the definition of the number a it follows that $h(0) = h'(0) = \cdots = h^{(a-1)}(0) = 0$, i.e.,

$$\sum_{j=0}^{k-1} w_{ij}\frac{t^j}{j!} = 0 \quad \text{for } i < a.$$

This holds for all $t \in (0, 1)$, and thus $w_{ij} = 0$ for $i < a$, which implies that $x^{-a}h(x) = x^{-a}G(x, t)$ is a polynomial for $x \leq t$.

Similarly, it follows from Lemma 4.15 (iii) that $(1 - x)^{-b} G(x, t)$ is a polynomial for $x \geq t$. □

The adjoint boundary value problem

Let

$$M_0^* = \{i \in \mathbb{N}_k : k - 1 - i \notin M_0\},$$
$$M_1^* = \{i \in \mathbb{N}_k : k - 1 - i \notin M_1\}.$$

The boundary value problem

$$g^{(k)} = f \quad \text{in } (0, 1);$$
$$g^{(i)}(0) = 0 \quad \text{for } i \in M_0^*,$$
$$g^{(j)}(1) = 0 \quad \text{for } j \in M_1^*$$

is called *adjoint* to BVP (4.4) (for $(a, b) = (0, 1)$).

The incidence matrix E^* of (M_0^*, M_1^*) can be constructed from the incidence matrix E of (M_0, M_1) so that we reverse each row of E and interchange 0's and 1's. For example, if

$$E = \begin{pmatrix} 1 & 1 & 1 & 0 & 1 & 0 \\ 0 & 1 & 0 & 1 & 0 & 0 \end{pmatrix},$$

then

$$E^* = \begin{pmatrix} 1 & 0 & 1 & 0 & 0 & 0 \\ 1 & 1 & 0 & 1 & 0 & 1 \end{pmatrix}.$$

It is easy to see that the pair (M_0^*, M_1^*) satisfies the Pólya condition if (M_0, M_1) does, and for the numbers a^*, b^*, c^*, d^* defined for (M_0^*, M_1^*) according to Definition 4.9 we have $a^* = c$, $b^* = d$, $c^* = a$ and $d^* = b$. The Green function for the adjoint problem is

$$G^*(x, t) = G(t, x) = \frac{(t - x)^{k-1}}{(k-1)!} \chi_{(0,t)}(x) + \sum_{i=0}^{k-1} \sum_{j=0}^{k-1} w_{ij} \frac{t^i}{i!} \frac{x^j}{j!}.$$

Corollary 4.17. *If $j < c$, then $w_{ij} = (-1)^{j+1} \delta_{i+j,k-1}$ (with δ the Kronecker symbol).*

Proof. For $t \in (0,1)$ fixed, set $h^*(x) = G^*(x,t)$. If $x < t$, then

$$h^*(x) = \frac{(t-x)^{k-1}}{(k-1)!} + \sum_{i=0}^{k-1}\sum_{j=0}^{k-1} w_{ij}\frac{t^i}{i!}\frac{x^j}{j!}.$$

Lemma 4.15 (ii) and the relation $a^* = c$ imply that $h^*(0) = h^{*'}(0) = \cdots = h^{*(c-1)}(0) = 0$, which leads to

$$\frac{(-1)^j t^{k-1-j}}{(k-1-j)!} + \sum_{i=0}^{k-1} w_{ij}\frac{t^i}{i!} = 0 \quad \text{for } j < c.$$

This polynomial in t vanishes for all $t \in (0,1)$, thus all coefficients are zero. Consequently,

$$\frac{(-1)^j}{i!}\delta_{i+j,k-1} + \frac{w_{ij}}{i!} = 0 \quad \text{for } j < c. \qquad \square$$

Corollary 4.18. $x^{-a}(1-t)^{-d}G(x,t)$ *is a polynomial for* $x \leq t$.

Proof. We know from Corollary 4.16 that $x^{-a}G(x,t)$ is a polynomial for $x \leq t$. The second assertion of this corollary, applied to the adjoint BVP, shows that $(1-t)^{-d}G(x,t)$ is also a polynomial if $x \leq t$, and the conclusion follows easily. $\qquad \square$

Now, we investigate the behaviour of $x^{-a}G(x,t)$ as $x \to 0$.

Lemma 4.19. *Let* $w(t) = \sum_{j=0}^{k-1} w_{aj}\frac{t^j}{j!}$. *Then for* $0 \leq i \leq k-1$ *we have*

(i) $w(t) \neq 0$ *for* $t \in (0,1)$,
(ii) $w^{(i)}(0) = 0$ *if and only if* $i \in M_0^*$ *when* $i < k-1-a$,
(iii) $w^{(k-1-a)}(0) \neq 0$,
(iv) $w^{(i)}(1) = 0$ *if and only if* $i \in M_1^*$.

Proof. Define

$$W(x,t) = x^{-a}\left(G(x,t) - \frac{(x-t)^{k-1}}{(k-1)!}\chi_{(0,x)}(t)\right) = \sum_{i=a}^{k-1}\sum_{j=0}^{k-1} w_{ij}\frac{x^{i-a}}{i!}\frac{t^j}{j!}$$

(we have shown in Corollary 4.17 that $w_{ij} = 0$ for $i < a$), thus $w(t) = a!W(0,t)$.

Since $W(x,t)$ is a polynomial, it is clear that

$$\frac{1}{a!}w^{(i)}(0) = \lim_{x \to 0}\left[\left(\frac{\partial}{\partial t}\right)^i W(x,t)|_{t=0}\right]$$

$$= \lim_{x \to 0}\left[x^{-a}\left(\frac{\partial}{\partial t}\right)^i G(x,t)|_{t=0} - \frac{(-1)^i x^{k-1-i-a}}{(k-1-i)!}\right].$$

For each fixed x, $G(x,t) = G^*(t,x)$ is a section of the Green function of the adjoint BVP, and consequently, if $i \in M_0^*$, then $(\frac{\partial}{\partial t})^i G(x,t)|_{t=0} = 0$. For such an i we have

$$\frac{1}{a!}w^{(i)}(0) = \lim_{x \to 0}\frac{(-1)^{i+1}x^{k-1-i-a}}{(k-1-i)!}$$

$$= \begin{cases} 0 & \text{if } i < k-1-a, \\ \dfrac{(-1)^{k-a}}{a!} & \text{if } i = k-1-a. \end{cases}$$

This proves assertion (iii) and the *if* part of (ii).

At the other endpoint we have for $i \in M_1^*$

$$\frac{1}{a!}w^{(i)}(1) = \lim_{x \to 0}\left[\left(\frac{\partial}{\partial t}\right)^i W(x,t)|_{t=1}\right]$$

$$= \lim_{x \to 0}\left[x^{-a}\left(\frac{\partial}{\partial t}\right)^i G(x,t)|_{t=1}\right] = 0$$

and thus, the *if* part of (iv) holds as well.

Let $T = \{i \in M_0^* : i < k-1-a\}$. The definition of c^* (which is a) shows that $k-1-a$ is the largest element of M_0^*. Thus T has exactly one element less than M_0^*.

Up to now we have shown that w satisfies the boundary conditions $w^{(i)}(0) = 0$ for $i \in T$ and $w^{(i)}(1) = 0$ for $i \in M_1^*$. The remaining part of the proof is similar to that of Lemma 4.15.

Let z_i^* be the number of zeros of $w^{(i)}$ in its (closed) support. Since $w^{(i)}$ is a polynomial, its support is the whole interval $[0,1]$

unless $w^{(i)} \equiv 0$. As we have seen, $w^{(k-1-a)}(0) \neq 0$ so $w^{(i)} \not\equiv 0$ for $i \leq k - 1 - a$.

For $i = 0$ we have the obvious inequality

$$z_0^* \geq \chi_T(0) + \chi_{M_1^*}(0). \tag{4.39}$$

If $w^{(i)} \not\equiv 0$ and $0 < i \leq k - 1$, then as in Lemma 4.15 we observe that the number of zeros of $w^{(i)}$ in $(0, 1)$ is at least $z_{i-1}^* - 1$ and, adding the boundary zeros, we have

$$z_i^* \geq z_{i-1}^* - 1 + \chi_T(i) + \chi_{M_1^*}(i). \tag{4.40}$$

If $w^{(i)} \equiv 0$, then we have $i > k - 1 - a$, so $i \notin T$. Also, both z_i^* and z_{i-1}^* are zero. It follows that (4.40) holds in this case as well, reducing to $0 \geq -1 + \chi_{M_1^*}(i)$.

Since the degree of w is at most $k - 1$ we must have

$$0 = z_k^*. \tag{4.41}$$

Adding inequality (4.39) and inequalities (4.40) for $1 \leq i \leq k - 1$ we obtain

$$z_0^* + z_1^* + \cdots + z_{k-1}^* \geq z_0^* + \cdots + z_{k-2}^* - (k-1) + \sum_{i=0}^{k-1}(\chi_T(i) + \chi_{M_1^*}(i))$$

and, by virtue of (4.41), this simplifies to

$$0 \geq -(k-1) + (|T| + |M_1^*|) = -(k-1) + (k-1) = 0.$$

Thus we have equality in (4.39) and in (4.40) for $1 \leq i \leq k - 1$.

Equality in (4.39) proves that w has no interior zeros which is assertion (i).

If $w^{(i)} \not\equiv 0$, equality in (4.40) means that $w^{(i)}$ has no boundary zeros except those accounted for by the terms $\chi_T(i)$ and $\chi_{M_1^*}(i)$. Hence the *only if* parts of (ii) and (iii) hold when $w^{(i)} \not\equiv 0$.

If $w^{(i)} \equiv 0$, then of course both $w^{(i)}(0) = 0$ and $w^{(i)}(1) = 0$. As we have seen, $i > k - 1 - a$ in this case so the *only if* part of (ii) holds vacuously. Equality in (4.40) when $w^{(i)} \equiv 0$ means that $\chi_{M_1^*}(1) = 1$, so $i \in M_1^*$ and the *only if* part of (iv) holds as well. $\square$

Corollary 4.20. *If $a + c = k$, then $w_{ac-1} \neq 0$, and if $a + c < k$, then $w_{ac} \neq 0$.*

Proof. If $a + c = k$, then Lemma 4.19 (iii) states that $w^{(c-1)}(0) \neq 0$. This means, due to the definition of w, that $w_{ac-1} \neq 0$.

As was mentioned above, $a + c = k - 1$ cannot hold (see (4.25)), so if $a + c < k$ we have $c < k - 1 - a$. Now by the definition of a^* (which is c) we have $c \notin M_0^*$, so Lemma 4.19 (ii) states $w^{(c)}(0) \neq 0$, which means that $w_{ac} \neq 0$. $\qquad\square$

Now we are able to derive the estimates of the Green function. First, let us introduce another boundary value problem which will reduce the number of estimates required.

The symmetric boundary value problem

For our pair (M_0, M_1), let

$$\widehat{M_0} = M_1, \quad \widehat{M_1} = M_0.$$

The boundary value problem

$$g^{(k)} = f \quad \text{in} \quad (0, 1);$$
$$g^{(i)}(0) = 0 \quad \text{for } i \in \widehat{M_0},$$
$$g^{(j)}(1) = 0 \quad \text{for } j \in \widehat{M_1}$$

is called *symmetric* to BVP (4.4) (for $(a, b) = (0, 1)$).

The incidence matrix $\widehat{E}$ of $(\widehat{M_0}, \widehat{M_1})$ simply means that we interchange the rows of the incidence matrix E of (M_0, M_1). Thus the Pólya condition still holds and it is easy to see that

$$\widehat{a} = b, \quad \widehat{b} = a, \quad \widehat{c} = d, \quad \widehat{d} = c$$

and indeed that

$$\widehat{A} = B, \quad \widehat{B} = A, \quad \widehat{C} = D, \quad \widehat{D} = C.$$

The Green function $\widehat{G}(x, t)$ of the symmetric BVP is given by

$$\widehat{G}(x, t) = G(1 - x, 1 - t).$$

Proof of Theorem 4.11. It is enough to prove this theorem under the restriction that $x + t \leq 1$ since if the estimates (4.29) are restricted to the Green function $\widehat{G}(x,t)$ of the symmetric BVP they become

$$|G(1-x, 1-t)| \approx x^b(1-x)^A t^D (1-t)^c \quad \text{for } 0 < x < t \leq 1-x,$$

$$|G(1-x, 1-t)| \approx x^B(1-x)^a t^d (1-t)^C \quad \text{for } 0 < t < x \leq 1-t.$$

Replacing x by $1-x$ and t by $1-t$ we obtain (4.29) for the original Green function $G(x,t)$ but restricted to $x + t \geq 1$.

A further restriction shows that it is enough to prove the second equivalence in (4.29) since the first is just the second applied to the adjoint BVP.

So to complete the proof we must establish the second relation in (4.29) under the restriction $x + t \leq 1$. To do this, it is enough to prove the following five statements:

(S1) $x^{-a}(1-x)^{-B} t^{-C}(1-t)^{-d}|G(x,t)|$ is continuous when $0 \leq x \leq t \leq 1-x$ except possibly at $(x,t) = (0,0)$. (Strictly speaking we mean that this function, defined on the *open* set, extends continuously to the closed set with perhaps one exceptional boundary point.)

(S2) $0 < x^{-a}(1-x)^{-B} t^{-C}(1-t)^{-d}|G(x,t)| < \infty$ for $0 < x \leq t \leq 1-x$.

(S3) $0 < \lim_{x \to 0} x^{-a}(1-x)^{-B} t^{-C}(1-t)^{-d}|G(x,t)| < \infty$ for $0 < t < 1$.

(S4) $0 < \lim_{(x,t) \to (0,1)} x^{-a}(1-x)^{-B} t^{-C}(1-t)^{-d}|G(x,t)| < \infty.$

(S5) $x^{-a}(1-x)^{-B} t^{-C}(1-t)^{-d}|G(x,t)|$ is bounded above and below (away from zero) in a neighbourhood of $(0,0)$ when $0 < x \leq t$.

On the set $0 \leq x \leq t \leq 1-x$, Corollary 4.18 shows that $x^{-a}(1-t)^{-d}G(x,t)$ is a polynomial, and the function $(1-x)^{-B} t^{-C}$ is continuous on this set except at $(0,0)$. The first statement follows.

The statements (S2) and (S3) follow by Lemma 4.15 (i) and Lemma 4.19 (i), respectively.

Lemma 4.19 (iv) and the definition of b^* (which is d) imply that

$$w(1) = w'(1) = \cdots = w^{(d-1)}(1) = 0, \quad w^{(d)}(1) \neq 0$$

so $(1-t)^{-d}w(t)$ is a polynomial in the variable $1-t$ with a non-zero constant term. Thus, for $x \leq t$, $x^{-a}(1-t)^{-d}G(x,t)$ is a polynomial in the variables x and $1-t$ with a non-zero constant term, and this proves (S4).

We prove (S5) in two cases. First suppose that $a+c < k$. In this case $C = c$, so it suffices to prove that $x^{-a}t^{-c}G(x,t)$ has a non-zero limit as $(x,t) \to (0,0)$ with $0 < x \leq t$. By Corollary 4.20, $w_{ac}/(a!c!)$ is non-zero. We estimate as follows:

$$\left| \frac{G(x,t)}{x^a t^c} - \frac{w_{ac}}{a!c!} \right| = \left| \left(\sum_{i=a}^{k-1} \sum_{j=0}^{k-1} w_{ij} \frac{x^{i-a}t^{j-c}}{i!j!} \right) - \frac{w_{ac}}{a!c!} \right|$$

$$\leq \sum_{i=a}^{k-1} \sum_{j=0}^{c-1} |w_{ij}| \frac{x^{i-a}}{i!} \frac{t^{j-c}}{j!} + \sum_{i=a+1}^{k-1} |w_{ic}| \frac{x^{i-a}}{i!c!}$$

$$+ \sum_{i=a}^{k-1} \sum_{j=c+1}^{k-1} |w_{ij}| \frac{x^{i-a}}{i!} \frac{t^{j-c}}{j!}.$$

By Corollary 4.17, $w_{ij} = (-1)^{j+1}\delta_{i+j,k-1}$ for $j < c$, so the last expression becomes

$$\sum_{i=a}^{k-1} \frac{x^{i-a}}{i!} \frac{t^{k-1-i-c}}{(k-1-i)!} + \sum_{i=a+1}^{k-1} |w_{ic}| \frac{x^{i-a}}{i!c!} + \sum_{i=a}^{k-1} \sum_{j=c+1}^{k-1} |w_{ij}| \frac{x^{i-a}}{i!} \frac{t^{j-c}}{j!}.$$

Now, $0 < x \leq t$ and $i - a \geq 0$, so in the first term, $x^{i-a}t^{k-1-i-c} \leq t^{k-1-a-c} \leq t$. The last inequality is justified because $0 < t < 1$ and $k - 1 - a - c \geq 1$. (Recall that $a + c = k - 1$ is impossible!) In the second term $x^{i-a} \leq x \leq t$ since $i \geq a+1$. In the third term we have $i - a \geq 0$ and $j \geq c+1$, so $x^{i-a}t^{j-c} \leq t$ as well. Thus the whole sum is dominated by a constant multiple of t and so it tends to zero if t does. This completes the case $a + c < k$.

In the second case, $a + c = k$, we do not have continuity at $(0,0)$ in general so the argument is more delicate. We note that $a > 0$ since otherwise $c = k$ would hold and $|M_0| = 0$, which is prohibited. Recall that in this case $C = c - 1$. We write

$$x^{-a}t^{-C}G(x,t) = \sum_{i=a}^{k-1}\sum_{j=0}^{k-1} w_{ij} \frac{x^{i-a}}{i!} \frac{t^{j-c+1}}{j!}$$

$$= \sum_{i=a}^{k-1}(-1)^{k-i}\frac{x^{i-a}}{i!}\frac{t^{k-i-c}}{(k-1-i)!}$$

$$+ \sum_{i=a}^{k-1}\sum_{j=c}^{k-1} w_{ij}\frac{x^{i-a}}{i!}\frac{t^{j-c+1}}{j!},$$

where we have again used Corollary 4.17 to simplify w_{ij} for $j < c$. The second term tends to zero with t since $x^{i-a}t^{j-c+1} \le t$ for each $i \ge a$ and $j \ge c$. It remains to show that the first term is bounded above and below in absolute value as $(x,t) \to (0,0)$.

Writing x as st for some $s \in [0,1]$ and using the hypothesis $a + c = k$, the first term simplifies to

$$\sum_{i=a}^{k-1}(-1)^{k-i}\frac{s^{i-a}}{i!(k-1-i)!}, \qquad (4.42)$$

which is continuous in s and hence bounded above in absolute value on $[0,1]$. To show that it is also bounded below in absolute value it is enough to prove that it has no zero in $[0,1]$. At $s = 0$ the value is

$$\frac{(-1)^{k-a}}{a!(k-1-a)!} \ne 0,$$

so we may complete the argument by showing that (4.42) is

$$(-1)^{k-a}s^{k-1-a}r_{k-1}(s),$$

where

$$r_n(s) = (-1)^{n-a}\sum_{i=a}^{n}\frac{(-s)^{i-n}}{i!(n-i)!}$$

is a *positive* function on $(0,1]$ for all $n \ge a$.

Indeed, for $n = a$ we have

$$r_a(s) = \frac{1}{a!} > 0.$$

Suppose $r_n(s) > 0$ when $0 < s \leq 1$ for some $n \geq a$. Then

$$r'_{n+1}(s) = -\frac{r_n(s)}{s^2} < 0,$$

so, for $0 < s \leq 1$, the functions r_n are decreasing and

$$r_n(s) \geq r_n(1) = \frac{(-1)^{n-a}}{n!} \sum_{i=a}^{n} (-1)^{n-i} \binom{n}{n-i}$$

$$= \frac{1}{n!} \binom{n-1}{n-a} > 0,$$

where the last equality is an identity for binomial coefficients. $\square$

4.4. Some Special Cases II

Some other boundary conditions

Up to now, we have given a complete characterization of weights u, v and parameters p, q for which the k-th order Hardy inequality (4.1) holds if we consider functions $g \in AC^{k-1}(M_0, M_1)$, i.e. satisfying boundary conditions of the form (4.2). The idea is based on the reduction of inequality (4.1) to a weighted norm inequality

$$\|Tf\|_{q,u} \leq C\|f\|_{p,v},$$

where Tf is expressed in terms of the Green function of the BVP (4.4),

$$(Tf)(x) = \int_a^x K_1(x,t)f(t)dt + \int_x^b K_2(x,t)f(t)dt,$$

with either $K_1(x,t) = \frac{1}{(k-1)!}(x-t)^{k-1}$, $K_2(x,t) \equiv 0$ (if $M_1 = \emptyset$) or $K_2(x,t) = \frac{1}{(k-1)!}(t-x)^{k-1}$, $K_1(x,t) \equiv 0$ (if $M_0 = \emptyset$ — see Sec. 4.2) or where either *non-negative* or *non-positive* kernels K_1, K_2 can be approximatively factorized (see Theorem 4.11), which then allows to use (twice) the "classical" first order Hardy inequality.

Of course, instead of (4.2), we could consider more general boundary conditions of the form

$$\sum_{j=0}^{k-1} \alpha_{ij} g^{(j)}(a) + \beta_{ij} g^{(j)}(b) = 0, \quad i = 0, 1, \ldots, k-1, \qquad (4.43)$$

and proceed in the same way, i.e., to solve the BVP $\{u^{(k)} = f$ in (a, b) and (4.43)$\}$ and investigate the corresponding Green function (if it exists); but even simple examples show that at least it is not guaranteed that the corresponding kernels do not change sign so that the argument with positive operators cannot be used.

Therefore, we will give only some examples for the case $k = 2$ where we consider the inequality

$$\left(\int_a^b |g(x)|^q u(x) dx \right)^{1/q} \le C \left(\int_a^b |g''(x)|^p v(x) dx \right)^{1/p} \qquad (4.44)$$

for functions $g \in AC^1$ satisfying

$$\alpha_{00} g(a) + \alpha_{01} g'(a) + \beta_{00} g(b) + \beta_{01} g'(b) = 0,$$
$$\alpha_{10} g(a) + \alpha_{11} g'(a) + \beta_{10} g(b) + \beta_{11} g'(b) = 0.$$

For simplicity, we will deal again with the special case $a = 0$, $b = 1$.

The general case, i.e., with general conditions (4.43), is dealt with in the recent paper [KKP1] by A. Kufner *et al.*, see also comment 4.10.10.

Robin-type boundary conditions

Consider inequality (4.44) (with $(a, b) = (0, 1)$) under the conditions

$$\alpha g(0) + \beta g'(0) = 0,$$
$$\gamma g(1) + \delta g'(1) = 0. \qquad (4.45)$$

The solution of the BVP

$$g'' = f \text{ in } (0, 1) \text{ and conditions } (4.45)$$

can be expressed as

$$g(x) = (Tf)(x) = \int_0^x K_1(x,t)f(t)dt + \int_x^1 K_2(x,t)f(t)dt, \quad (4.46)$$

where

$$K_1(x,t) = \frac{1}{\Delta}(\alpha t - \beta)(\gamma + \delta - \gamma x), \quad 0 < t < x,$$

$$\hspace{8.5cm} (4.47)$$

$$K_2(x,t) = \frac{1}{\Delta}(\alpha x - \beta)(\gamma + \delta - \gamma t), \quad x < t < 1,$$

provided that

$$\Delta := \beta\gamma - \alpha(\gamma + \delta) \neq 0. \quad (4.48)$$

The kernels K_1, K_2 are not necessarily both non-negative or both non-positive but they are *factorized*. Thus we can use the results of Chap. 2 (see Theorem 2.3) and obtain immediately the following *necessary and sufficient conditions* for the validity of the inequality

$$\left(\int_0^1 |(Tf)(x)|^q u(x)dx\right)^{1/q} \leq C \left(\int_0^1 |f(x)|^p v(x)dx\right)^{1/p}$$

which is equivalent to (4.44) (for $(a,b) = (0,1)$) under the conditions (4.45):

(i) For the case $1 < p \leq q < \infty$,

$$\sup_{0<x<1} \left(\int_x^1 |\gamma + \delta - \gamma t|^q u(t)dt\right)^{1/q}$$

$$\times \left(\int_0^x |\alpha t - \beta|^{p'} v^{1-p'}(t)dt\right)^{1/p'} < \infty,$$

$$\sup_{0<x<1} \left(\int_0^x |\alpha t - \beta|^q u(t)dt\right)^{1/q}$$

$$\times \left(\int_x^1 |\gamma + \delta - \gamma t|^{p'} v^{1-p'}(t)dt\right)^{1/p'} < \infty.$$

(ii) For the case $0 < q < p < \infty$, $1 < p < \infty$,

$$\left(\int_0^1 \left(\int_x^1 |\gamma + \delta - \gamma t|^q u(t) dt \right)^{r/p} \right.$$

$$\times \left(\int_0^x |\alpha t - \beta|^{p'} v^{1-p'}(t) dt \right)^{r/p'} |\gamma + \delta - \gamma x|^q u(x) dx \right)^{1/r} < \infty,$$

$$\left(\int_0^1 \left(\int_0^x |\alpha t - \beta|^q u(t) dt \right)^{r/p} \right.$$

$$\times \left(\int_x^1 |\gamma + \delta - \gamma t|^{p'} v^{1-p'}(t) dt \right)^{r/p'} |\alpha x - \beta|^q u(x) dx \right)^{1/r} < \infty$$

with $\frac{1}{r} = \frac{1}{q} - \frac{1}{p}$.

A counterexample

If condition (4.48) is violated, then the Hardy inequality (4.44) under boundary conditions (4.45) becomes meaningless. Indeed: for $\alpha = \gamma = 1$, $\beta = 0$ and $\delta = -1$, i.e., for the boundary conditions

$$g(0) = 0, \quad g(1) - g'(1) = 0$$

we have $\Delta = \beta\gamma - \alpha(\gamma + \delta) = 0$ and the function $g(x) = x$ satisfies these conditions but $g''(x) \equiv 0$ and so the right-hand side in (4.44) is zero while the left hand side is not.

Another example

Consider again inequality (4.44) (with $(a, b) = (0, 1)$), but now with the conditions

$$\alpha g(0) + \beta g(1) = 0,$$

$$\gamma g'(0) + \delta g'(1) = 0. \tag{4.49}$$

The solution of the BVP

$$g'' = f \text{ in } (0, 1) \text{ and conditions } (4.49)$$

can again be expressed in the form (4.46), but now with

$$K_1(x,t) = Cx + (B-1)t - BC, \quad 0 < t < x < 1,$$
$$K_2(x,t) = (C-1)x + Bt - BC, \quad 0 < x < t < 1 \tag{4.50}$$

with $B = \frac{\beta}{\alpha+\beta}$, $C = \frac{\gamma}{\gamma+\delta}$ provided

$$\Delta := (\alpha + \beta)(\gamma + \delta) \neq 0. \tag{4.51}$$

(i) Again, the Hardy inequality becomes meaningless if condition (4.51) is violated. As a counterexample we can use the function $g(x) = (\alpha + \beta)x - \beta$ with $\alpha + \beta \neq 0$, which satisfies (4.49) (the second condition with $\gamma = 1$, $\delta = -1$, i.e., $\gamma + \delta = 0$) and again, $g''(x) \equiv 0$.

(ii) The kernels K_1, K_2 from (4.50) cannot be in general factorized or approximated by a product of a function of x and a function of t, moreover, they can *change sign* in the corresponding triangle $0 < t < x < 1$ or $0 < x < t < 1$. Consequently, the approaches described above cannot be used in the general case. However, in some special cases we can give necessary and sufficient conditions. E.g., let us suppose that $B = 0$ (i.e., $\beta = 0$ in (4.49)). Then we have

$$K_1(x,t) = Cx - t, \quad K_2(x,t) = (C-1)x.$$

If $0 < C < 1$ then $K_1(x,t)$ changes its sign in the triangle $0 < t < x < 1$. But for $C > 1$, we have $K_1(x,t) > 0$ for $0 < t < x < 1$ and, moreover, this kernel satisfies conditions (2.25) and (2.26) of Chap. 2. Thus, for the operator T defined by

$$(Tf)(x) = \int_0^x (Cx - t)f(t)dt + (C-1)x \int_x^1 f(t)dt \tag{4.52}$$

with $C > 1$, we can use the positivity argument, consider T for $f \geq 0$ and obtain the following *triple* of necessary and sufficient conditions for the validity of inequality (4.44) (for $(a,b) = (0,1)$) under conditions (4.49) with our choice of $B(= 0)$ and $C(> 1)$, i.e., under the conditions

$$g(0) = 0, \quad \gamma g'(0) + (1 - \gamma)g'(1) = 0, \quad \gamma > 1:$$

for $1 < p \le q < \infty$,

$$\sup_{0 < x < 1} \left(\int_x^1 (\gamma x - t)^q u(t) dt \right)^{1/q} \left(\int_0^x v^{1-p'}(t) dt \right)^{1/p'} < \infty,$$

$$\sup_{0 < x < 1} \left(\int_x^1 u(t) dt \right)^{1/q} \left(\int_0^x (\gamma t - x)^{p'} v^{1-p'}(t) dt \right)^{1/p'} < \infty, \quad (4.53)$$

$$\sup_{0 < x < 1} \left(\int_0^x t^q u(t) dt \right)^{1/q} \left(\int_x^1 v^{1-p'}(t) dt \right)^{1/p'} < \infty.$$

Recall that the first two conditions in (4.53) are necessary and sufficient for the validity of the inequality

$$\left(\int_0^1 \left(\int_0^x (\gamma x - t) f(t) dt \right)^q u(x) dx \right)^{1/q} \le C_1 \left(\int_0^1 f^p(x) v(x) dx \right)^{1/p}$$

for $f \ge 0$, i.e., they cover the first integral in (4.52), while the third condition in (4.53) is necessary and sufficient for

$$\left(\int_0^1 \left(x \int_x^1 f(t) dt \right)^q u(x) dx \right)^{1/q}$$

$$\le C_2 \left(\int_0^1 f^p(x) v(x) dx \right)^{1/p}, \quad f \ge 0$$

and covers the rest of the operator T.

4.5. Reducing the Conditions

Remark 4.21. While in the case of the first order Hardy inequalities we have only *one* (necessary and sufficient) condition, in the case of a higher order we have a *pair* of conditions (and sometimes even more, see (4.53)). In some special cases — i.e., for some special weights — we are able to reduce this pair to *only one* condition. Let us describe an approach which, unfortunately, works only for $(a, b) = (0, \infty)$ and for the case $M_1 = \emptyset$ or $M_0 = \emptyset$, provided that, moreover, $1 < p \le q < \infty$.

(i) Let us suppose $M_0 = \emptyset$, $M_1 = \mathbb{N}_k$. As we have seen (see Theorem 4.3 and Remark 4.4), the Hardy inequality

$$\left(\int_0^\infty |g(x)|^q u(x) dx \right)^{1/q} \le C \left(\int_0^\infty |g^{(k)}(x)|^p v(x) dx \right)^{1/p} \qquad (4.54)$$

holds for all functions $g \in AC^{k-1} (\emptyset, \mathbb{N}_k)$, i.e., functions satisfying

$$g(\infty) = g'(\infty) = \cdots = g^{(k-1)}(\infty) = 0$$

if and only if (for $1 < p \le q < \infty$) the following two functions are bounded on $(0, \infty)$:

$$B_1(x) = \left(\int_0^x (x-t)^{(k-1)q} u(t) dt \right)^{1/q} \left(\int_x^\infty v^{1-p'}(t) dt \right)^{1/p'},$$

$$\qquad (4.55)$$

$$B_2(x) = \left(\int_0^x u(t) dt \right)^{1/q} \left(\int_x^\infty (t-x)^{(k-1)p'} v^{1-p'}(t) dt \right)^{1/p'}.$$

(ii) Analogously, inequality (4.54) holds for all functions $g \in AC^{k-1}(\mathbb{N}_k, \emptyset)$ (i.e., satisfying $g^{(i)}(0) = 0$ for $i = 0, 1, \ldots, k-1$) if and only if the following two functions are bounded on $(0, \infty)$:

$$\widetilde{B}_1(x) = \left(\int_x^\infty (t-x)^{q(k-1)} u(t) dt \right)^{1/q} \left(\int_0^x v^{1-p'}(t) dt \right)^{1/p'},$$

$$\qquad (4.56)$$

$$\widetilde{B}_2(x) = \left(\int_x^\infty u(t) dt \right)^{1/q} \left(\int_0^x (x-t)^{p'(k-1)} v^{1-p'}(t) dt \right)^{1/p'}.$$

(iii) Now, we will show that for some classes of weight functions u, v, the boundedness of only one of the functions B_1, B_2 (or $\widetilde{B}_1, \widetilde{B}_2$) is necessary and sufficient.

Theorem 4.22. (i) *Denote*

$$U(x) = \int_0^x (x-t)^{q(k-1)} u(t) dt \qquad (4.57)$$

and suppose that there exists a positive constant c_0 such that

$$U(2x) \le c_0 U(x) \text{ for every } x \in (0, \infty). \qquad (4.58)$$

Then inequality (4.54) holds for every $g \in AC^{k-1}(\emptyset, \mathbb{N}_k)$ if and only if the function $B_2(x)$ from (4.55) is bounded.

(ii) *Denote*

$$V(x) = \int_0^x (x-t)^{p'(k-1)} v^{1-p'}(t) dt \qquad (4.59)$$

and suppose that there exists a positive constant c_0 such that

$$V(2x) \le c_0 V(x) \text{ for every } x \in (0, \infty). \qquad (4.60)$$

Then inequality (4.54) holds for every $g \in AC^{k-1}(\mathbb{N}_k, \emptyset)$ if and only if the function $\widetilde{B}_1(x)$ from (4.56) is bounded.

Proof. (a) *Necessity.* If inequality (4.54) holds, then *both* the functions B_1, B_2 (or $\widetilde{B}_1, \widetilde{B}_2$) are bounded, and thus, so is B_2 (or $\widetilde{B}_1$).

(b) *Sufficiency.* It follows from (4.58) that

$$\int_0^x u(t) dt = \int_0^x x^{q(k-1)} u(t) x^{-q(k-1)} dt$$

$$\ge \int_0^x (x-t)^{q(k-1)} u(t) dt \cdot x^{-q(k-1)}$$

$$\ge \frac{1}{c_0} \int_0^{2x} (2x-t)^{q(k-1)} u(t) dt \cdot x^{-q(k-1)}.$$

Further,

$$\int_x^\infty (t-x)^{p'(k-1)} v^{1-p'}(t) dt$$

$$= \int_x^\infty \left(\frac{t}{x} - 1\right)^{p'(k-1)} v^{1-p'}(t) x^{p'(k-1)} dt$$

$$\geq \int_{2x}^{\infty} \left(\frac{t}{x} - 1\right)^{p'(k-1)} v^{1-p'}(t)dt \cdot x^{p'(k-1)}$$

$$\geq \int_{2x}^{\infty} v^{1-p'}(t)dt \cdot x^{p'(k-1)}$$

since in the third integral we have $2x < t < \infty$, i.e., $\frac{t}{x} > 2$ and $\frac{t}{x} - 1 > 1$.

Consequently, we have

$$B_2(x) = \left(\int_0^x u(t)dt\right)^{1/q} \left(\int_x^{\infty} (t-x)^{p'(k-1)} v^{1-p'}(t)dt\right)^{1/p'}$$

$$\geq c_0^{-q} \left(\int_0^{2x} (2x-t)^{q(k-1)} u(t)dt\right)$$

$$\times x^{-(k-1)} \left(\int_{2x}^{\infty} v^{1-p'}(t)dt\right)^{1/p'} x^{k-1}$$

$$= c_0^{-q} B_1(2x)$$

and thus, the boundedness of B_2 on $(0, \infty)$ implies also the boundedness of B_1 and so, the validity of (4.54). So, we proved case (i), and the proof for case (ii) follows analogously, showing that $\widetilde{B}_2(2x) \leq c\widetilde{B}_1(x)$. $\qquad\square$

4.6. Overdetermined Classes ($k=1$)

Up to now, we have investigated the Hardy inequality (4.1) for functions $g \in AC^{k-1}(M_0, M_1)$ where the index sets satisfied

$$|M_0| + |M_1| = k. \tag{4.61}$$

In this case necessary and sufficient conditions for (4.1) to hold are (almost) fully described. Now, we will investigate (4.1) for functions $g \in AC^{k-1}(M_0, M_1)$ where

$$|M_0| + |M_1| > k, \tag{4.62}$$

which means that in the incidence matrix are more 1's that 0's. Again we suppose that the Pólya condition is satisfied (see Sec. 4.1). The class $AC^{k-1}(M_0, M_1)$ with M_0, M_1 satisfying (4.62) will be called *overdetermined*.

We will see that the solution of the problem to find necessary and sufficient conditions for the validity of (4.1) on overdetermined classes is less satisfactory than in the *well-determined* case (4.61): the full answer is known only in some special cases and the conditions are very cumbersome. But first, let us start with the first order Hardy inequality (1.25).

The case k = 1

Here, the only overdetermined class is the class of functions $g \in AC^0$ satisfying

$$g(a) = g(b) = 0 \tag{4.63}$$

and as was mentioned in Chap. 1, the inequality

$$\left(\int_a^b |g(x)|^q u(x) dx \right)^{1/q} \le C \left(\int_a^b |g'(x)|^p v(x) dx \right)^{1/p} \tag{4.64}$$

holds for such functions g if and only if, for $1 < p \le q < \infty$,

$$\sup_{(c,d)\subset(a,b)} \left[\left(\int_c^d u(t) dt \right)^{1/q} \right.$$

$$\left. \times \min \left\{ \left(\int_a^c v^{1-p'}(t) dt \right)^{1/p'} , \left(\int_d^b v^{1-p'}(t) dt \right)^{1/p'} \right\} \right] < \infty \tag{4.65}$$

(see (1.33) in Remark 1.4).

Example 4.23. If we choose $(a, b) = (-\infty, \infty)$, then inequality (4.64) holds with $u(x) = e^{\alpha x^2}$, $\alpha > 0$, $v(x) = e^{\alpha x^2 p/q}$ if g satisfies $g(-\infty) = 0$ *and* $g(+\infty) = 0$, but it does *not hold* if g satisfies only one of these two conditions (see [OK, Example 6.13]).

Remark 4.24. The result just mentioned cannot be used if $p > q$ and also if *both* integrals appearing in the minimum in (4.65) are infinite. In the sequel, both disadvantages will be removed.

For simplicity, we consider again the case

$$(a, b) = (0, 1).$$

Our approach in the foregoing parts was mostly based on the reduction of the Hardy inequality (4.1) to a weighted norm inequality, acting with some operator $T : L^p(v) \to L^q(u)$. In what follows, we will use operators acting on certain special subspaces of $L^p(v)$. Therefore, we will start with a useful lemma.

Let us recall that an operator T is called *positive* if it maps non-negative functions to non-negative functions.

Lemma 4.25. *Let $1 < p < \infty$, $0 < q < \infty$, and let u, v be weight functions on $(0, 1)$. Suppose that there is a number $z \in (0, 1)$ such that*

$$\int_0^z v^{1-p'}(x)dx \approx \int_z^1 v^{1-p'}(x)dx < \infty \ \ or$$

$$\int_0^z v^{1-p'}(x)dx = \int_z^1 v^{1-p'}(x)dx = \infty, \qquad (4.66)$$

and set

$$H = \left\{ h : \int_0^z h(x)dx = \int_z^1 h(x)dx \right\}. \qquad (4.67)$$

Let T be a positive linear operator. Then T maps $L^p(v) \cap H$ into $L^q(u)$ if and only if T maps $L^p(v)$ into $L^q(u)$.

Proof. The *if* part is trivial. To prove the *only if* part suppose that $T : H \cap L^p(v) \to L^q(u)$. Since $L^1 \cap L^p(v)$ is dense in $L^p(v)$ it is enough to show that $T : L^1 \cap L^p(v) \to L^q(u)$.

Fix $g \in L^1 \cap L^p(v)$ and suppose, without loss of generality, that

$$\int_0^z |g(x)|dx \leq \int_z^1 |g(x)|dx.$$

(i) Suppose that the first condition in (4.66) holds:

$$\int_0^z v^{1-p'}(x)dx \approx \int_z^1 v^{1-p'}(x)dx < \infty,$$

and set

$$h(x) := \alpha v^{1-p'}(x) \left(\int_z^1 |g(t)|dt - \int_0^z |g(t)|dt \right) \chi_{(0,z)}(x),$$

where $1/\alpha = \int_0^z v^{1-p'}(x)dx$. Obviously $h \geq 0$, and it can be easily shown that $|g| + h \in H$. Furthermore, using Hölder's inequality and (4.66), we have

$$\|h\|_{p,v} = \alpha \left(\int_0^z v^{(1-p')p}(x)v(x)dx \right)^{1/p} \left(\int_z^1 |g(t)|dt - \int_0^z |g(t)|dt \right)$$

$$\leq \alpha^{1/p'} \int_z^1 |g(t)|dt$$

$$\leq \alpha^{1/p'} \left(\int_z^1 v^{1-p'}(t)dt \right)^{1/p'} \left(\int_0^1 |g(t)|^p v(t)dt \right)^{1/p}$$

$$\leq \widetilde{C} \|g\|_{p,v}.$$

Since $g \leq |g| + h$, the positivity of T implies

$$\|Tg\|_{q,u} \leq \|T(|g| + h)\|_{q,u} \leq C\| |g| + h\|_{p,v}$$

$$\leq C\|g\|_{p,v} + C\|h\|_{p,v} \leq (1 + \widetilde{C})C\|g\|_{p,v}$$

and case (i) is proved.

(ii) Suppose that $\int_0^z v^{1-p'}(x)dx = \int_z^1 v^{1-p'}(x)dx = \infty$. For each positive integer n set

$$h_n(x) := \alpha_n v_n^{1-p'}(x) \left(\int_z^1 |g(t)|dt - \int_0^z |g(t)|dt \right) \chi_{(0,z)}(x),$$

where $v_n^{1-p'}(x) = v^{1-p'}(x) \cdot \chi_{S_n}(x)$ with $S_n = \{x \in (0,1) : v^{1-p'}(x) < n\}$, and $1/\alpha_n = \int_0^z v_n^{1-p'}(x)dx$. Again, $h_n \geq 0$, $|g| +$

202 *Weighted Inequalities of Hardy Type*

$h_n \in H$ and $g \le |g| + h_n$ so that for every n we have

$$\|Tg\|_{q,u} \le \|T(|g| + h_n)\|_{q,u} \le C\|g\|_{p,v} + \|h_n\|_{p,v}.$$

Now $v_n^{1-p'}$ is zero where $v \ne v_n$, so

$$\|h_n\|_{p,v} = \alpha_n \left(\int_0^z v_n^{(1-p')p}(x)v(x)dx \right)^{1/p}$$

$$\times \left(\int_z^1 |g(t)|dt - \int_0^z |g(t)|dt \right)$$

$$\le \alpha_n^{1/p'} \int_0^1 |g(t)|dt.$$

For $n \to \infty$ we have that $\alpha_n \to 0$ so finally $\|Tg\|_{q,u} \le C\|g\|_{p,v}$, which completes case (ii) and the proof. $\qquad\square$

Definition 4.26. For fixed $z \in (0,1)$, let $S = S_1 + S_2$, where

$$(S_1 f)(x) := \left(\int_0^x f(t)dt \right) \chi_{(0,z)}(x);$$

$$(S_2 f)(x) := \left(\int_x^1 f(t)dt \right) \chi_{(z,1)}(x). \tag{4.68}$$

Note that S_1 and S_2, and hence S, are positive operators.

Lemma 4.27. *Suppose that g satisfies*

$$g' = f \quad in \quad (0,1), \quad g(0) = g(1) = 0, \tag{4.69}$$

and set $h(x) = (\chi_{(0,z)}(x) - \chi_{(z,1)}(x))f(x)$. Then $g = Sh$.

Proof. Since $g(0) = g(1) = 0$ we have

$$g(x) = \int_0^x f(t)dt = -\int_x^1 f(t)dt$$

and hence, for $z \in (0,1)$ arbitrary but fixed,

$$g(x) = \chi_{(0,z)}(x) \int_0^x f(t)dt - \chi_{(z,1)}(x) \int_x^1 f(t)dt$$

$$= (S_1 f)(x) - (S_2 f)(x) = (Sh)(x),$$

since $h(x) = (\chi_{(0,z)}(x) - \chi_{(z,1)}(x))f(x)$ means that

$$h(x) = \begin{cases} f(x) & \text{for } x \in (0,z), \\ -f(x) & \text{for } x \in (z,1). \end{cases}$$ $\qquad\square$

The main result for $k = 1$ reads as follows:

Theorem 4.28. *Let* $0 < q < \infty$, $1 < p < \infty$. *Suppose* u, v *are weight functions and condition* (4.66) *is satisfied with some* $z \in (0,1)$. *Then the first order Hardy inequality*

$$\left(\int_0^1 |g(x)|^q u(x)dx \right)^{1/q} \le C \left(\int_0^1 |g'(x)|^p v(x)dx \right)^{1/p} \qquad (4.70)$$

holds for all $g \in AC^0$ *satisfying*

$$g(0) = g(1) = 0 \qquad (4.71)$$

if and only if either
   (i) $1 < p \le q < \infty$ *and*

$$\sup_{0<x<z} \left(\int_x^z u(t)dt \right)^{1/q} \left(\int_0^x v^{1-p'}(t)dt \right)^{1/p'} < \infty,$$

$$\sup_{z<x<1} \left(\int_z^x u(t)dt \right)^{1/q} \left(\int_x^1 v^{1-p'}(t)dt \right)^{1/p'} < \infty, \qquad (4.72)$$

or

(ii) $0 < q < p < \infty$, $1 < p < \infty$ *and*

$$\left(\int_0^z \left(\int_x^z u(t)dt \right)^{r/p} \left(\int_0^x v^{1-p'}(t)dt \right)^{r/p'} u(x)dx \right)^{1/r} < \infty,$$

$$\left(\int_z^1 \left(\int_z^x u(t)dt \right)^{r/p} \left(\int_x^1 v^{1-p'}(t)dt \right)^{r/p'} u(x)dx \right)^{1/r} < \infty$$

$$(4.73)$$

with $\frac{1}{r} = \frac{1}{q} - \frac{1}{p}$.

Proof. We begin by showing that (4.70) holds if and only if S : $L^p(v) \to L^q(u)$ with S the operator from Definition 4.26.

Suppose first that $S : L^p(v) \to L^q(u)$ and that g satisfies (4.69). Now, with $h = (\chi_{(0,z)} - \chi_{(z,1)})f$, we use Lemma 4.27 and the boundedness of S to get

$$\|g\|_{q,u} = \|Sh\|_{q,u} \le C\|h\|_{p,v} = C\|f\|_{p,v},$$

which is (4.70) since $f = g'$.

Conversely, suppose that (4.70) holds for g satisfying (4.71). According to Lemma 4.25, it is enough to prove that $\|Sh\|_{q,u} \le C\|h\|_{p,v}$ for functions $h \in L^p(v)$ satisfying $\int_0^z h(x)dx = \int_z^1 h(x)dx$ in order to conclude that $S : L^p(v) \to L^q(u)$. Fix such an h and define f and g by

$$f(x) = (\chi_{(0,z)}(x) - \chi_{(z,1)}(x))h(x), \quad g(x) = \int_0^x f(t)dt.$$

Since $g(1) = \int_0^1 f(t)dt = \int_0^z h(t)dt - \int_z^1 h(t)dt = 0$, it is clear that $\|g\|_{q,u} \le C\|g'\|_{p,v} = C\|f\|_{p,v}$. Thus, using Lemma 4.27,

$$\|Sh\|_{q,u} = \|g\|_{q,u} \le C\|f\|_{p,v} = C\|h\|_{p,v}.$$

To complete the proof, we show that the boundedness of S is equivalent to the conditions in (4.72) or (4.73).

Since S is the sum of two positive operators, it is bounded if and only if both S_1 and S_2 are bounded. The boundedness of

$S_1 : L^p(v) \to L^q(u)$ means that there exists a constant C such that

$$\left(\int_0^1 \left| \left(\int_0^x f(t)dt \right) \chi_{(0,z)}(x) \right|^q u(x)dx \right)^{1/q}$$

$$\leq C \left(\int_0^1 |f(x)|^p v(x)dx \right)^{1/p}$$

for all functions f on $(0,1)$. Since the left hand side does not depend on the values of f on $(z,1)$, the above inequality is clearly equivalent to the inequality

$$\left(\int_0^z \left| \int_0^x f(t)dt \right|^q u(x)dx \right)^{1/q} \leq C \left(\int_0^z |f(x)|^p v(x)dx \right)^{1/q}$$

for all functions on $(0,z)$. But this is the Hardy inequality on the interval $(0,z)$, which holds if and only if the first condition in (4.72) (for $p \leq q$) or in (4.73) (for $p > q$) holds.

A similar analysis shows that the boundedness of S_2 reduces to the conjugate Hardy inequality on the interval $(z,1)$, which yields the second condition in (4.72) or in (4.73). This completes the proof. □

Remark 4.29. (i) As follows from the proof, we have derived conditions of the validity of (4.70) under (4.71) by *glueing together* the Hardy inequalities on $(0,z)$ (for g satisfying $g(0) = 0$) and on $(z,1)$ (for g satisfying $g(1) = 0$). This is evident also from conditions (4.72) and (4.73), which can be rewritten as

$$A(0,z;u,v) < \infty,$$

$$\tilde{A}(z,1;u,v) < \infty, \tag{4.72$'$}$$

or

$$A^*(0,z;u,v) < \infty,$$

$$\tilde{A}^*(z,1;u,v) < \infty. \tag{4.73$'$}$$

(ii) Condition (4.66) is crucial for our considerations. Of course, the first part of this condition could be replaced by equality, claiming that there is a $z \in (0, 1)$ such that

$$\int_0^z v^{1-p'}(t)dt = \int_z^1 v^{1-p'}(t)dt \quad (< \infty). \qquad (4.74)$$

Let us note that for some weights v it is not possible to find a z satisfying (4.66) (e.g. if one of the integrals in (4.74) is finite and the other infinite). In this case, one has to use condition (4.65). As an example, let us mention the weight $v(x)$ defined as

$$v(x) = \begin{cases} 1 & \text{for } x \in \left(0, \dfrac{1}{2}\right], \\ 2x - 1 & \text{for } x \in \left[\dfrac{1}{2}, 1\right) \end{cases}$$

and the parameter $p = 2$.

4.7. Overdetermined Classes ($k > 1$)

We will deal with the special overdetermined case

$$|M_0| + |M_1| = k + 1 \qquad (4.75)$$

with the following particular choice:

$$M_i = \{M_i^{k-1}, \ k - 1\}, \qquad (4.76)$$

where $M_i^{k-1} \subset \mathbb{N}_{k-1} = \{0, 1, \ldots, k-2\}$, $i = 0, 1$, $|M_0^{k-1}| + |M_1^{k-1}| = k - 1$ and the couple (M_0^{k-1}, M_1^{k-1}) satisfies the Pólya condition. Consequently, BVP

$$g^{(k-1)} = h \quad \text{in } (0, 1), \quad g^{(i)}(0) = 0 \quad \text{for } i \in M_0^{k-1},$$
$$g^{(j)}(1) = 0 \quad \text{for } j \in M_1^{k-1} \qquad (4.77)$$

has a unique solution,

$$g(x) = \int_0^1 G^{k-1}(x, t)h(t)dt. \qquad (4.78)$$

Now, the solution of BVP

$$g^{(k)} = f, \quad g^{(i)}(0) = 0 \quad \text{for } i \in M_0,$$
$$g^{(j)}(1) = 0 \quad \text{for } j \in M_1 \tag{4.79}$$

can be combined from the BVP (4.77) and from the (overdetermined) BVP

$$h' = f \quad \text{in} \quad (0,1), \quad h(0) = h(1) = 0 \tag{4.80}$$

whose solution is, according to Lemma 4.27, given by

$$h = S_1 f - S_2 f$$

with S_1, S_2 from Definition 4.26, provided there is a $z \in (0,1)$ such that (4.66) is satisfied.

If we denote

$$F(x) = (\chi_{(0,z)}(x) - \chi_{(z,1)}(x))f(x) = \begin{cases} f(x) & \text{for } x \in (0,z), \\ -f(x) & \text{for } x \in (z,1), \end{cases}$$

we have $h = SF$, where $S = S_1 + S_2$, and the solution of BVP (4.79) is given by

$$g(x) = \int_0^1 G^{k-1}(x,t)(SF)(t)dt$$

$$= \int_0^1 G^{k-1}(x,t)\left[\chi_{(0,z)}(t)\int_0^t F(s)ds + \chi_{(z,1)}(t)\int_t^1 F(s)ds\right]dt$$

$$= \int_0^z \left(\int_s^z G^{k-1}(x,t)dt\right) F(s)ds$$

$$+ \int_z^1 \left(\int_z^s G^{k-1}(x,t)dt\right) F(s)ds, \tag{4.81}$$

where we have used Fubini's theorem.

Let us denote the last expression by

$$(Tf)(x).$$

Then the following assertion will be useful:

Lemma 4.30. *Let $0 < q < \infty$ and $1 < p < \infty$, let u, v be weight functions on $(0,1)$ and let $z \in (0,1)$ satisfy (4.66). Suppose that M_0^{k-1}, M_1^{k-1} are subsets of $\mathbb{N}_{k-1}$, $|M_0^{k-1}| + |M_1^{k-1}| = k - 1$ and (M_0^{k-1}, M_1^{k-1}) satisfies the Pólya condition. Then the Hardy inequality*

$$\|g\|_{q,u} \le C \|g^{(k)}\|_{p,v} \tag{4.82}$$

holds for functions $g \in AC^{k-1}(M_0, M_1)$ with $M_i = M_i^{k-1} \cup \{k - 1\}$, $i = 0, 1$, if and only if $T : L^p(v) \to L^q(u)$.

Proof. Suppose $T : L^p(v) \to L^q(u)$ and f and g satisfy (4.79). Set $F = (\chi_{(0,z)} - \chi_{(z,1)})f$. Since f and $g^{(k-1)} = h$ satisfy (4.80) we may apply Lemma 4.27 to get $g^{(k-1)} = SF$, where S is the operator from Definition 4.26, and due to (4.78) we finally obtain

$$g(x) = \int_0^1 G^{k-1}(x, t)(SF)(t)dt = (TF)(x).$$

Thus

$$\|g\|_{q,u} = \|TF\|_{q,u} \le C \|F\|_{p,v} = C \|f\|_{p,v},$$

which is (4.82) since $f = g^{(k)}$.

Conversely, suppose that (4.82) holds. Since T is a positive operator (recall that, due to Theorem 4.11, G^{k-1} does not change sign on $(0,1) \times (0,1)$), Lemma 4.25 shows that it is enough to prove that $\|Th\|_{q,u} \le C\|h\|_{p,v}$ for functions $h \in L^p(w)$ satisfying (4.67) in order to conclude that $T : L^p(v) \to L^q(u)$. Fix such an h and define f and g by

$$f(x) = (\chi_{(0,z)}(x) - \chi_{(z,1)}(x))h(x), \quad g(x) = \int_0^1 G^{k-1}(x, t)(Sh)(t)dt.$$

Similarly as above we obtain that $g(x) = (Th)(x)$. The definition of g shows that Sh and g satisfy BVP (4.77) so that g satisfies the boundary conditions

$$g^{(i)}(0) = 0 \quad \text{for } i \in M_0^{k-1}, \quad g^{(j)}(1) = 0 \quad \text{for } j \in M_1^{k-1}.$$

We also have $g^{(k-1)} = Sh$ so that using Definition 4.26 we obtain

$$g^{(k-1)}(0) = (S_1 h)(0) = 0, \quad g^{(k-1)}(1) = (S_2 h)(1) = 0.$$

Finally, differentiation yields $g^{(k)} = f$ and we have shown that g and f satisfy BVP (4.79). Thus

$$\|Th\|_{q,u} = \|g\|_{q,u} \le C\|g^{(k)}\|_{p,v} = C\|f\|_{p,v} = C\|h\|_{p,v},$$

which completes the proof. □

The case $M_0 = \mathbb{N}_k$, $M_1 = \{k-1\}$

We start with this simpler case since the necessary and sufficient conditions are more transparent. Here, $g \in AC^{k-1}(M_0, M_1)$ has to satisfy

$$g^{(i)}(0) = 0 \quad \text{for } i = 0, 1, \ldots, k-1, \quad g^{(k-1)}(1) = 0, \qquad (4.83)$$

i.e., we have $M_0^{k-1} = \{0, 1, \ldots, k-2\} = \mathbb{N}_{k-1}$, $M_1^{k-1} = \emptyset$. According to Sec. 4.2, the Green function $G^{k-1}(x, t)$ has the simple form

$$G^{k-1}(x, t) = \frac{1}{(k-2)!}(x - t)^{k-2}\chi_{(0,x)}(t).$$

However, since we assume throughout this section that the weight function v satisfies condition (4.66), there exists a $z \in (0, 1)$ such that the operator T defined in (4.81) has now the form

$$(Th)(x) = \int_0^z \left[\int_s^z \frac{(x-t)^{k-2}}{(k-2)!}\chi_{(0,x)}(t)dt \right] h(s)ds$$
$$+ \int_z^1 \left[\int_z^s \frac{(x-t)^{k-2}}{(k-2)!}\chi_{(0,x)}(t)dt \right] h(s)ds.$$

If $x < z$ the second term vanishes and, performing the inner integration in the first term, we have

$$(Th)(x) = \int_0^x \frac{(x-s)^{k-1}}{(k-1)!}h(s)ds.$$

If $x > z$, then some calculations yield

$$(Th)(x) = \int_0^z \frac{(x-s)^{k-1} - (x-z)^{k-1}}{(k-1)!} h(s) ds$$

$$+ \int_z^x \frac{(x-z)^{k-1} - (x-s)^{k-1}}{(k-1)!} h(s) ds$$

$$+ \frac{(x-z)^{k-1}}{(k-1)!} \int_x^1 h(s) ds.$$

Thus $(k-1)!(Th)(x) = (T_1 h)(x) + (T_2 h)(x) + (T_3 h)(x) + (T_4 h)(x)$, where

$$(T_1 h)(x) = \chi_{(0,z)}(x) \int_0^x (x-s)^{k-1} h(s) ds;$$

$$(T_2 h)(x) = \chi_{(z,1)}(x) \int_0^z [(x-s)^{k-1} - (x-z)^{k-1}] h(s) ds;$$

$$(T_3 h)(x) = \chi_{(z,1)}(x) \int_z^x [(x-z)^{k-1} - (x-s)^{k-1}] h(s) ds;$$

$$(T_4 h)(x) = \chi_{(z,1)}(x)(x-z)^{k-1} \int_x^1 h(s) ds.$$

T_1 is the Riemann-Liouville operator on $(0, z)$ (see Remark 2.8 (iii)). Using the Binomial formula

$$(x-s)^{k-1} - (x-z)^{k-1} = [(x-z) + (z-s)]^{k-1} - (x-z)^{k-1}$$

$$= \sum_{j=1}^{k-1} \binom{k-1}{j} (x-z)^{k-j-1} (z-s)^j,$$

T_2 can be simplified to

$$(T_2 h)(x) = \chi_{(z,1)}(x) \sum_{j=1}^{k-1} \binom{k-1}{j} (x-z)^{k-j-1} \int_0^z (z-s)^j h(s) ds.$$

Using the estimate

$$(s-z)(x-z)^{k-2} \le (x-z)^{k-1} - (x-s)^{k-1} \le k(s-z)(x-z)^{k-2}$$

for $0 < z < s < x < 1$, we can write

$$(T_3 h)(x) \approx \chi_{(z,1)}(x)(x - z)^{k-2} \int_z^x (s - z)h(s)ds,$$

where the last expression is the Hardy operator on $(z, 1)$ applied to the function $(s - z)h(s)$. Finally, T_4 is the conjugate Hardy operator on $(z, 1)$.

Now, we are able to describe the necessary and sufficient conditions:

Theorem 4.31. *Let u, v be weights on $(0,1)$ and let $z \in (0, 1)$ satisfy (4.66). Then the Hardy inequality (4.82) holds for all functions $g \in AC^{k-1}$ satisfying (4.83) if and only if either*
 (i) $1 < p \le q < \infty$ *and*

$$\sup_{0 < x < z} \left(\int_x^z (t - x)^{q(k-1)} u(t)dt \right)^{1/q} \left(\int_0^x v^{1-p'}(t)dt \right)^{1/p'} < \infty, \quad (4.84)$$

$$\sup_{0 < x < z} \left(\int_x^z u(t)dt \right)^{1/q} \left(\int_0^x (x - t)^{p'(k-1)} v^{1-p'}(t)dt \right)^{1/p'} < \infty, \quad (4.85)$$

$$\sup_{z < x < 1} \left(\int_x^1 (t - z)^{q(k-2)} u(t)dt \right)^{1/q}$$

$$\times \left(\int_z^x (t - z)^{p'} v^{1-p'}(t)dt \right)^{1/p'} < \infty, \quad (4.86)$$

$$\sup_{z < x < 1} \left(\int_z^x (t - z)^{q(k-1)} u(t)dt \right)^{1/q} \left(\int_x^1 v^{1-p'}(t)dt \right)^{1/p'} < \infty, \quad (4.87)$$

$$\sup_{j=1,\ldots,k-1} \left(\int_z^1 (t - z)^{q(k-j-1)} u(t)dt \right)^{1/q}$$

$$\times \left(\int_0^z (z - t)^{p'j} v^{1-p'}(t)dt \right)^{1/p'} < \infty; \quad (4.88)$$

or

(ii) $1 < q < p < \infty$, (4.88) *holds again and, with* $\frac{1}{r} = \frac{1}{q} - \frac{1}{p}$,

$$\left(\int_0^z \left(\int_x^z (t-x)^{q(k-1)} u(t) dt \right)^{r/q} \right.$$

$$\left. \times \left(\int_0^x v^{1-p'}(t) dt \right)^{r/q'} v^{1-p'}(t) dt \right)^{1/r} < \infty, \qquad (4.84')$$

$$\left(\int_0^z \left(\int_x^z u(t) dt \right)^{r/p} \right.$$

$$\left. \times \left(\int_0^x (x-t)^{p'(k-1)} v^{1-p'}(t) dt \right)^{r/p'} u(x) dx \right)^{1/r} < \infty, \qquad (4.85')$$

$$\left(\int_z^1 \left(\int_x^1 (t-z)^{q(k-2)} u(t) dt \right)^{r/q} \left(\int_z^x (t-z)^{p'} v^{1-p'}(t) dt \right)^{r/q'} \right.$$

$$\left. \times (t-z)^{p'} v^{1-p'}(x) dx \right)^{1/r} < \infty, \qquad (4.86')$$

$$\left(\int_z^1 \left(\int_z^x (t-z)^{q(k-1)} u(t) dt \right)^{r/q} \right.$$

$$\left. \times \left(\int_x^1 v^{1-p'}(t) dt \right)^{r/q'} v^{1-p'}(x) dx \right)^{1/r} < \infty. \qquad (4.87')$$

Proof. By Lemma 4.30 it is enough to prove the equivalence of the conditions (4.84)–(4.87'), (4.88) and the boundedness of $T : L^p(v) \to L^q(u)$.

But T (more precisely, $(k-1)!T$) is the sum of positive operators T_1, T_2, T_3 and T_4, and thus their separate boundedness is necessary and sufficient for the boundedness of their sum.

Now, T_1 is bounded if and only if (4.84) and (4.85) (for $p \le q$) or (4.84'), (4.85') (for $p > q$) hold (see Theorem 4.3). The Hardy operator T_3 is bounded if and only if (4.86) (or (4.86′)) holds, and similarly for the Hardy operator T_4 and condition (4.87) (or (4.87′)). Finally, using Hölder's inequality it can be shown that T_2 is bounded if and only if (4.88) holds (see Theorem 4.12). $\qquad\qquad\square$

Remark 4.32. (i) Note that we have supposed $q \in (1,\infty)$ and not $q \in (0,\infty)$ as e.g. in Lemma 4.30. The reason is that the results for the (Rieman-Liouville) operator T_1 are available only for $q > 1$.

(ii) Obviously, it is easy to formulate a similar theorem for the case $M_0 = \{k - 1\}$, $M_1 = \mathbb{N}_k$, i.e., for $g \in AC^{k-1}$ satisfying

$$g^{(k-1)}(0) = 0, \quad g^{(i)}(1) = 0 \quad \text{for } i = 0, 1, \ldots, k - 1.$$

One has only to use the fact that the Green function $G'^{k-1}(x,t)$ has the simple form

$$\frac{1}{(k - 2)!}(t - x)^{k-2}\chi_{(x,1)}(t)$$

and then proceed similarly as on pp. 209–212.

(iii) In contrast to the *well-determined* case where only two or at most three necessary and sufficient conditions appeared, here we have a collection of *five* conditions even in the very simple case $|M_0| = k$, $|M_1| = 1$. As we will just see, the situation becomes more cumbersome if we allow, roughly speaking, $|M_1| > 1$.

The case $\mathsf{M_i = M_i^k - 1 \cup \{k - 1\}}$, $\mathsf{M_i^{k-1} \neq \emptyset}$, $\mathsf{i = 0,1}$

Let us suppose that (M_0^{k-1}, M_1^{k-1}) satisfies the Pólya condition, $M_i^k \subset \mathbb{N}_{k-1}$, $|M_0^{k-1}| + |M_1^{k-1}| = k - 1$, but contrary to the cases mentioned after Lemma 4.30 and in Remark 4.32 (ii), let us assume that M_i^{k-1} is non-empty for $i = 0, 1$. Denote again by a, b, c, d, A, B, C, D the numbers introduced in Definition 4.9, but this time for the couple (M_0^{k-1}, M_1^{k-1}).

For the Green function $G^{k-1}(x,t)$ of BVP (4.77) we then have, according to Theorem 4.11, estimates (4.29) and if we use these estimates in the expression for T in (4.81),

$$(Th)(x) = \int_0^z \left(\int_s^z G^{k-1}(x,t)dt \right) h(s)ds$$

$$+ \int_z^1 \left(\int_z^s G^{k-1}(x,t)dt \right) h(s)ds,$$

then due to the estimates

$$\int_s^z G^{k-1}(x,t)dt \approx x^a(1-x)^B \int_s^z t^C(1-t)^d dt \quad \text{if } x \le s;$$

$$\int_s^z G^{k-1}(x,t)dt \approx x^A(1-x)^b \int_s^x t^C(1-t)^D dt$$

$$+ x^a(1-x)^B \int_s^z t^C(1-t)^d dt \quad \text{if } s < x \le z;$$

$$\int_s^z G^{k-1}(x,t)dt \approx x^A(1-x)^b \int_s^z t^c(1-t)^D dt \quad \text{if } z < x;$$

$$\int_z^s G^{k-1}(x,t)dt \approx x^a(1-x)^B \int_z^s t^C(1-t)^d dt \quad \text{if } x \le z;$$

$$\int_z^s G^{k-1}(x,t)dt \approx x^A(1-x)^b \int_z^x t^C(1-t)^D dt$$

$$+ x^a(1-x)^B \int_x^s t^C(1-t)^d \quad \text{if } z < x \le s;$$

$$\int_z^s G^{k-1}(x,t)dt \approx x^A(1-x)^b \int_z^s t^c(1-t)^D dt \quad \text{if } s < x,$$

we obtain that

$$(Th)(x) \approx (T_1 h)(x) + \cdots + (T_8 h)(x),$$

where

$$(T_1 h)(x) = \chi_{(0,z)}(x) x^a (1-x)^B \int_x^z \left(\int_s^z t^C (1-t)^d dt \right) h(s) ds,$$

$$(T_2 h)(x) = \chi_{(0,z)}(x) x^A (1-x)^b \int_0^x \left(\int_s^x t^c (1-t)^D dt \right) h(s) ds,$$

$$(T_3 h)(x) = \chi_{(0,z)}(x) x^a (1-x)^B \int_0^x \left(\int_x^z t^C (1-t)^d dt \right) h(s) ds,$$

$$(T_4 h)(x) = \chi_{(0,z)}(x) x^a (1-x)^B \int_z^1 \left(\int_z^s t^C (1-t)^d dt \right) h(s) ds,$$

$$(T_5 h)(x) = \chi_{(z,1)}(x) x^A (1-x)^b \int_z^x \left(\int_z^s t^c (1-t)^D dt \right) h(s) ds,$$

$$(T_6 h)(x) = \chi_{(z,1)}(x) x^a (1-x)^B \int_z^1 \left(\int_x^s t^C (1-t)^d dt \right) h(s) ds,$$

$$(T_7 h)(x) = \chi_{(z,1)}(x) x^A (1-x)^b \int_x^1 \left(\int_z^x t^c (1-t)^D dt \right) h(s) ds,$$

$$(T_8 h)(x) = \chi_{(z,1)}(x) x^A (1-\dot{x})^b \int_0^z \left(\int_s^z t^c (1-t)^D dt \right) h(s) ds.$$

Here T_1 is the conjugate Hardy operator on $(0, z)$, T_2 is a Hardy-type operator as investigated in Chap. 2, with kernel $k(x,s) = \int_s^x t^c (1-t)^D dt$ (see Example 2.7 (iv)), T_3 is a Hardy operator on $(0, z)$ etc. Consequently, we can formulate our main result:

Theorem 4.33. *Let u, v be weights on $(0,1)$ and suppose that there is a number $z \in (0,1)$ such that (4.66) is satisfied. Let M_0, M_1 be of the form $M_i = M_i^{k-1} \cup \{k-1\}$, where M_0^{k-1}, M_1^{k-1} are nonempty subsets of $\mathbb{N}_{k-1}$ satisfying the Pólya condition and such that $|M_0^{k-1}| + |M_1^{k-1}| = k - 1$. Then the Hardy inequality (4.82) holds for all functions $g \in AC^{k-1}(M_0, M_1)$ if and only if either*

(i) $1 < p \leq q < \infty$ and

$$\sup_{0<x<z} \left(\int_0^x t^{aq}(1-t)^{Bq} u(t) dt \right)^{1/q} \left(\int_x^z H_1^{p'}(x,t) v^{1-p'}(t) dt \right)^{1/p'} < \infty$$

(4.89)

with $H_1(z,t) = \int_t^z s^C (1-s)^d ds$;

$$\sup_{0<x<z} \left(\int_x^z H_2^q(x,t) t^{Aq}(1-t)^{bq} u(t) dt \right)^{1/q} \left(\int_0^x v^{1-p'}(t) dt \right)^{1/p'} < \infty,$$

(4.90)

$$\sup_{0<x<z} \left(\int_x^z t^{Aq}(1-t)^{bq} u(t) dt \right)^{1/q} \left(\int_0^x H_2^{p'}(t,x) v^{1-p'}(t) dt \right)^{1/p'} < \infty$$

with $H_2(x,t) = \int_x^t s^c (1-s)^D ds$;

$$\sup_{0<x<z} \left(\int_x^z H_1^q(t,z) t^{aq}(1-t)^{Bq} u(t) dt \right)^{1/q} \left(\int_0^x v^{1-p'}(t) dt \right)^{1/p'} < \infty;$$

(4.91)

$$\left(\int_0^z t^{aq}(1-t)^{Bq} u(t) dt \right)^{1/q} \left(\int_z^1 H_1^{p'}(t,z) v^{1-p'}(t) dt \right)^{1/p'} < \infty;$$

(4.92)

$$\sup_{z<x<1} \left(\int_x^1 t^{Aq}(1-t)^{bq} u(t) dt \right)^{1/q}$$

$$\times \left(\int_z^x H_2^{p'}(t,z) v^{1-p'}(t) dt \right)^{1/p'} < \infty;$$

(4.93)

$$\sup_{z<x<1} \left(\int_z^x H_1^q(x,t) t^{aq}(1-t)^{Bq} u(t) dt \right)^{1/q}$$

$$\times \left(\int_x^1 v^{1-p'}(t) dt \right)^{1/p'} < \infty,$$

$$\sup_{z<x<1} \left(\int_z^x t^{aq}(1-t)^{Bq} u(t)dt \right)^{1/q}$$

$$\times \left(\int_x^1 H_1^{p'}(t,x)v^{1-p'}(t)dt \right)^{1/p'} < \infty; \qquad (4.94)$$

$$\sup_{z<x<1} \left(\int_z^x H_2^q(t,z)t^{Aq}(1-t)^{bq} u(t)dt \right)^{1/q}$$

$$\times \left(\int_x^1 v^{1-p'}(t)dt \right)^{1/p'} < \infty; \qquad (4.95)$$

$$\left(\int_z^1 t^{Aq}(1-t)^{bq} u(t)dt \right)^{1/q} \left(\int_0^z H_2^{p'}(z,t)v^{1-p'}(t)dt \right)^{1/p'} < \infty, \qquad (4.96)$$

(ii) or $1 < q < p < \infty$, (4.92) and (4.96) hold again and, with $\frac{1}{r} = \frac{1}{q} - \frac{1}{p}$,

$$\left(\int_0^z \left(\int_0^x t^{aq}(1-t)^{Bq} u(t)dt \right)^{r/q} \right.$$

$$\times \left(\int_x^z H_1^{p'}(z,t)v^{1-p'}(t)dt \right)^{r/q'} H_1^{p'}(z,x)v^{1-p'}(x)dx \right)^{1/r} < \infty; \qquad (4.89')$$

$$\left(\int_0^z \left(\int_x^z H_2^q(x,t)t^{Aq}(1-t)^{bq} u(t)dt \right)^{r/q} \right.$$

$$\times \left(\int_0^x v^{1-p'}(t)dt \right)^{r/q'} v^{1-p'}(x)dx \right)^{1/r} < \infty, \qquad (4.90')$$

$$\left(\int_0^z \left(\int_x^z t^{Aq}(1-t)^{bq} u(t)dt \right)^{r/p} \right.$$

$$\times \left(\int_0^x H_2^{p'}(t,x)v^{1-p'}(t)dt \right)^{r/p'} x^{Aq}(1-x)^{bq}u(x)dx \right)^{1/r} < \infty;$$

$$\left(\int_0^z \left(\int_x^z H_1^q(t,z)t^{aq}(1-t)^{Bq}u(t)dt \right)^{r/q} \right.$$

$$\left. \times \left(\int_0^x v^{1-p'}(t)dt \right)^{r/q'} v^{1-p'}(x)dx \right)^{1/r} < \infty; \tag{4.91'}$$

$$\left(\int_z^1 \left(\int_x^1 t^{Aq}(1-t)^{bq}u(t)dt \right)^{r/p} \right.$$

$$\left. \times \left(\int_z^x H_2^{p'}(t,z)v^{1-p'}(t)dt \right)^{r/p'} x^{Aq}(1-x)^{bq}u(x)dx \right)^{1/r} < \infty; \tag{4.93'}$$

$$\left(\int_z^1 \left(\int_z^x H_1^q(x,t)t^{aq}(1-t)^{Bq}u(t)dt \right)^{r/q} \right.$$

$$\left. \times \left(\int_x^1 v^{1-p'}(t)dt \right)^{r/q'} v^{1-p'}(x)dx \right)^{1/r} < \infty, \tag{4.94'}$$

$$\left(\int_z^1 \left(\int_z^x t^{aq}(1-t)^{Bq}u(t)dt \right)^{r/p} \right.$$

$$\left. \times \left(\int_x^1 H_1^{p'}(t,x)v^{1-p'}(t)dt \right)^{r/p'} x^{aq}(1-x)^{Bq}u(x)dx \right)^{1/r} < \infty;$$

$$\left(\int_z^1 \left(\int_z^x H_2^q(t,z)t^{Aq}(1-t)^{bq}u(t)dt \right)^{r/q} \right.$$

$$\left. \times \left(\int_x^1 v^{1-p'}(t)dt \right)^{r/q'} v^{1-p'}(x)dx \right)^{1/r} < \infty. \tag{4.95'}$$

Sketch of proof. Again we apply Lemma 4.30, according to which the Hardy inequality is equivalent to the boundedness of $T : L^p(v) \rightarrow L^q(u)$. But T is equivalent to a sum of *positive* operators $T_1, \ldots, T_8$ so that we may examine each T_i individually.

Since T_1 is a conjugate Hardy operator on $(0, z)$, it is bounded if and only if (4.89) (for $p \leq q$) or (4.89') (for $p > q$) holds. T_2 is bounded, according to Theorem 2.17 and Example 2.7 (iv), if and only if the two conditions (4.90) or (4.90'), respectively, hold. The boundedness of the Hardy operator T_3 is equivalent to (4.91) or (4.91'), respectively, and the boundedness of T_4 is equivalent to (4.92) due to Hölder's inequality.

In the same way as $T_1, \ldots, T_4$ give rise to the first five conditions, $T_5, \ldots, T_8$ give rise to the last five. $\qquad\qquad\square$

Example 4.34. Suppose M_0, M_1 are as in Theorem 4.33 and set

$$u(t) = t^\alpha (1 - t)^\beta, \quad v^{1-p'}(t) = t^\gamma (1 - t)^\delta.$$

Then the k-th order Hardy inequality (4.82) holds provided

$$\gamma + 1 > 0, \quad \delta + 1 > 0, \quad \alpha + 1 + aq > 0, \quad \beta + 1 + bq > 0,$$

where a, b depend on M_0^{k-1}, M_1^{k-1} as in Definition 4.9.

In fact, if we set $z = \frac{1}{2}$, then (4.66) holds so that it remains to verify the conditions of Theorem 4.33. We know that either $A = a$ or $A = a - 1$ and similarly either $B = b$ or $B = b - 1$.

Suppose $A = a$ and $B = b$. Verifying the conditions of Theorem 4.33, we proceed as follows:

(i) Each condition involves integrals over subintervals of $(0, z)$ or $(z, 1)$, so extend the range of integration to either $(0, z)$ or $(z, 1)$.

(ii) Use the fact that (positive or negative) powers of $1 - x, 1 - t, 1 - s$ are bounded above on $(0, z)$ and powers of x, t, s are bounded above on $(z, 1)$.

(iii) Use the restrictions on $\alpha, \beta, \gamma, \delta$ to evaluate the remaining integrals.

We illustrate this procedure by showing that the second condition in (4.90) is satisfied:

$$\sup_{0<x<z} \left(\int_x^z t^{Aq}(1-t)^{bq} t^\alpha (1-t)^\beta dt \right)^{1/q}$$

$$\times \left(\int_0^x \left(\int_t^x s^c(1-s)^D ds \right)^{p'} t^\gamma (1-t)^\delta dt \right)^{1/p'}$$

$$\leq K \sup_{0<x<z} \left(\int_0^z t^{Aq+\alpha} dt \right)^{1/q} \left(\int_0^z \left(\int_0^z s^c ds \right)^{p'} t^\gamma dt \right)^{1/p'},$$

which is finite because $c \geq 0$, $\gamma + 1 > 0$ and $Aq + \alpha + 1 = aq + \alpha + 1 > 0$. ($K$ is the constant arising from step (ii); it depends on $bq + \beta$ and δ.)

Similarly we can verify the remaining conditions even in the case $p > q$. If $A = a - 1$ or $B = b - 1$, we have to modify the estimates on $\alpha, \beta, \gamma, \delta$. It can be shown that these restrictions are also necessary for (4.82) to hold.

4.8. Overdetermined Classes (Another Approach)

Definition 4.35. A couple (M_0, M_1) of subsets of $\mathbb{N}_k = \{0, 1, \ldots, k - 1\}$ will be called *standard* if it satisfies the Pólya condition (see Sec. 4.1) and if

$$|M_0| + |M_1| = k.$$

Recall that for standard couples we have established necessary and sufficient conditions of the validity of the k-th order Hardy inequality (4.30) for functions $g \in AC^{k-1}(M_0, M_1)$ and that the Green function $G(x, t)$ of the BVP

$$g^{(k)} = f \quad \text{in } (0,1),$$

$$g^{(i)}(0) = 0 \quad \text{for } i \in M_0,$$

$$g^{(j)}(1) = 0 \quad \text{for } j \in M_1 \tag{4.97}$$

is uniquely determined. The investigation of the Hardy inequality (4.1) is then equivalent to the investigation of the weighted norm inequality

$$\|Tf\|_{q,u} \leq C\|f\|_{p,v} \tag{4.98}$$

for $f \geq 0$, where

$$(Tf)(x) = \int_0^1 G(x,t)f(t)dt. \tag{4.99}$$

Up to now, we have investigated overdetermined problems for the special case

$$M_i = M_i^{k-1} \cup \{k-1\}, \quad i = 0, 1,$$

where (M_0^{k-1}, M_1^{k-1}) was a *standard couple*, but with respect to a Hardy inequality of order $k - 1$. In this section we will deal with *general* overdetermined couples (M_0, M_1), i.e., couples satisfying

$$|M_0| + |M_1| > k,$$

which are constructed as follows: We add some *new boundary conditions* to the conditions described by a standard couple $(\widehat{M}_0, \widehat{M}_1)$. (Obviously, since $(\widehat{M}_0, \widehat{M}_1)$ satisfies the Pólya condition, the same is true for (M_0, M_1).) We will show that necessary and sufficient conditions of the validity of (4.30) for $g \in AC^{k-1}(M_0, M_1)$ are the *same* as the corresponding conditions for $g \in AC^{k-1}(\widehat{M}_0, \widehat{M}_1)$ *provided the weights u and v satisfy some additional assumptions.*

A particular case

We will explain our idea on the particular standard couple

$$\widehat{M}_0 = \mathbb{N}_k, \quad \widehat{M}_1 = \emptyset.$$

Necessary and sufficient conditions of the validity of (4.30) for $g \in AC^{k-1}(\widehat{M}_0, \widehat{M}_1)$ are described in Theorem 4.3 (see (4.8) and/or (4.9)).

To the boundary conditions described by $\widehat{M}_0, \widehat{M}_1$, i.e.,

$$g(0) = g'(0) = \cdots = g^{(k-1)}(0) = 0, \tag{4.100}$$

we add conditions

$$g^{(j)}(1) = 0 \quad \text{for } j \in M, \tag{4.101}$$

where M is a nonempty subset of $\mathbb{N}_k$. Then we have the situation described above, with

$$M_0 = \widehat{M}_0, \quad M_1 = M.$$

For the operator T from (4.99) we have

$$(Tf)(x) = \frac{1}{(k-1)!} \int_0^x (x-t)^{k-1} f(t) dt, \quad x \in (0,1).$$

The function $g = Tf$ obviously satisfies conditions (4.100) and the equation $g^{(k)} = f$ in $(0,1)$. Moreover, since

$$g^{(j)}(x) = \frac{(-1)^j}{(k-j-1)!} \int_0^x (x-t)^{k-j-1} f(t) dt, \quad j = 0, 1, \ldots, k-1,$$

conditions (4.101) lead to the assumptions

$$\int_0^1 (1-t)^{k-j-1} f(t) dt = 0 \quad \text{for } j \in M. \tag{4.102}$$

Let us denote by F_M the set of all functions $f \in L^p(v)$ which satisfy (4.102). Furthermore, suppose that the weight function v satisfies

$$\int_0^1 (1-t)^{(k-j-1)p'} v^{1-p'}(t) dt < \infty \quad \text{for } j \in M, \tag{4.103}$$

and denote by V_M the linear hull of the functions

$$\varphi_j(t) = (1-t)^{k-j-1} v^{-1}(t), \quad j \in M.$$

Obviously, due to condition (4.103), $\varphi_j \in L^{p'}(v)$ and V_M is a finite-dimensional subspace of $L^{p'}(v)$ (of dimension $m = |M|$).

If we define a duality $\langle \cdot, \cdot \rangle_v$ between $L^p(v)$ and $L^{p'}(v)$ by

$$\langle g, h \rangle_v = \int_0^1 g(t)h(t)v(t)dt, \quad g \in L^p(v), \ h \in L^{p'}(v)^1,$$

then assumptions (4.102) can be rewritten as

$$\langle f, \varphi_j \rangle_v = 0 \quad \text{for } j \in M.$$

Thus, if we denote by $V_M^\perp$ the *orthogonal complement* of V_M, i.e. the set of all $g \in L^p(v)$ such that $\langle g, \varphi_j \rangle_v = \int_0^1 g(t)(1-t)^{k-j-1}dt = 0$ for $j \in M$, we have shown that

$$F_M = V_M^\perp$$

and F_M is a closed subspace of $L^p(v)$ of finite codimension m.

So we have proved the following assertion.

Lemma 4.36. *Let M be a nonempty subset of $\{0, 1, \ldots, k-1\}$. Then the Hardy inequality (4.30) for the overdetermined class $AC^{k-1}(\mathbb{N}_k, M)$ is equivalent to the weighted norm inequality*

$$\|Tf\|_{q,u} \le c\|f\|_{p,v} \quad for \ all \quad f \in F_M \qquad (4.104)$$

with $F_M \subset L^p(v)$ determined by (4.102). Moreover, if the weight function v satisfies (4.103), then $F_M = V_M^\perp$.

Remark 4.37. (i) Thus, the investigation of the Hardy inequality under the overdetermined conditions (4.100), (4.101) can be reduced to the investigation of inequality (4.104) on the subset $V_M^\perp$ of $L^p(v)$ provided v satisfies (4.103).

(ii) Obviously, conditions (4.103) may be replaced by a *single* condition

$$\int_0^1 (1-t)^{(k-j_0-1)p'} v^{1-p'}(t)dt < \infty \quad \text{with} \quad j_0 = \max\{j : j \in M\}.$$
$$(4.105)$$

(iii) If M is empty, then we have no additional conditions on v and the subset $V_M^\perp$ coincides with the whole space $L^p(v)$. This

---

[1]Notice that this duality concept is different from that introduced in Sec. 1.1.

means: To investigate (4.30) under the standard conditions (4.100) is equivalent to investigating (4.104) on the *whole* space $L^p(v)$.

The main result reads — for the particular choice $\widehat{M_0} = \mathbb{N}_k$, $\widehat{M_1} = \emptyset$ — as follows.

Theorem 4.38. *Let* $1 < p < \infty$, $0 < q < \infty$. *Let* M *be a nonempty subset of* $\mathbb{N}_k$, $j_0 = \max\{j : j \in M\}$ *and suppose that the weight functions* u, v *satisfy conditions* (4.105) *and*

$$\int_0^1 \left(\int_0^x (x-t)^{k-1}(1-t)^{(k-j_0-1)(p'-1)}v^{1-p'}(t)dt \right)^q u(x)dx < \infty.$$

(4.106)

Then the k-th order Hardy inequality (4.30) *holds for g from the overdetermined class* $AC^{k-1}(\mathbb{N}_k, M)$ *if and only if it holds for* $g \in AC^{k-1}(\mathbb{N}_k, \emptyset)$ *(i.e., satisfying the standard conditions* (4.100)).

Proof. The *if* part is obvious since $AC^{k-1}(\mathbb{N}_k, M)$ is a subset of $AC^{k-1}(\mathbb{N}_k, \emptyset)$.

To prove the *only if* part, let us first suppose that M contains only *one* element, $M = \{j_0\}$. Then there is only one function φ : $\varphi(t) = (1-t)^{k-j_0-1}v^{-1}(t)$, $\varphi \in L^p(v)$ due to (4.105), and the set V_M is one-dimensional.

Define a function ψ by

$$\psi(t) = C_0(1-t)^{(k-j_0-1)(p'-1)}v^{1-p'}(t)$$

with a suitable constant $C_0 > 0$. Condition (4.105) guarantees that

$$\psi \in L^p(v) \quad \text{and} \quad \langle \psi, \varphi \rangle_v > 0$$

since

$$\int_0^1 \psi^p(t)v(t)dt = C_0^p \int_0^1 (1-t)^{(k-j_0-1)p'}v^{1-p'}(t)dt < \infty$$

and

$$\langle \psi, \varphi \rangle_v = \int_0^1 \psi(t)\varphi(t)v(t)dt = C_0 \int_0^1 (1-t)^{(k-j_0-1)p'}v^{1-p'}(t)dt.$$

If we choose C_0 such that $\langle \psi, \varphi \rangle_v = 1$, we can write every $g \in L^p(v)$ in the form

$$g = f + C\psi, \quad f \in V_M^\perp. \tag{4.107}$$

Indeed, if we put $f = g - \langle g, \varphi \rangle_v \psi$, we have $\langle f, \varphi \rangle_v = \langle g, \varphi \rangle_v - \langle g, \varphi \rangle_v \langle \psi, \varphi \rangle_v = 0$, i.e., $f \in V_M^\perp$, and (4.107) follows with $C = \langle g, \varphi \rangle_v$.

Condition (4.106) is nothing else then the assertion

$$T\psi \in L^q(u)$$

and thus T maps the one-dimensional subset $\{c\psi\}$, $c \in \mathbb{R}$, of $L^p(v)$ continuously into $L^q(u)$. Since T maps $V_M^\perp$ continuously into $L^q(u)$ if and only if (4.30) holds for $AC^{k-1}(\mathbb{N}_k, M)$ (due to Lemma 4.36), and since according to (4.107),

$$L^p(v) = V_M^\perp \oplus \{c\psi\} = F_M \oplus \{c\psi\},$$

T maps the *whole* space $L^p(v)$ into $L^q(u)$. But then, according to Remark 4.52 (iii), inequality (4.30) holds for $g \in AC^{k-1}(\mathbb{N}_k, \emptyset)$, and this completes the proof for $M = \{j_0\}$.

If M contains more than one element from the set $\{0, 1, \ldots, k-1\}$, then we can proceed similarly, using the concept of biorthogonality (see, e.g., T. Kato [Ka1, Theorem 1.22]). We denote again $j_0 = \max\{j : j \in M\}$ and suppose that (4.105) and (4.106) hold. Then for the functions $\varphi_j(t) = (1-t)^{k-j-1} v^{-1}(t)$, $j \in M$, introduced on p. 222, with $\varphi_j \in L^{p'}(v)$, there exist functions $\psi_i \in L^p(v)$ such that $\langle \psi_i, \varphi_j \rangle_v = \delta_{ij}$, $i, j \in M$, and we can write, in analogy to (4.107), every function $g \in L^p(v)$ in the form

$$g = f + h,$$

where f belongs to $V_M^\perp$ and h belongs to the linear hull Ψ_0 of the ψ_i's, i.e., to a finite-dimensional subspace of $L^p(v)$. It can be shown that the functions ψ_i can be expressed as linear combinations of the functions

$$(1-t)^{(k-j-1)(p'-1)} v^{1-p'}(t), \quad j \in M.$$

Condition (4.106) implies that $T\psi_i \in L^q(u)$, $i \in M$, and thus $T : \Psi \to L^q(u)$. The conclusion now follows as in the case $M = \{j_0\}$. $\square$

Example 4.39. Consider inequality (4.30) under the conditions

$$g(0) = g'(0) = \cdots = g^{(k-1)}(0) = 0, \quad g(1) = 0.$$

We have the situation just described, with $M = \{j_0\} = \{0\}$, and thus, according to Theorem 4.38, inequality (4.30) holds if and only if conditions (4.8) (for $p \leq q$) or (4.9) (for $p > q$) are satisfied, provided that, in addition,

$$\int_0^1 (1-t)^{(k-1)p'} v^{1-p'}(t) dt < \infty$$

and

$$\int_0^1 \left(\int_0^x (x-t)^{k-1} (1-t)^{(k-1)(p'-1)} v^{1-p'}(t) dt \right)^q u(x) dx < \infty.$$

Another example

Consider inequality (4.30) under the conditions

$$g(0) = g'(0) = \cdots = g^{(k-1)}(0) = 0, \quad g^{(k-1)}(1) = 0.$$

Now we have $M = \{k-1\}$, and thus inequality (4.30) holds if and only if conditions (4.8) or (4.9) are satisfied, provided that, in addition,

$$\int_0^1 v^{1-p'}(t) dt < \infty$$

and

$$\int_0^1 \left(\int_0^x (x-t)^{k-1} v^{1-p'}(t) dt \right)^q u(x) dx < \infty.$$

Exactly this case was investigated on pp. 209–212. Compare the results with Theorem 4.31.

A general standard couple

The approach described for the special case $\widehat{M}_0 = \mathbb{N}_k$, $\widehat{M}_1 = \emptyset$ can be used for any standard couple $\widehat{M}_0$, $\widehat{M}_1$. Let us describe shortly how to proceed:

A function $g \in AC^{k-1}(\widehat{M_0}, \widehat{M_1})$ can be expressed with help of an integral operator T:

$$g(x) = (Tf)(x) = \int_0^1 G(x,t)f(t)dt, \quad f \in L^p(v), \qquad (4.108)$$

where the kernel G — the Green function — is known. If our *overdetermined* couple M_0, M_1 is determined by *additional* conditions

$$g^{(\alpha)}(0) = 0, \quad g^{(\beta)}(1) = 0,$$

with $\alpha \in A$, $\beta \in B$, $A, B \subset \mathbb{N}_k$, $A \cap \widehat{M_0} = \emptyset$, $B \cap \widehat{M_1} = \emptyset$ (so that $M_0 = \widehat{M_0} \cup A$, $M_1 = \widehat{M_1} \cup A$), we simply use these conditions in (4.108) and obtain additional conditions on f:

$$\int_0^1 \frac{\partial^\alpha G}{\partial x^\alpha}(0,t)f(t)dt = 0, \quad \int_0^1 \frac{\partial^\beta G}{\partial x^\beta}(1,t)f(t)dt = 0, \qquad (4.109)$$

$\alpha \in A$, $\beta \in B$. Supposing additionally that the weight function v satisfies

$$\int_0^1 \left| \frac{\partial^\alpha G}{\partial x^\alpha}(0,t) \right|^{p'} v^{1-p'}(t)dt < \infty, \quad \int_0^1 \left| \frac{\partial^\beta G}{\partial x^\beta}(1,t) \right|^{p'} v^{1-p'}(t)dt < \infty,$$

conditions (4.109) describe a closed subspace $F(A, B)$ of $L^p(v)$ (with finite codimension) and we consider the weighted norm inequality

$$\|Tf\|_{q,u} \leq C\|f\|_{p,v}$$

for f from $F(A, B)$, which is the *orthogonal complement* of a certain subspace of $L^{p'}(v)$.

Finally, we try to find conditions on u and v which allow to extend the last inequality to the whole space $L^p(v)$.

Without going into details, we explain our approach by a simple example.

Example 4.40. We consider the third order Hardy inequality

$$\left(\int_0^1 |g(x)|^q u(x)dx\right)^{1/q} \leq C \left(\int_0^1 |g'''(x)|^p v(x)dx\right)^{1/p} \quad (4.110)$$

for functions $g \in AC^2(M_0, M_1)$ with $M_0 = \{0, 1\}$, $M_1 = \{0, 2\}$, i.e., satisfying

$$g(0) = 0, \quad g(1) = 0, \quad g'(0) = 0, \quad g''(1) = 0. \quad (4.111)$$

We start with the *standard* conditions

$$g(0) = 0, \quad g'(0) = 0, \quad g(1) = 0,$$

i.e., take $\widehat{M_0} = \{0, 1\}$, $\widehat{M_1} = \{0\}$, and the necessary and sufficient conditions of the validity of inequality (4.110) for $g \in AC^2(\widehat{M_0}, \widehat{M_1})$ have, according to Theorem 4.13, the form (for $1 < p \leq q < \infty$)

$$\sup_{0 < x < 1} \left(\int_x^1 t^q (1-t)^q u(t)dt\right)^{1/q}$$

$$\times \left(\int_0^x t^{p'} (1-t)^{p'} v^{1-p'}(t)dt\right)^{1/p'} < \infty, \quad (4.112)$$

$$\sup_{0 < x < 1} \left(\int_0^x t^{2q} u(t)dt\right)^{1/q} \left(\int_x^1 (1-t)^{2p'} v^{1-p'}(t)dt\right)^{1/p'} < \infty.$$

In this case we have

$$g(x) = (Tf)(x) = \frac{1}{2} \int_0^x t(2x^2 - x^2 t - 2x + t)f(t)dt$$

$$- \frac{1}{2}x^2 \int_x^1 (1-t)^2 f(t)dt$$

and the additional condition $g''(1) = 0$ leads to the following condition on f:

$$\int_0^1 t(2-t)f(t)dt = 0. \quad (4.113)$$

Thus inequality (4.110) holds for g satisfying (4.111) if the inequality $\|Tf\|_{q,u} \leq C\|f\|_{p,v}$ holds for all $f \in L^p(v)$ satisfying (4.113). But condition (4.113) can be rewritten in the form $\langle f, \varphi \rangle_v = 0$ with $\varphi(t) = t(2 - t)v^{-1}(t)$ provided φ belongs to $L^{p'}(v)$, which means that v satisfies the assumption

$$\int_0^1 t^{p'} v^{1-p'} dt < \infty. \tag{4.114}$$

Then, similarly as in the proof of Theorem 4.38, we can construct a function $\psi \in L^p(v)$, $\psi(t) = C_0 t^{p'-1}(2-t)^{p'-1} v^{1-p'}(t)$, and decompose the space $L^p(v)$. If, moreover,

$$\int_0^1 \left(x(1 - x) \int_0^x t^{p'}(1 - t)v^{1-p'}(t)dt \right.$$

$$\left. + \, x^2 \int_x^1 (1 - t)^2 t^{p'-1} v^{1-p'}(t)dt \right)^q u(x)dx < \infty, \tag{4.115}$$

then $T\psi \in L^q(u)$ and we can conclude that conditions (4.112) are necessary and sufficient for (4.110) with g satisfying the *overdetermined* conditions (4.111) provided u and v satisfy assumptions (4.114) and (4.115).

(Note that in (4.114) and (4.115) we used the fact that $G(x, t) \approx x(1 - x)t(1 - t)$ for $0 < t < x$ according to Theorem 4.11, formula (4.29), and that $1 < 2 - t < 2$ for $t \in (0, 1)$.)

Remark 4.41. If an overdetermined couple (M_0, M_1) is given, we sometimes can choose *different* standard couples $\widehat{M_0}, \widehat{M_1}$ and obtain different necessary and sufficient conditions in dependence on the particular additional conditions on u and v.

So, in Example 4.40, we had $M_0 = \{0, 1\}$ and $M_1 = \{0, 2\}$ and we have chosen $\widehat{M_0} = \{0, 1\}$, $\widehat{M_1} = \{0\}$, i.e. $A = \emptyset$, $B = \{2\}$ in the notation on pp. 226–228. But it is also possible to take $\widehat{M_0} = \{0, 1\}$, $\widehat{M_1} = \{2\}$ which is again a standard couple, and now $A = \emptyset$, $B = \{0\}$. Then we have

$$g(x) = (Tf)(x) = \frac{1}{2} \int_0^x t(t - 2x)f(t)dt - \frac{1}{2}x^2 \int_x^1 f(t)dt$$

and the necessary and sufficient conditions for $g \in AC^2(\widehat{M_0}, \widehat{M_1})$ have now — instead of (4.112) — the form (for $1 < p \leq q < \infty$)

$$\sup_{0<x<1} \left(\int_x^1 t^q u(t) dt \right)^{1/q} \left(\int_0^x t^{p'} v^{1-p'}(t) dt \right)^{1/p'} < \infty,$$

$$\sup_{0<x<1} \left(\int_0^x t^{2q} u(t) dt \right)^{1/q} \left(\int_x^1 v^{1-p'}(t) dt \right)^{1/p'} < \infty.$$

The additional condition $g(1) = 0$ leads again to condition (4.113) on f, and consequently, we can again choose φ and ψ as in Example 4.40. This leads to condition (4.114) again, but assumption (4.115), which guarantees that $T\psi \in L^q(u)$, now reads

$$\int_0^1 \left(x \int_0^x t^{p'} v^{1-p'}(t) dt + x^2 \int_x^1 t^{p'-1} v^{1-p'}(t) dt \right)^q u(x) dx < \infty.$$

In Example 4.40 it is also possible to choose $\widehat{M_0} = \{1\}$, $\widehat{M_1} = \{0, 2\}$. In this case, we have

$$g(x) = (Tf)(x) = \int_0^x t(1-x) f(t) dt + \frac{1}{2} \int_x^1 (2t - x^2 - t^2) f(t) dt$$

and, according to Theorem 4.12, there is only one necessary and sufficient condition for the Hardy inequality (4.110) with $g \in AC^2(\widehat{M_0}, \widehat{M_1})$:

$$\left(\int_0^1 (1-x)^q u(x) dx \right)^{1/q} \left(\int_0^1 t^{p'} v^{1-p'}(t) dt \right)^{1/p'} < \infty.$$

The additional condition $g(0) = 0$ leads again to assumption (4.113) about f while assumption (4.115) is replaced by

$$\int_0^1 \left((1-x) \int_0^x t^{p'} v^{1-p'}(t) dt + (1-x) \int_x^1 t^{p'} v^{1-p'}(t) dt \right)^q u(x) dx$$

$$= \left(\int_0^1 (1-x)^q u(x) dx \right) \left(\int_0^1 t^{p'} v^{1-p'}(t) dt \right)^q < \infty.$$

In view of (4.114), the last two conditions can be replaced by the single condition

$$\int_0^1 (1-x)^q u(x)dx < \infty,$$

which together with (4.114) guarantees the validity of (4.110) for $g \in AC^2(M_0, M_1)$.

The case $M_0 = M_1 = \mathbb{N}_k$

Now, let us consider the *maximal overdetermined* class, namely, the functions $g \in AC^{k-1}$ satisfying

$$g^{(i)}(0) = g^{(i)}(1) = 0 \quad \text{for } i = 0, 1, \ldots, k-1. \tag{4.116}$$

In this case we can modify the foregoing approaches.

We choose $z \in (0,1)$ arbitrary but fixed and introduce operators $S_{1,z}$, $S_{2,z}$, defined by

$$(S_{1,z}f)(x) = \frac{1}{(k-1)!} \int_0^x (x-t)^{k-1} f(t)dt, \quad x \in (0,z),$$

$$(S_{2,z}f)(x) = \frac{1}{(k-1)!} \int_x^1 (t-x)^{k-1} f(t)dt, \quad x \in (z,1). \tag{4.117}$$

If we define

$$g(x) = \chi_{(0,z)}(x)(S_{1,z}f)(x) + (-1)^k \chi_{(z,1)}(x)(S_{2,z}f)(x)$$
$$=: (T_z f)(x), \tag{4.118}$$

then g satisfies conditions (4.116) and we have $g^{(k)} = f$ on $(0,z) \cup (z,1)$. If f satisfies the assumptions

$$\int_0^1 t^i f(t)dt = 0, \quad i = 0, 1, \ldots, k-1, \tag{4.119}$$

then $g^{(i)}(z+0) = g^{(i)}(z-0)$, $0 \le i \le k-1$, and we have immediately the following assertion:

Lemma 4.42. *The k-th order Hardy inequality (4.30) with overdetermined conditions (4.116) is equivalent to the weighted norm inequality*

$$\|T_z f\|_{q,u} \leq C \|f\|_{p,v} \tag{4.120}$$

for all $f \in L^p(v)$ *satisfying conditions (4.119), where* T_z *is given by (4.118),* $z \in (0,1)$ *fixed.*

So, similarly as in the foregoing sections, we have reduced the Hardy inequality (4.30) under conditions (4.116) to the special weighted norm inequality (4.120) on a *subset* of $L^p(v)$. Now, we will try to extend T_z to the whole space $L^p(v)$, and for this purpose, we introduce the functions

$$\varphi_i(t) = t^i v^{-1}(t), \quad i = 0, 1, \ldots, k-1.$$

The additional assumption

$$\int_0^1 v^{1-p'}(t) dt < \infty \tag{4.121}$$

guarantees that $\varphi_i \in L^{p'}(v)$, and conditions (4.119) can be rewritten as $\langle f, \varphi_i \rangle_v = 0$. Analogously as in the proof of Theorem 4.38, we introduce functions $\psi_i \in L^p(v)$ such that $\langle \psi_i, \varphi_j \rangle_v = \delta_{ij}$, in terms of the functions

$$\kappa_i(t) = t^{i(p'-1)} v^{1-p'}(t).$$

Condition (4.121) guarantees that $\kappa_i \in L^p(v)$, and the additional assumptions

$$\int_0^z \left(\int_0^x (x-t)^{k-1} t^{i(p'-1)} v^{1-p'}(t) dt \right)^q u(x) dx < \infty,$$

$$\int_z^1 \left(\int_x^1 (x-t)^{k-1} t^{i(p'-1)} v^{1-p'}(t) dt \right)^q u(x) dx < \infty \tag{4.122}$$

guarantee that $T_z \psi_i \in L^q(u)$. Consequently, T_z maps the whole space $L^p(v)$ continuously into $L^q(u)$ provided the Hardy inequality (4.30) holds for g satisfying (4.116) and the weight functions u, v satisfy (4.121) and (4.122).

Now let us look for necessary conditions. Assume that (4.120) holds for all $f \in L^p(v)$.

(i) If k is *even*, then T_z is the sum of positive operators $S_{1,z}$ and $S_{2,z}$, and we have

$$|S_{i,z}f| \le S_{i,z}|f| \le (S_{1,z} + S_{2,z})|f| = T_z|f|$$

and

$$\|S_{i,z}f\|_{q,u} \le \|T_z\,|f|\,\|_{q,u} \le C\|\,|f|\,\|_{p,v} = C\|f\|_{p,v} \qquad (4.123)$$

for every $f \in L^p(v)$ and $i = 1, 2$. Consequently, the necessary conditions

$$\sup_{0<x<z} \left(\int_x^z (t-x)^{(k-1)q} u(t)dt \right)^{1/q} \left(\int_0^x v^{1-p'}(t)dt \right)^{1/p'} < \infty,$$
$$\hspace{10cm} (4.124)$$
$$\sup_{0<x<z} \left(\int_x^z u(t)dt \right)^{1/q} \left(\int_0^x (x-t)^{(k-1)p'} v^{1-p'}(t)dt \right)^{1/p'} < \infty$$

(for $i = 1$, i.e., on $(0, z)$) and

$$\sup_{z<x<1} \left(\int_z^x (x-t)^{(k-1)q} u(t)dt \right)^{1/q} \left(\int_x^1 v^{1-p'}(t)dt \right)^{1/p'} < \infty,$$
$$\hspace{10cm} (4.125)$$
$$\sup_{z<x<1} \left(\int_z^x u(t)dt \right)^{1/q} \left(\int_x^1 (t-x)^{(k-1)p'} v^{1-p'}(t)dt \right)^{1/p'} < \infty$$

(for $i = 2$, i.e., on $(z, 1)$) have to be satisfied.

(ii) If k is *odd*, then T_z is the difference of positive operators $S_{1,z}$ and $S_{2,z}$. For $f \in L^p(v)$, define

$$h(x) = \begin{cases} f(x) & \text{for } x \in (0, z), \\ -f(x) & \text{for } x \in (z, 1). \end{cases}$$

Then $h \in L^p(v)$, $\|h\|_{p,v} = \|f\|_{p,v}$ and $\|S_{i,z}f\|_{q,u} = \|S_{i,z}h\|_{q,u}$ since $S_{1,z}$ and $S_{2,z}$ are concentrated on $(0, z)$ and $(z, 1)$, respectively. Since $T_z f = \chi_{(0,z)} S_{1,z} h + \chi_{(z,1)} S_{2,z} h$, we again obtain (4.123) and the necessary conditions (4.124) and (4.125).

On the other hand, it follows from (4.124) that

$$\left(\int_0^z |(S_{1,z}f)(x)|^q u(x)dx \right)^{1/q}$$

$$\leq C \left(\int_0^z |f(x)|^p v(x)dx \right)^{1/p} \leq C\|f\|_{p,v} \qquad (4.126)$$

and it follows from (4.125) that

$$\left(\int_z^1 |(S_{2,z}f)(x)|^q u(x)dx \right)^{1/q}$$

$$\leq C \left(\int_z^1 |f(x)|^p v(x)dx \right)^{1/p} \leq C\|f\|_{p,v}, \qquad (4.127)$$

and consequently, conditions (4.124) and (4.125) are also sufficient for (4.120) to hold on $L^p(v)$.

So, we have proved the following assertion.

Theorem 4.43. *Let $z \in (0,1)$ be arbitrary but fixed. Let $1 < p \leq q < \infty$. Then the Hardy inequality (4.30) holds for $g \in AC^{k-1}(M_0, M_1)$ with $M_0 = M_1 = \mathbb{N}_k$ if and only if conditions (4.124) and (4.125) are satisfied provided the weight functions u and v satisfy assumptions (4.121) and (4.122).*

Obviously, an analogous assertion holds also for $1 < q < p < \infty$ with (4.124) and (4.125) replaced by the appropriate conditions expressed in terms of A and $\widetilde{A}, A^*, \widetilde{A}^*$ (see Remark 4.4). Thus, (4.124) could be replaced by

$$A(0, z; (t-x)^{(k-1)q}u(t), v(t)) < \infty,$$

$$A^*(0, z; u(t), (x-t)^{(1-k)p}v(t)) < \infty,$$

and analogously for (4.125).

Remark 4.44. Let us consider the special case $k = 1$. We have shown in Theorem 4.28 that for v satisfying (4.121), conditions (4.72) are necessary and sufficient for (4.30) to hold with $g \in AC^{k-1}(\mathbb{N}_k, \mathbb{N}_k)$ (of course, $k = 1$!). Since conditions (4.72) are exactly the conditions (4.124) and (4.125) for $k = 1$, it seems that the second assumption on u and v, namely (4.122), is *not relevant*. But this is not the case since — for $k = 1$ — condition (4.122) is satisfied automatically.

To show it, assume without loss of generality that (for $z \in (0, 1)$ fixed)

$$\int_0^z v^{1-p'}(t)dt \leq \int_z^1 v^{1-p'}(t)dt.$$

Define f by

$$f(x) = \begin{cases} v^{1-p'}(x) + c & \text{for } x \in (0, z), \\ v^{1-p'}(x) & \text{for } x \in (z, 1), \end{cases}$$

where the constant c is chosen so that

$$\int_0^z f(x)dx = \int_z^1 f(x)dx,$$

and define h by

$$h(x) = \begin{cases} f(x) & \text{for } x \in (0, z), \\ -f(x) & \text{for } x \in (z, 1). \end{cases}$$

Then $\int_0^1 h(x)dx = 0$, i.e., h satisfies (4.119) (note that $k = 1$, i.e., $i = 0$).

According to Lemma 4.42, we have $\|T_z h\|_{q,u} \leq C \|h\|_{p,v}$, and consequently, we have (4.126) and (4.127). But due to the positivity of $S_{1,z}$, we have from (4.126) that

$$\int_0^z \left(\int_0^x v^{1-p'}(t)dt \right)^q u(x)dx = \int_0^z |(S_{1,z} v^{1-p'})(x)|^q u(x)dx$$

$$\leq \int_0^z |S_{1,z}(v^{1-p'} + c)(x)|^q u(x)dx$$

$$= \int_0^z |(S_{1,z} f)(x)|^q u(x)dx$$

$$\leq C \|f\|_{p,v}^q < \infty,$$

which is the first condition in (4.122) (we have $k = 1$, i.e. $i = 0$) while the second follows analogously from (4.127).

4.9. Again the Interval (0,∞)

Introduction

In contrast to the results of Chaps. 1 and 2 where the role of the interval (a, b) was not important and the results could be derived from the corresponding assertions for the generic interval $(0, \infty)$, the situation for higher order Hardy inequalities is substantially different. We tried to illustrate it on pp. 171–174; here we will mention some interesting results concerning the inequality

$$\left(\int_0^\infty |g(x)|^q u(x) dx \right)^{1/q} \le C \left(\int_0^\infty |g^{(k)}(x)|^p v(x) dx \right)^{1/p} \quad (4.128)$$

with parameters p, q such that $1 < p, q < \infty$ and under various choices of the sets M_0, M_1. These results are due to V.D. Stepanov and M. Nasyrova; among other, the difference between the *well-determined* ($|M_0| + |M_1| = k$) and overdetermined ($|M_0| + |M_1| > k$) cases in a certain sense disappears.

The results concern mainly the conditions at the "right endpoint" ∞. Therefore, let us recall a result which was already mentioned in Remark 4.4 and which is (in a more general setting) due to V.D. Stepanov [St1].

Theorem 4.45. *Let $1 < p, q < \infty$. Then inequality (4.128) holds for all functions g satisfying*

$$g(\infty) = g'(\infty) = \cdots = g^{(k-1)}(\infty) = 0 \quad (4.129)$$

if and only if

$$\mathcal{A} := \max(\mathbb{A}_{k,0}, \mathbb{A}_{k,1}) < \infty, \quad (4.130)$$

where

$$
\mathbb{A}_{k,0} := \begin{cases}
\sup_{t>0} \left(\int_0^t (t-x)^{(k-1)q} u(x)dx \right)^{1/q} \\
\quad \times \left(\int_t^\infty v^{1-p'}(x)dx \right)^{1/p'} \quad \text{for } p \le q, \\
\left(\int_0^\infty \left(\int_0^t (t-x)^{(k-1)q} u(x)dx \right)^{r/q} \right. \\
\quad \left. \left(\int_t^\infty v^{1-p'}(x)dx \right)^{r/q'} v^{1-p'}(t)dt \right)^{1/r} \quad \text{for } p > q;
\end{cases}
$$

$$
\mathbb{A}_{k,1} := \begin{cases}
\sup_{t>0} \left(\int_0^t u(x)dx \right)^{1/q} \left(\int_t^\infty (x-t)^{(k-1)p'} v^{1-p'}(x)dx \right)^{1/p'} \\
\hspace{8cm} \text{for } p \le q \\
\left(\int_0^\infty \left(\int_0^t u(x)dx \right)^{r/p} \right. \\
\quad \left. \left(\int_t^\infty (x-t)^{(k-1)p'} v^{1-p'}(x)dx \right)^{r/p'} u(t)dt \right)^{1/r} \\
\hspace{8cm} \text{for } p > q
\end{cases}
$$

with $1/r = 1/q - 1/p$. *Moreover, the best constant* C *in* (4.128) *satisfies*

$$
C \approx \mathcal{A}.
$$

The key result reads as follows:

Theorem 4.46. *Let* $1 < p, q < \infty$. *Then inequality* (4.128) *holds for all functions* g *satisfying*

$$
g(\infty) = 0 \tag{4.131}
$$

if and only if it holds for g *satisfying* (4.129), *i.e., if and only if* $\mathcal{A} < \infty$, *with* $\mathcal{A}$ *defined by* (4.130).

Proof. (i) The necessity of condition (4.130) follows from Theorem 4.45: If (4.128) holds for g satisfying (4.131), then it holds also for g satisfying (4.129), and hence $\mathcal{A} < \infty$.

(ii) To prove the sufficiency of (4.130) assume that $\|g^{(k)}\|_{p,v} < \infty$, $g(\infty) = 0$ and $\mathcal{A} < \infty$. Denote $g^{(k)} = F$ and define $\tilde{g}$ by

$$\tilde{g}(x) = \frac{1}{(k-1)!} \int_x^\infty (z-x)^{k-1} F(z)dz, \; x > 0.$$

Then $\tilde{g}^{(k)}(x) = F(x) = g^{(k)}(x)$, and since $\mathcal{A} < \infty$, we have by Hölder's inequality that

$$|\tilde{g}(x)| \le \frac{1}{(k-1)!} \|F\|_{p,v} \left(\int_x^\infty (z-x)^{(k-1)p'} v^{1-p'}(z)dz \right)^{1/p'}$$

and $\tilde{g}(x) \to 0$ for $x \to \infty$, i.e. $\tilde{g}(\infty) = 0$. But also $g(\infty) = 0$, and hence, $g = \tilde{g}$. Moreover, $\tilde{g}(\infty) = \tilde{g}'(\infty) = \cdots = \tilde{g}^{(k-1)}(\infty) = 0$ and consequently, due to Theorem 4.45, inequality (4.128) holds for $\tilde{g} = g$. □

Remark 4.47. (i) It follows from the foregoing theorem that in the case of the interval $(0, \infty)$, only **one** zero condition at infinity for the **least** derivative is important. V.D. Stepanov calls this phenomenon the *heuristic principle*. From Theorem 4.46 it follows that necessary and sufficient conditions of the validity of inequality (4.128) for functions g satisfying (4.131) [i.e., with the second line of the incidence matrix of the form $1, 0, 0, \ldots, 0, 0$] are the same as for functions satisfying

$$g(\infty) = g'(\infty) = 0 \quad (2^{\text{nd}} \text{ line } 1, 1, 0, \ldots, 0, 0)$$

or

$$g(\infty) = g'(\infty) = g''(\infty) = 0 \quad (2^{\text{nd}} \text{ line } 1, 1, 1, 0, \ldots, 0, 0)$$

etc., up to

$$g(\infty) = g'(\infty) = \cdots = g^{(k-2)}(\infty) = 0 \quad (2^{\text{nd}} \text{ line } 1, 1, 1, 1, \ldots, 1, 0)$$

or finally (Theorem 4.46)

$$g(\infty) = g'(\infty) = \cdots = g^{(k-1)}(\infty) = 0 \quad (2^{\text{nd}} \text{ line } 1, 1, 1, \ldots, 1, 1).$$

Instead of *starting* with (4.131), we can also start with functions g satisfying

$$g^{(j)}(\infty) = 0 \quad (2^{\text{nd}} \text{ line } 0, \ldots, 0, 1, 0, \ldots, 0, 0)$$

with a fixed j, $0 < j < k - 1$, and add the condition $g^{(j+1)}(\infty) = 0$ or the conditions $g^{(j+1)}(\infty) = g^{(j+2)}(\infty) = 0$ etc. up to the set of conditions $g^{(s)}(\infty) = 0$ for $s = j, j + 1, \ldots, k - 1$.

(ii) This heuristic principle was studied and used by M. Nasyrova and V.D. Stepanov [NS1] to the case $k = 2$ and $p = q = 2$, and then extended by M. Nasyrova [N1] (see also [N2]) to the whole scale of parameters, $1 < p, q < \infty$. In this last paper, the case $k = 2$ is fully described and for $k > 2$ some special choices of the *boundary conditions* are investigated. Here, we will shortly deal with the case $k = 2$.

The second order Hardy inequality

Let us deal with the inequality

$$\|g\|_{q,u} \le C \|g''\|_{p,v} \tag{4.132}$$

on the interval $(0, \infty)$. This inequality can be investigated under one of the following nontrivial boundary conditions on g (we list them together with the corresponding incidence matrices):

(i) $\quad g(\infty) = 0 \qquad \begin{pmatrix} 0 & 0 \\ 1 & 0 \end{pmatrix}$

(ii) $\quad g(0) = g(\infty) = 0 \quad \begin{pmatrix} 1 & 0 \\ 1 & 0 \end{pmatrix}$

(iii) $\quad g(0) = g'(0) = 0 \quad \begin{pmatrix} 1 & 1 \\ 0 & 0 \end{pmatrix}$

(iv) $g(\infty) = g'(\infty) = 0$ $\begin{pmatrix} 0 & 0 \\ 1 & 1 \end{pmatrix}$

(v) $g(0) = g'(\infty) = 0$ $\begin{pmatrix} 1 & 0 \\ 0 & 1 \end{pmatrix}$

(vi) $g'(0) = g(\infty) = 0$ $\begin{pmatrix} 0 & 1 \\ 1 & 0 \end{pmatrix}$

(vii) $g(0) = g'(0) = g(\infty) = 0$ $\begin{pmatrix} 1 & 1 \\ 1 & 0 \end{pmatrix}$

(viii) $g(0) = g'(0) = g'(\infty) = 0$ $\begin{pmatrix} 1 & 1 \\ 0 & 1 \end{pmatrix}$

(ix) $g(0) = g(\infty) = g'(\infty) = 0$ $\begin{pmatrix} 1 & 0 \\ 1 & 1 \end{pmatrix}$

(x) $g'(0) = g(\infty) = g'(\infty) = 0$ $\begin{pmatrix} 0 & 1 \\ 1 & 1 \end{pmatrix}$

(xi) $g(0) = g'(0) = g(\infty) = g'(\infty) = 0$ $\begin{pmatrix} 1 & 1 \\ 1 & 1 \end{pmatrix}$

The case (iii) is solved by Theorem 4.3, the case (iv) by Theorem 4.45 and the case (v) by Theorem 4.8. The case (i) is solved since it is equivalent to (iv) due to Theorem 4.46. The heuristic principle (Theorem 4.46) indicates that

(ii) is equivalent to (ix),

(vi) is equivalent to (x),

(vii) is equivalent to (xi),

and in M. Nasyrova [N2], these equivalences are proved and criteria of validity of (4.132) in all cases are given.

Example 4.48. In Example 4.7, the case (vi), i.e., $g'(0) = g(\infty) = 0$ was considered for the special weights $v(x) = x^{\alpha}$, $u(x) = x^{\alpha-2p}$ and for $p = q$. The criteria mentioned in the foregoing subsection show that inequality (4.132), i.e., inequality (4.17) holds for $\alpha > 2p - 1$.

Remark 4.49. Some new results concerning higher order Hardy-type inequalities (even with three weights involved) are proved and discussed in the paper [KP1] by A. Kalybay and L.E. Persson, see also the doctoral thesis of A. Kalybay [Kal1].

Remark 4.50. We consider again the standard Hardy inequality

$$\left(\int_a^b |g(x)|^q u(x) dx \right)^{1/q} \leq C \left(\int_a^b |g'(x)|^p v(x) dx \right)^{1/p}$$

with $g(a) = 0$ or $g(b) = 0$. Instead of generalizing this to the case with higher derivatives on the right-hand side as in this chapter it is natural to consider multi-dimensional generalizations of the type

$$\left(\int_\Omega |g(x)|^q u(x) dx \right)^{1/q} \leq C \left(\int_\Omega |\nabla g|^p v(x) dx \right)^{1/p}, \qquad (4.133)$$

where $\nabla g = \text{grad } g$ and $\Omega \subset \mathbb{R}^n, n = 2, 3, \ldots$ and with suitable boundary conditions. It seems not to be an easy task to generalize (4.133) to a higher order setting like in the one-dimensional case in this chapter. For the power weighted case a number of interesting results of type (4.133) are known, see e.g. Comments and Remarks 4.10.11.

4.10. Comments and Remarks

4.10.1. The Pólya condition mentioned in Sec. 4.1 is a very particular case of a general Pólya condition appearing in the theory of Birkhoff interpolation. For details, see R.A. Lorentz [Lor1].

4.10.2. Theorem 4.3 is in fact due to V. Stepanov [St1]. It is a consequence of his more general results concerning Riemann-Liouville operators.

4.10.3. The results summarized in Theorem 4.8 are due to A. Kufner and H.P. Heinig [KH1].

4.10.4. Necessary and sufficient conditions for the validity of the k-th order Hardy inequality in the general (well-determined) case (see Sec. 4.3) have been investigated by A. Kufner and A. Wannebo [KW1] first for $k = 2$ and $k = 3$. Then A. Kufner [Ku2] gave a proof for general $k \in \mathbb{N}$ if $M_0 \cap M_1 = \emptyset$ and formulated a conjecture for general index sets M_0, M_1. Finally, G. Sinnamon [Si4] proved this conjecture. In Sec. 4.3, we follow his approach.

4.10.5. The special cases described in Sec. 4.4 can be found in A. Kufner [Ku6].

4.10.6. The reduction of conditions (Sec. 4.5) was proposed by A. Kufner in [Ku3].

4.10.7. Condition (4.65) was derived by P. Gurka in an unpublished paper for the case $p \le q$. Then B. Opic modified his approach and extended the results also to the case $p > q$. For details, see [OK, Sec. 8]. The approach to overdetermined classes for the case when the weight function v satisfies (4.66) for $k = 1$ (Theorem 4.28) as well as for the special overdetermined classes if $k > 1$ (Sec. 4.7) is essentially due to G. Sinnamon and can be found in A. Kufner and G. Sinnamon [KSin1]. The idea of "splitting the interval $(0, 1)$ by some z" was suggested by R. Oinarov.

4.10.8. The approach to general overdetermined problems described in Sec. 4.8 is due to A. Kufner and H. Leinfelder [KL1]. Partial results can be found in A. Kufner and C.G. Simader [KSi1]. See also M. Nasyrova and V.D. Stepanov [NS2], where the maximal overdetermined case ($|M_0| + |M_1| = 2k$), $k = 2$, $p = q = 2$, is characterized.

4.10.9. The higher order Hardy inequality on $(0, \infty)$ has also been dealt with in T. Kilgore [Ki1].

4.10.10. The case of the k-th order Hardy inequality (4.1) with general boundary conditions (4.43) was recently solved by A. Kufner *et al.* in [KKP1]. Conditions on the coefficients α_{ij} and β_{ij} are given so that the corresponding boundary value problem, i.e., $u^{(k)} = f$ in

(a, b) and conditions (4.43), is uniquely solvable. Let the determinant

$$\Delta := \begin{vmatrix} \alpha_{1,1} + \beta_{1,1} & \cdots & \sum_{j=1}^{m} \frac{(m-1)!}{(m-j)!}[\alpha_{1,j}a^{m-j} + \beta_{1,j}b^{m-j}] & \cdots \xrightarrow{j:} & \sum_{j=1}^{k} \frac{(k-1)!}{(k-j)!}[\alpha_{1,j}a^{k-j} + \beta_{1,j}b^{k-j}] \\ \downarrow i: & \cdots & \vdots & \cdots & \downarrow i: \\ \alpha_{i,1} + \beta_{i,1} & \cdots & \sum_{j=1}^{m} \frac{(m-1)!}{(m-j)!}[\alpha_{i,j}a^{m-j} + \beta_{i,j}b^{m-j}] & \cdots \xrightarrow{j:} & \sum_{j=1}^{k} \frac{(k-1)!}{(k-j)!}[\alpha_{i,j}a^{k-j} + \beta_{i,j}b^{k-j}] \\ \downarrow i: & \cdots & \vdots & \cdots & \downarrow i: \\ \alpha_{k,1} + \beta_{k,1} & \cdots & \sum_{j=1}^{m} \frac{(m-1)!}{(m-j)!}[\alpha_{k,j}a^{m-j} + \beta_{k,j}b^{m-j}] & \cdots \xrightarrow{j:} & \sum_{j=1}^{k} \frac{(k-1)!}{(k-j)!}[\alpha_{k,j}a^{k-j} + \beta_{k,j}b^{k-j}] \end{vmatrix}$$

be nonzero. The unique solution of the boundary value problem then has the form

$$u(x) = \sum_{m=1}^{k} c_m x^{m-1} - \int_x^b \frac{(x-t)^{k-1}}{(k-1)!} f(t)\, dt$$

with coefficients $c_1, c_2, \ldots, c_k$, where the constants c_m are the solutions of the system

$$\sum_{m=1}^{k} c_m \left[\sum_{j=1}^{m} \frac{(m-1)!}{(m-j)!}[\alpha_{i,j}a^{m-j} + \beta_{i,j}b^{m-j}] \right]$$

$$= \sum_{j=1}^{k} \alpha_{i,j} \int_a^b \frac{(a-t)^{k-j}}{(k-j)!} g(t)\, dt,$$

$i = 1, \ldots, k$, whose determinant Δ is defined above.

The corresponding Green function has the form

$$G(x,t) = \sum_{n=1}^{k} P_n(x)t^{n-1} - \frac{(x-t)^{k-1}}{(k-1)!}\chi_{(x,b)}(t),$$

where $P_n(x) = \sum_{m=1}^{k} a_{n,m}x^{m-1}$, $n = 1,\ldots,k$, are polynomials of order $\leq k-1$.

In [KKP1] conditions are given (sufficient and also necessary and sufficient) for the k-th order Hardy inequality to hold under the conditions (4.43). These conditions are expressed in terms of the Green function.

4.10.11. With concrete power weights in (4.133) a great number of interesting results is known, e.g. the following Hardy-Sobolev inequality:

$$\left(\int_{\mathbb{R}^n} |f(x)|^q |x|^{\frac{(n-2)}{2}q-n}dx\right)^{1/q} \leq C \left(\int_{\mathbb{R}^n} |\nabla f(x)|^2 dx\right)^{1/2},$$

where $2 \leq q \leq 2n/(n-3)$, $n \geq 3$. Also the sharp constant C is known.

Another important early development was done in 1984 by L. Caffarelli *et al.* [CKN1]. The following inequality is sometimes referred to as the Caffarelli-Kohn-Nirenberg inequality (see e.g. [DEL1]):

$$\left(\int_{\mathbb{R}^n} |f(x)|^q |x|^{-bq}dx\right)^{1/q} \leq C \left(\int_{\mathbb{R}^n} |\nabla f(x)|^2 |x|^{-2a}dx\right)^{1/2},$$

where $a \leq b \leq a+1$ if $n \geq 3$, $a < b \leq a+1$ if $n = 2$, $a+1/2 < b \leq a+1$ if $n = 1$ and $a < (n-2)/2$. The exponent

$$q = \frac{2n}{n-2+2(b-a)}$$

is obtained by a standard dilation argument.

Just to mention one important recent paper related to this inequality we refer to J. Dolbeault *et al.* [DEL1]. In particular, the authors here characterized the so-called *symmetry breaking region* in the Caffarelli-Kohn-Nirenberg inequalities. As a consequence they can identify the optimal functions and sharp constants in the symmetry region. This solves a longstanding conjecture. We also refer to the paper [DETT1] and the references in the papers we have cited above. Except the authors mentioned above several authors have done important contributions related to inequality (4.133) for the power weighted case, e.g. concerning refinements, best constants, various domains Ω, etc.

5

Fractional Order Hardy Inequalities

5.1. Introduction

As mentioned in Chap. 1, now we will investigate *fractional order Hardy inequalities*, i.e., inequalities of the form

$$\|g\|_{q,u} \le C \|g^{(\lambda)}\|_{p,v}, \quad 0 < \lambda < 1, \tag{5.1}$$

and

$$\|g^{(\lambda)}\|_{q,u} \le C \|g'\|_{p,v}, \quad 0 < \lambda < 1, \tag{5.2}$$

where

$$\|g^{(\lambda)}\|_{r,w} := \left(\int_a^b \int_a^b \frac{|g(x) - g(y)|^r}{|x-y|^{1+\lambda r}} w(x,y) dx dy \right)^{1/r}. \tag{5.3}$$

Here $w(x,y)$ is a weight function defined in $(a,b) \times (a,b)$, $-\infty \le a < b \le +\infty$, and $g^{(\lambda)}$ denotes the (formal) fractional derivative of order λ, $0 < \lambda < 1$.

The following result was mentioned in Chap. 3 (see p. 128):

Proposition 5.1. *Let* $1 < p < \infty$, $0 < \lambda < 1$, $\lambda \neq 1/p$. *Then, for every* $g \in C_0^\infty(0, \infty)$,

$$\left(\int_0^\infty \left| \frac{g(x)}{x^\lambda} \right|^p dx \right)^{1/p} \leq C \left(\int_0^\infty \int_0^\infty \frac{|g(x) - g(y)|^p}{|x - y|^{1+\lambda p}} dx dy \right)^{1/p}.$$
(5.4)

Remark 5.2. Inequality (5.4) is a special case of inequality (5.1): We take $q = p$, $u(x) = x^{-\lambda p}$ and $v(x, y) \equiv 1$. It was derived independently by G.N. Jakovlev [J1] and P. Grisvard [Gr1], but it might have been known earlier (see 5.6.1 in Sec. 5.6). Below, in Theorem 5.9, we generalize inequality (5.4).

The following result is a counterpart of inequality (5.4) and a special case of inequality (5.2). It was derived in Chap. 3 (see Example 3.18) but we will give here a different proof.

Theorem 5.3. *Let* $1 < p < \infty$, $0 < \lambda < 1$. *Then, for every* $g \in AC(0, \infty)$,

$$\left(\int_0^\infty \int_0^\infty \frac{|g(x) - g(y)|^p}{|x - y|^{1+\lambda p}} dx dy \right)^{1/p} \leq C \left(\int_0^\infty |g'(x)|^p x^{(1-\lambda)p} dx \right)^{1/p},$$
(5.5)

where $C = 2^{1/p} \lambda^{-1} (p(1 - \lambda))^{-1/p}$ *is the best possible constant.*

Proof. Using Fubini's theorem and the symmetry of the integrand, we rewrite the left-hand side in (5.5) into the form

$$\left(\int_0^\infty \int_0^\infty \frac{|g(x) - g(y)|^p}{|x - y|^{1+\lambda p}} dx dy \right)^{1/p}$$

$$= \left(\int_0^\infty \int_0^x \frac{|g(x) - g(y)|^p}{|x - y|^{1+\lambda p}} dy dx + \int_0^\infty \int_x^\infty \frac{|g(x) - g(y)|^p}{|x - y|^{1+\lambda p}} dy dx \right)^{1/p}$$

$$= \left(2 \int_0^\infty \int_x^\infty \frac{|g(x) - g(y)|^p}{|x - y|^{1+\lambda p}} dy dx \right)^{1/p}$$

$$= 2^{1/p} \left(\int_0^\infty \left(\int_x^\infty |x - y|^{-1-\lambda p} \left| \int_x^y g'(t) dt \right|^p dy \right) dx \right)^{1/p}. \quad (5.6)$$

Now we use the Hardy inequality for the interval (x, ∞) with a fixed $x > 0$:

$$\int_x^\infty |x - y|^{-1-\lambda p} \left| \int_x^y g'(t) dt \right|^p dy$$

$$\leq \lambda^{-p} \int_x^\infty |x - y|^{-1-\lambda p + p} |g'(y)|^p dy$$

(see Chap. 1; the necessary and sufficient condition for its validity is satisfied since $\lambda > 0$). Using this inequality and Fubini's theorem in (5.6) we obtain

$$\int_0^\infty \int_0^\infty \frac{|g(x) - g(y)|^p}{|x - y|^{1+\lambda p}} dx dy$$

$$\leq 2\lambda^{-p} \int_0^\infty \left(\int_x^\infty |x - y|^{-1-\lambda p + p} |g'(y)|^p dy \right) dx$$

$$= 2\lambda^{-p} \int_0^\infty |g'(y)|^p \left(\int_0^y (y - x)^{-1-\lambda p + p} dx \right) dy$$

$$= \frac{2\lambda^{-p}}{p - \lambda p} \int_0^\infty |g'(y)|^p y^{p - \lambda p} dy.$$

Thus, due to the fact that $\lambda < 1$, inequality (5.5) holds. The constant C is the best possible because of the sharpness of the constant in the classical Hardy inequality. $\qquad\square$

Remark 5.4. If *both* the inequalities (5.4) and (5.5) hold, we have a *refinement* of the classical Hardy inequality (1.25):

$$\|g\|_{p,u} \leq C \|g'\|_{p,v}$$

with $(a, b) = (0, \infty)$, $u(x) = x^{-\lambda p}$, $v(x) = x^{(1-\lambda)p}$. Of course, here the assumptions on g are stronger.

Remark 5.5. This chapter is organized as follows: In Sec. 5.2 we present and prove some (unweighted) fractional order Hardy inequalities of type (5.1) for the case $a = 0$, $b = \infty$ (see Proposition 5.1). Our key lemma (Lemma 5.6) is of independent interest and explains partly the restriction $\lambda \neq 1/p$ in Proposition 5.1. Section 5.3 deals with some inequalities of the type (5.1) and (5.2) for the general (weighted) case, with $b = \infty$ as well as $b < \infty$. In Sec. 5.4, some relations between fractional order Hardy inequalities and interpolation theory are discussed. Further results and generalizations are dealt with in Sec. 5.5.

5.2. An Elementary Approach. The Unweighted Case

In this section we consider the standard case $a = 0$, $b = \infty$, and the "unweighted" modification, i.e., $w(x, y) \equiv 1$ in (5.3) (cf. Proposition 5.1). Nonetheless, it will be obvious that similar results can be derived also for $b < \infty$.

Our key lemma reads as follows:

Lemma 5.6. *Let $1 \leq p < \infty$, $\alpha \in \mathbb{R} \setminus \{0\}$. Assume that $\int_0^x g(t)dt$ exists for every $x > 0$ and that either*

$$\alpha > 0 \quad and \quad \lim_{x \to 0} \frac{1}{x} \int_0^x g(t)dt = 0 \tag{5.7}$$

or

$$\alpha < 0 \quad and \quad \lim_{x \to \infty} \frac{1}{x} \int_0^x g(t)dt = 0. \tag{5.8}$$

Then

$$\left(\int_0^\infty \left| \frac{g(x)}{x^\alpha} \right|^p \frac{dx}{x} \right)^{1/p} \leq C(\alpha) \left(\int_0^\infty \left| \frac{g(x) - \frac{1}{x} \int_0^x g(t)dt}{x^\alpha} \right|^p \frac{dx}{x} \right)^{1/p} \tag{5.9}$$

with $C(\alpha) = 1 + 1/|\alpha|$.

Remark 5.7. (i) In particular, Lemma 5.6 implies that inequality (5.9) holds for all $\alpha \neq 0$ and any locally integrable function g with compact support in $(0, \infty)$.

(ii) Inequality (5.9) *does not* hold for $\alpha = 0$ with any finite constant $C = C(0)$ even if we assume that the limit conditions in (5.7) or (5.8) are satisfied. As a counterexample, we can take for g the function $g_a(x) = \chi_{(1,a)}(x)$, $1 < a < \infty$, and let $a \to \infty$.

Proof of Lemma 5.6. Define

$$h(x) := g(x) - \frac{1}{x} \int_0^x g(t)dt. \tag{5.10}$$

For $0 < x_0 < x_1 < \infty$, integration by parts yields

$$\begin{aligned}
\int_{x_0}^{x_1} \frac{h(t)}{t} dt &= \int_{x_0}^{x_1} \frac{g(t)}{t} dt - \int_{x_0}^{x_1} \frac{1}{t^2} \int_0^t g(s)ds\,dt \\
&= \int_{x_0}^{x_1} \frac{g(t)}{t} dt + \left[\frac{1}{t} \int_0^t g(s)ds \right]_{x_0}^{x_1} - \int_{x_0}^{x_1} \frac{g(t)}{t} dt \\
&= \frac{1}{x_1} \int_0^{x_1} g(t)dt - \frac{1}{x_0} \int_0^{x_0} g(t)dt. \tag{5.11}
\end{aligned}$$

Putting $x_1 = x$ and letting $x_0 \to 0$, we obtain that in case (5.7)

$$\int_0^x \frac{h(t)}{t} dt = \frac{1}{x} \int_0^x g(t)dt,$$

and substituting this into (5.10), we obtain that

$$g(x) = h(x) + \int_0^x \frac{h(t)}{t} dt. \tag{5.12}$$

Analogously, putting $x_0 = x$ in (5.11) and letting $x_1 \to \infty$, we obtain that in the case (5.8)

$$\int_x^\infty \frac{h(t)}{t} dt = -\frac{1}{x} \int_0^x g(t)dt$$

and substituting this into (5.10), we obtain that

$$g(x) = h(x) - \int_x^\infty \frac{h(t)}{t} dt. \tag{5.13}$$

Now, the estimate (5.9) under assumption (5.7) or (5.8) follows from (5.12) and (5.13), respectively, by virtue of the Minkowski and Hardy inequalities. Namely, in case (5.7), we have from (5.12) that

$$\left(\int_0^\infty \left| \frac{g(x)}{x^\alpha} \right|^p \frac{dx}{x} \right)^{1/p}$$

$$\leq \left(\int_0^\infty \left| \frac{h(x)}{x^\alpha} \right|^p \frac{dx}{x} \right)^{1/p} + \left(\int_0^\infty \left| x^{-\alpha} \int_0^x \frac{h(t)}{t} dt \right|^p \frac{dx}{x} \right)^{1/p}$$

$$\leq \left(1 + \frac{1}{\alpha} \right) \left(\int_0^\infty \left| \frac{h(x)}{x^\alpha} \right|^p \frac{dx}{x} \right)^{1/p}$$

$$= \left(1 + \frac{1}{\alpha} \right) \left(\int_0^\infty \left| \frac{g(x) - \frac{1}{x} \int_0^x g(t) dt}{x^\alpha} \right|^p \frac{dx}{x} \right)^{1/p},$$

while in case (5.8), we have from (5.13) that

$$\left(\int_0^\infty \left| \frac{g(x)}{x^\alpha} \right|^p \frac{dx}{x} \right)^{1/p}$$

$$\leq \left(\int_0^\infty \left| \frac{h(x)}{x^\alpha} \right|^p \frac{dx}{x} \right)^{1/p} + \left(\int_0^\infty \left| x^{-\alpha} \int_x^\infty \frac{h(t)}{t} dt \right|^p \frac{dx}{x} \right)^{1/p}$$

$$\leq \left(1 + \frac{1}{|\alpha|} \right) \left(\int_0^\infty \left| \frac{h(x)}{x^\alpha} \right|^p \frac{dx}{x} \right)^{1/p}$$

$$= \left(1 + \frac{1}{|\alpha|} \right) \left(\int_0^\infty \left| \frac{g(x) - \frac{1}{x} \int_0^x g(t) dt}{x^\alpha} \right|^p \frac{dx}{x} \right)^{1/p}. \qquad \square$$

Remark 5.8. Inequality (5.9) does *not* hold for $0 < p < 1$ with any positive constant C. To see this we take $h(x) = h_\varepsilon(x) = \chi_{(1,1+\varepsilon)}(x)$ with $\varepsilon > 0$. After calculating the corresponding function $g(x) = g_\varepsilon(x)$ by (5.12) for the case $\alpha > 0$ and by (5.13) for the case $\alpha < 0$, i.e.

$$g_\varepsilon(x) = (1 + \ln x)\chi_{(1,1+\varepsilon)}(x) + \ln(1 + \varepsilon)\chi_{(1+\varepsilon,\infty)}(x)$$

and

$$g_\varepsilon(x) = -\ln(1 + \varepsilon)\chi_{(0,1)}(x) + (1 + \ln x - \ln(1 + \varepsilon))\chi_{(1,1+\varepsilon)}(x),$$

respectively, inserting these functions g_ε into (5.9) and letting $\varepsilon \to 0$, we find that the corresponding "constants" $C = C(\varepsilon)$ tend to infinity.

Now we are ready to prove the following sharpened version of Proposition 5.1:

Theorem 5.9. *Let $1 \leq p < \infty$, $0 < \lambda < 1$, $\lambda \neq 1/p$. Assume that $\int_0^x g(t)dt$ exists for every $x > 0$ and that either*

$$\frac{1}{p} < \lambda < 1 \quad \text{and} \quad \lim_{x \to 0} \frac{1}{x} \int_0^x g(t)dt = 0 \tag{5.14}$$

or

$$0 < \lambda < \frac{1}{p} \quad \text{and} \quad \lim_{x \to \infty} \frac{1}{x} \int_0^x g(t)dt = 0. \tag{5.15}$$

Then

$$\left(\int_0^\infty \left| \frac{g(x)}{x^\lambda} \right|^p dx \right)^{1/p} \leq C_{\lambda,p} \left(\int_0^\infty \int_0^\infty \frac{|g(x) - g(y)|^p}{|x - y|^{1+\lambda p}} dx dy \right)^{1/p} \tag{5.16}$$

with

$$C_{\lambda,p} = 2^{-1/p} \left(1 + \frac{p}{|\lambda p - 1|} \right). \tag{5.17}$$

Proof. Hölder's inequality yields

$$\left| g(x) - \frac{1}{x} \int_0^x g(t)dt \right|^p = \left| \frac{1}{x} \int_0^x (g(x) - g(t))dt \right|^p$$

$$\leq \frac{1}{x} \int_0^x |g(x) - g(t)|^p dt.$$

Therefore, putting $\alpha = \lambda - \frac{1}{p}$ (and hence, $\alpha p + 2 = 1 + \lambda p > 0$) and using Fubini's theorem and the symmetry of the integrand, we obtain that

$$\left(\int_0^\infty \left| \frac{g(x) - \frac{1}{x} \int_0^x g(t)dt}{x^\lambda} \right|^p dx \right)^{1/p}$$

$$= \left(\int_0^\infty \left| \frac{g(x) - \frac{1}{x} \int_0^x g(t)dt}{x^\alpha} \right|^p \frac{dx}{x} \right)^{1/p}$$

$$\leq \left(\int_0^\infty \int_0^x \frac{|g(x) - g(t)|^p}{x^{\alpha p + 2}} dt dx \right)^{1/p}$$

$$\leq \left(\int_0^\infty \int_0^x \frac{|g(x) - g(t)|^p}{|x - t|^{\alpha p + 2}} dt dx \right)^{1/p}$$

$$= \left(\frac{1}{2} \int_0^\infty \int_0^\infty \frac{|g(x) - g(t)|^p}{|x - t|^{1 + \lambda p}} dt dx \right)^{1/p}.$$

The estimate now follows from Lemma 5.6 since

$$\int_0^\infty \left| \frac{g(x)}{x^\lambda} \right|^p dx = \int_0^\infty \left| \frac{g(x)}{x^\alpha} \right|^p \frac{dx}{x}. \qquad \square$$

Remark 5.10. (i) The restrictions on the parameters in Theorem 5.9 are essential. Indeed: If either $\lambda \geq 1$ or $\lambda \leq 0$, then the integral on the right hand side in (5.16) diverges e.g. for each non-zero function g from $C_0^\infty(0, \infty)$. Moreover, if $\lambda = 1/p$, $1 < p < \infty$, then by inserting into (5.16) the function

$$g_\varepsilon(x) = \frac{x - \varepsilon}{\varepsilon} \chi_{(\varepsilon, 2\varepsilon)}(x) + \chi_{(2\varepsilon, 1/2)}(x) + 2(1 - x)\chi_{(1/2, 1)}(x), \quad \varepsilon > 0,$$

and letting $\varepsilon \to 0$ we find (after some tedious but straightforward calculations) that the corresponding constants $C_{\lambda, p}(\varepsilon)$ tend to infinity.

(ii) As already mentioned in Remark 5.8, the basic inequality (5.9) does not hold for $0 < p < 1$. Therefore it is a surprising fact that inequality (5.16) holds also for the case $0 < p < 1$ provided that g and λ satisfy the assumptions of Theorem 5.9. This can be seen in the following way: Choose parameters p_1 and λ_1, $p_1 \geq 1$, $0 < \lambda_1 < 1$, such that $\lambda_1 p_1 = \lambda p$, apply Theorem 5.9 with λ_1 and p_1 instead of λ and p, respectively, to the function $|g(x)|^{p/p_1}$ and use the elementary inequality

$$\big| |g(x)|^{p/p_1} - |g(y)|^{p/p_1} \big| \leq |g(x) - g(y)|^{p/p_1}$$

(cf. A. Kufner and H. Triebel [KT1]).

Inequality (5.4) (i.e., inequality (5.16)) appears in literature sometimes in a slightly modified form (see the references mentioned in Remark 5.2). To cover also these cases, let us close this section with the following equivalent form of the assertion of Theorem 5.9 for the case $1/p < \lambda < 1$.

Theorem 5.11. *Let* $1 \le p < \infty$, $1/p < \lambda < 1$. *Assume that* $\int_0^x g(t)dt$ *exists for every* $x > 0$ *and that*

$$\lim_{x \to 0} \frac{1}{x} \int_0^x g(t)dt = \mu, \quad \mu \in \mathbb{R}.$$

Then

$$\int_0^\infty \left| \frac{g(x) - \mu}{x^\lambda} \right|^p dx \le C_{\lambda,p} \left(\int_0^\infty \int_0^\infty \frac{|g(x) - g(y)|^p}{|x - y|^{1+\lambda p}} dx dy \right)^{1/p}$$

with $C_{\lambda,p}$ *from* (5.17).

Remark 5.12. If the right-hand side is finite, then g is equivalent to a continuous function $\tilde{g}$ on $[0, \infty)$ and $\mu = \tilde{g}(0)$.

5.3. The General Weighted Case

Now we consider the case when the two-variables weight $w(x, y)$ is not identically equal to 1. Also the interval (a, b) can be arbitrary — finite or infinite.

First, let us note that by adopting the methods used in Sec. 5.2, fractional order Hardy inequalities can be obtained also for some weighted cases. For instance, if we choose $w(x, y) = x^\gamma$, $\gamma \in \mathbb{R}$, we can easily see that by using the methods of the proof of Theorem 5.9, we obtain the following generalization.

Theorem 5.13. *Let* $1 \le p < \infty$, $\lambda \ge -1/p$ *and* $\gamma \ne \lambda p - 1$. *Assume that* $\int_0^x g(t)dt$ *exists for every* $x > 0$ *and that either*

$$\gamma < \lambda p - 1 \quad and \quad \lim_{x \to 0} \frac{1}{x} \int_0^x g(t)dt = 0 \tag{5.18}$$

or

$$\gamma > \lambda p - 1 \quad and \quad \lim_{x \to \infty} \frac{1}{x} \int_0^x g(t)dt = 0. \tag{5.19}$$

Then

$$\left(\int_0^\infty \left| \frac{g(x)}{x^\lambda} \right|^p x^\gamma dx \right)^{1/p}$$

$$\leq C_{\lambda,p} \left(\int_0^\infty \int_0^\infty \frac{|g(x) - g(y)|^p}{|x - y|^{1+\lambda p}} x^\gamma dx dy \right)^{1/p} \qquad (5.20)$$

with

$$C_{\lambda,p} = 2^{-1/p} \left(1 + \frac{p}{|\lambda p - \gamma - 1|} \right).$$

Remark 5.14. Due to a certain symmetry of the right hand side of (5.20), we can replace there x^γ by y^γ. Consequently, we obtain an inequality of the type (5.1) with the two-variables weight

$$v(x,y) = \alpha x^\gamma + \beta y^\delta, \quad \gamma, \delta \neq \lambda p - 1,$$

on the right-hand side, provided $\alpha > 0$, $\beta > 0$ and γ, δ are either both less than or both greater than $\lambda p - 1$.

In the foregoing results the weights have been *power functions*. Now we deal also with more general weights. The next theorem is given without proof since the argument is essentially the same as in Theorem 5.17 below.

Theorem 5.15. *Let* $1 < p < \infty$, $\beta \geq 0$. *Assume that*

$$\lim_{x \to \infty} \frac{1}{x} \int_0^x g(t) dt = 0. \qquad (5.21)$$

Moreover, let $u(x)$ *and* $w(x)$ *be weight functions on* $(0, \infty)$ *satisfying*

$$\widetilde{A} := \sup_{x>0} \left(\int_0^x u(t) dt \right)^{1/p} \left(\int_x^\infty w^{1-p'}(t) dt \right)^{1/p'} < \infty, \qquad (5.22)$$

and denote $B = \widetilde{A}^p p^p (p - 1)^{1-p}$. *If*

$$v(x) = x^{\beta-1} u(x) + x^{\beta-1-p} w(x),$$

then

$$\int_0^\infty |g(x)|^p u(x) dx \leq C \int_0^\infty \int_0^x \frac{|g(x) - g(y)|^p}{|x - y|^\beta} v(x) dy dx \qquad (5.23)$$

with

$$C = 2^{p-1} \max(1, B). \qquad (5.24)$$

Remark 5.16. (i) We can apply Theorem 5.15 for power weights (i.e., weights of the form x^α, $\alpha \in \mathbb{R}$) to obtain various modifications of Theorems 5.9 and 5.13. Note that the estimate (5.23) is better than the corresponding estimate (5.20) since in the inner integral on the right hand side, the integration is taken over $(0, x)$ instead of $(0, \infty)$.

(ii) Also notice that condition (5.22) is a (necessary and sufficient) condition for the validity of the conjugate Hardy inequality with weights u and w (see condition (0.24) with $a = 0$, $b = \infty$).

(iii) Due to assumption (5.21), Theorem 5.15 corresponds to the second part of Theorem 5.13 (under assumption (5.19)). The counterpart corresponding to (5.18) reads as follows.

Theorem 5.17. *Let $1 < p < \infty$, $\beta \geq 0$. Assume that*

$$\lim_{x \to 0} \frac{1}{x} \int_0^x g(t)dt = 0.$$

Moreover, let $u(x)$ and $w(x)$ be weight functions on $(0, \infty)$ satisfying

$$A := \sup_{x>0} \left(\int_x^\infty u(t)dt \right)^{1/p} \left(\int_0^x w^{1-p'}(t)dt \right)^{1/p'} < \infty, \qquad (5.25)$$

and denote $B = A^p \, p^p (p-1)^{1-p}$. If

$$v(x) = x^{\beta-1} u(x) + x^{\beta-1-p} w(x),$$

then inequality (5.23) holds with C from (5.24).

Proof. As in the proof of Lemma 5.6, define

$$h(x) := g(x) - \frac{1}{x} \int_0^x g(t)dt$$

and find that

$$g(x) = h(x) + \int_0^x \frac{h(t)}{t} dt$$

— see (5.10) and (5.12). Hence

$$\int_0^\infty |g(x)|^p u(x)dx$$

$$\leq 2^{p-1}\left(\int_0^\infty |h(x)|^p u(x)dx + \int_0^\infty \left|\int_0^x \frac{h(t)}{t}dt\right|^p u(x)dx\right)$$

$$\leq 2^{p-1}\left(\int_0^\infty |h(x)|^p u(x)dx + B\int_0^\infty \left|\frac{h(x)}{x}\right|^p w(x)dx\right),$$

where we have used the Hardy inequality

$$\int_0^\infty \left|\int_0^x f(t)dt\right|^p u(x)dx \leq B\int_0^\infty |f(x)|^p w(x)dx$$

for $f(t) = \frac{h(t)}{t}$, which is satisfied due to condition (5.25). Thus we have that

$$\int_0^\infty |g(x)|^p u(x)dx \leq 2^{p-1}\max(1,B)\int_0^\infty |h(x)|^p W(x)dx$$

$$= C\int_0^\infty \left|g(x) - \frac{1}{x}\int_0^x g(y)dy\right|^p W(x)dx$$

with $W(x) = u(x) + x^{-p}w(x) = x^{1-\beta}v(x)$. This implies inequality (5.23) with C from (5.24) since by Hölder's inequality,

$$\left|g(x) - \frac{1}{x}\int_0^x g(y)dy\right|^p \leq \left(\frac{1}{x}\int_0^x |g(x) - g(y)|dy\right)^p$$

$$\leq \frac{1}{x}\int_0^x |g(x) - g(y)|^p dy$$

$$\leq x^{\beta-1}\int_0^x \frac{|g(x) - g(y)|^p}{|x-y|^\beta}dy$$

due to the fact that for $\beta \geq 0$, $x^\beta/|x-y|^\beta \geq 1$ with $0 < y < x$. □

Remark 5.18. By applying Theorems 5.15 and 5.17 with $\beta = 1+\lambda p$, $u(x) = x^{\gamma-\lambda p}$ and $w(x) = x^{\gamma-\lambda p+p}$ we find that for $\lambda \geq -1/p$ with

$1 < p < \infty$,

$$\int_0^\infty \left| \frac{g(x)}{x^\lambda} \right|^p x^\gamma dx \leq C \int_0^\infty \int_0^\infty \frac{g(x) - g(y)|^p}{|x - y|^{1+\lambda p}} |x - y|^\gamma dx dy$$

provided that either

$$\gamma < \lambda p - 1 \quad \text{and} \quad \lim_{x \to 0} \frac{1}{x} \int_0^x g(t) dt = 0$$

or

$$\lambda p - 1 < \gamma \leq \lambda p + 1 \quad \text{and} \quad \lim_{x \to \infty} \frac{1}{x} \int_0^x g(t) dt = 0.$$

Hence, this result reflects the possibility to investigate the more general inequality

$$\int_0^\infty |g(x)|^p u(x) dx$$

$$\leq C \int_0^\infty \int_0^\infty |g(x) - g(y)|^p w(|x - y|) dx dy \qquad (5.26)$$

which is a special case of (5.1) for the two-variables weight function $v(x, y)$ of the type

$$v(x, y) = w(|x - y|).$$

Inequality (5.26) and its modification for the interval $(0, b)$ with $b \leq \infty$ was investigated by V.I. Burenkov and W.D. Evans [BE1]. Let us give one of their results.

Theorem 5.19. *Let $1 \leq p < \infty$, $0 < B \leq \infty$ and let w be a weight function on $(0, B)$ such that*

$$u(x) := \int_x^B w(t) dt < \infty \qquad (5.27)$$

for every $x \in (0, B)$. Suppose that there exists a constant $c \in (1, 2)$ such that

$$u(x) \leq cu(2x), \quad x \in (0, B/2). \qquad (5.28)$$

Then, for every $b \in (0, B)$ and all $g \in L^p(0, b)$,

$$\left(\int_0^b |g(x)|^p u(x) dx \right)^{1/p} \leq C \left\{ \left(u(b) \int_0^b |g(x)|^p \right)^{1/p} \right.$$

$$\left. + \left(\int_0^b \int_0^b |g(x) - g(y)|^p w(|x - y|) dx dy \right)^{1/p} \right\} \quad (5.29)$$

with C independent of g and b.

In particular, for all $g \in L^p(0, B)$,

$$\left(\int_0^B |g(x)|^p u(x) dx \right)^{1/p}$$

$$\leq C \left(\int_0^B \int_0^B |g(x) - g(y)|^p w(|x - y|) dx dy \right)^{1/p} . \quad (5.30)$$

Proof. Let U be the inverse of u from (5.27) so that $U(u(x)) = u(U(x)) = x$. Since w is a weight function, both u and U are continuous and strictly decreasing.

Define the function δ by

$$\delta(x) := U(cu(x)), \quad x \in (0, b)$$

with c the constant from (5.28). Then

$$\delta(x) \leq \frac{1}{2} x \quad (5.31)$$

since, due to (5.28), $\delta(x) = U(cu(x)) \leq U(u(x/2)) = x/2$ and

$$u(\delta(x)) = u(U(cu(x))) = cu(x).$$

Consequently,

$$\int_{\delta(x)}^x w(t) dt = \int_{\delta(x)}^\infty w(t) dt - \int_x^\infty w(t) dt$$

$$= u(\delta(x)) - u(x) = (c - 1)u(x). \quad (5.32)$$

Let $\varepsilon \in (0, b/2)$. Since $|g(x)| \le |g(y)| + |g(x) - g(y)|$, the Minkowski inequality yields

$$J := \left(\int_\varepsilon^{b/2} \int_{2x}^b |g(x)|^p w(y-x) dy dx \right)^{1/p}$$

$$\le \left(\int_\varepsilon^{b/2} \int_{2x}^b |g(y)|^p w(y-x) dy dx \right)^{1/p}$$

$$+ \left(\int_\varepsilon^{b/2} \int_{2x}^b |g(x) - g(y)|^p w(y-x) dy dx \right)^{1/p}$$

$$=: I_1 + I_2$$

(notice that $x \in (\varepsilon, b/2)$ implies $2x < b$, and $y \in (2x, b)$ implies $y > x$, i.e. $|x - y| = y - x$). Substitution $y - x = t$ and (5.27) yield

$$J = \left(\int_\varepsilon^{b/2} |g(x)|^p \int_x^{b-x} w(t) dt dx \right)^{1/p}$$

$$= \left(\int_\varepsilon^{b/2} |g(x)|^p (u(x) - u(b-x)) dx \right)^{1/p}$$

while Fubini's theorem and substitution $y - x = t$ yield, together with (5.31) and (5.32), that

$$I_1 = \left(\int_{2\varepsilon}^b |g(y)|^p \int_\varepsilon^{y/2} w(y-x) dx dy \right)^{1/p}$$

$$= \left(\int_{2\varepsilon}^b |g(y)|^p \int_{y/2}^{y-\varepsilon} w(t) dt dy \right)^{1/p}$$

$$\le \left(\int_\varepsilon^b |g(y)|^p \int_{\delta(y)}^y w(t) dt dy \right)^{1/p}$$

$$= \left(\int_\varepsilon^b |g(y)|^p (c-1) u(y) dy \right)^{1/p}.$$

Due to the symmetry of the integrand, we have

$$I_2 \le \left(\int_0^b \int_0^y |g(x) - g(y)|^p w(|x - y|) dx dy \right)^{1/p}$$

$$\le \left(\frac{1}{2} \int_0^b \int_0^b |g(x) - g(y)|^p w(|x - y| dx dy \right)^{1/p} =: C_0 \quad (5.33)$$

and hence finally, since $u(x) \le u(b/2) \le cu(b)$ for $x \in (b/2, b)$,

$$\left(\int_\varepsilon^{b/2} |g(x)|^p (u(x) - u(b - x)) dx \right)^{1/p}$$

$$\le (c - 1)^{1/p} \left[\left(\int_\varepsilon^{b/2} |g(x)|^p u(x) dx \right)^{1/p} \right.$$

$$\left. + \left(\int_{b/2}^b |g(x)|^p dx \right)^{1/p} (cu(b))^{1/p} \right] + C_0.$$

For $x \in (0, b/2)$, we have $b - x > b/2$ and hence $u(b - x) \le u(b/2) \le cu(b)$ due to (5.28). Thus

$$\left(\int_\varepsilon^{b/2} |g(x)|^p u(x) dx \right)^{1/p}$$

$$\le \left(\int_\varepsilon^{b/2} |g(x)|^p (u(x) - u(b - x)) dx \right)^{1/p}$$

$$+ \left(\int_\varepsilon^{b/2} |g(x)|^p u(b - x) dx \right)^{1/p}$$

$$\le (c - 1)^{1/p} \left(\int_\varepsilon^{b/2} |g(x)|^p u(x) dx \right)^{1/p}$$

$$+ (c(c-1)u(b))^{1/p} \left(\int_{b/2}^{b} |g(x)|^p dx \right)^{1/p} + C_0$$

$$+ (cu(b))^{1/p} \left(\int_{0}^{b/2} |g(x)|^p dx \right)^{1/p}$$

and consequently

$$(1 - (c-1)^{1/p}) \left(\int_{\varepsilon}^{b/2} |g(x)|^p u(x) dx \right)^{1/p}$$

$$\leq (cu(b))^{1/p} \left(\int_{0}^{b} |g(x)|^p dx \right)^{1/p}$$

$$+ \left(\frac{1}{2} \right)^{1/p} \left(\int_{0}^{b} \int_{0}^{b} |g(x) - g(y)|^p w(|x - y|) dx dy \right)^{1/p}. \quad (5.34)$$

Moreover,

$$\left(\int_{\varepsilon}^{b} |g(x)|^p u(x) dx \right)^{1/p} \leq \left(\int_{\varepsilon}^{b/2} |g(x)|^p u(x) dx \right)^{1/p}$$

$$+ (cu(b))^{1/p} \left(\int_{b/2}^{b} |g(x)|^p dx \right)^{1/p} \quad (5.35)$$

and since $1 - (c-1)^{1/p} > 0$, it follows from (5.34) and (5.35) that

$$\left(\int_{\varepsilon}^{b} |g(x)|^p u(x) dx \right)^{1/p} \leq C \left\{ \left(u(b) \int_{0}^{b} |g(x)|^p dx \right)^{1/p} \right.$$

$$+ \left. \left(\int_{0}^{b} \int_{0}^{b} |g(x) - g(y)|^p w(|x - y|) dx dy \right)^{1/p} \right\}$$

with C independent of g, b and ε.

Now, (5.29) follows by letting $\varepsilon \to 0+$, and inequality (5.30) follows from (5.29) by letting b tend to B. □

Now we shall present a fractional order Hardy inequality with a general two-variables weight $v(x, y)$. We shall work in the interval $(0, b)$ with b finite or infinite.

Theorem 5.20. *Let $1 < p < \infty$, $\lambda \geq -1/p$. Let $v(x, y)$ be a weight function on $(0, b) \times (0, b)$ with $0 < b \leq \infty$ and suppose that $\int_0^x v^{1-p'}(x, t)dt < \infty$ for every $x < b$ and $\int_0^y v^{1-p'}(t, y)dt < \infty$ for every $y < b$.*
(i) Denote

$$w(x) = \left(\frac{1}{x} \int_0^x v^{1-p'}(x, y)dy \right)^{1-p}$$

and

$$u(x) = w(x)x^{-\lambda p}.$$

If

$$A_p := \sup_{0 < x < b} \left(\int_x^b \frac{w(t)}{t^{p(\lambda+1)}}dt \right)^{1/p} \left(\int_0^x w^{1-p'}(t)t^{\lambda p'}dt \right)^{1/p'} < \infty \tag{5.36}$$

and

$$K := \frac{A_p p}{(p-1)^{1/p'}} < 1,$$

then for $g \in L^p(u)$ we have

$$\left(\int_0^b |g(x)|^p u(x)dx \right)^{1/p}$$

$$\leq \frac{1}{1-K} \left(\int_0^b \int_0^b \frac{|g(x) - g(y)|^p}{|x - y|^{1+\lambda p}} v(x, y)dxdy \right)^{1/p}. \tag{5.37}$$

(ii) *Denote*

$$\widetilde{w}(y) = \left(\frac{1}{y} \int_0^y v^{1-p'}(x,y)dx\right)^{1-p}$$

and

$$\widetilde{u}(y) = \widetilde{w}(y)y^{-\lambda p}.$$

If

$$\widetilde{A}_p := \sup_{0<y<b} \left(\int_y^b \frac{\widetilde{w}(t)}{t^{p(\lambda+1)}}dt\right)^{1/p} \left(\int_0^y \widetilde{w}^{1-p'}(t)t^{\lambda p'}dt\right)^{1/p'} < \infty$$

$$(5.38)$$

and

$$\widetilde{K} := \frac{\widetilde{A}_p p}{(p-1)^{1/p'}} < 1,$$

then inequality (5.37) *holds for* $g \in L^p(u)$ *with* u *and* K *replaced by* $\widetilde{u}$ *and* $\widetilde{K}$, *respectively.*

Proof. First note that

$$\int_0^b \int_0^b \frac{|g(x)-g(y)|^p}{|x-y|^{1+\lambda p}}v(x,y)dxdy$$

$$= \int_0^b \int_0^x \frac{|g(x)-g(y)|^p}{|x-y|^{1+\lambda p}}v(x,y)dydx$$

$$+ \int_0^b \int_x^b \frac{|g(x)-g(y)|^p}{|x-y|^{1+\lambda p}}v(x,y)dydx$$

$$\geq \int_0^b \int_0^x \frac{|g(x)-g(y)|^p}{x^{1+\lambda p}}v(x,y)dydx$$

$$+ \int_0^b \int_x^b \frac{|g(x)-g(y)|^p}{y^{1+\lambda p}}v(x,y)dydx$$

$$=: I_1 + I_2.$$

Hölder's inequality yields

$$\left| \int_0^x (g(x) - g(y))dy \right|^p$$

$$\leq \int_0^x |g(x) - g(y)|^p v(x,y)dy \left(\int_0^x v^{1-p'}(x,y)dy \right)^{p-1}$$

$$= \frac{x^{p-1}}{w(x)} \int_0^x |g(x) - g(y)|^p v(x,y)dy$$

and consequently

$$I_1 \geq \int_0^b \frac{w(x)x^{1-p}}{x^{1+\lambda p}} \left| \int_0^x (g(x) - g(y))dy \right|^p dx$$

$$= \int_0^b \frac{w(x)}{x^{\lambda p}} \left| g(x) - \frac{1}{x}\int_0^x g(y)dy \right|^p dx.$$

Using now the Minkowski and Hardy inequalities (the latter due to the fact that $A_p < \infty$) we obtain that

$$\left(\int_0^b |g(x)|^p u(x)dx \right)^{1/p}$$

$$= \left(\int_0^b |g(x)|^p w(x)x^{-\lambda p}dx \right)^{1/p}$$

$$\leq \left(\int_0^b \frac{w(x)}{x^{\lambda p}} \left| g(x) - \frac{1}{x}\int_0^x g(y)dy \right|^p dx \right)^{1/p}$$

$$+ \left(\int_0^b \frac{w(x)}{x^{\lambda p+p}} \left| \int_0^x g(y)dy \right|^p dx \right)^{1/p}$$

$$\leq I_1^{1/p} + K \left(\int_0^b |g(x)|^p u(x)dx \right)^{1/p},$$

and since $K < 1$, inequality (5.37) follows.

Part (ii) can be proved completely analogously, starting from the integral I_2 instead of I_1. □

Remark 5.21. (i) The proof of Theorem 5.20 shows also that the inequality

$$(1 - K) \left(\int_0^b |g(x)|^p u(x) dx \right)^{1/p} + (1 - \widetilde{K}) \left(\int_0^b |g(x)|^p \widetilde{u}(x) dx \right)^{1/p}$$

$$\leq \left(\int_0^b \int_0^b \frac{|g(x) - g(y)|^p}{|x - y|^{1 + \lambda p}} v(x, y) dx dy \right)^{1/p}$$

holds whenever conditions (5.36) and (5.38) are satisfied simultaneously (and $K, \widetilde{K} < 1$).

(ii) In the step of the proof of Theorem 5.20 where the Hardy inequality was applied, we used the fact that the constant in this inequality can be estimated by $A_p k(p, p)$ according to formula (0.26). In our case, $k(p, p) = p^{1/p} p'^{1/p'} = p(p - 1)^{-1/p'}$, which leads to the formula for the constant K.

Example 5.22. If we take $v(x, y) = x^\gamma$ and $b = \infty$ in Theorem 5.20 and replace $g(x)$ by $g(x) - g(0)$, we obtain the inequality

$$\left(\int_0^\infty |g(x) - g(0)|^p x^{\gamma - \lambda p} dx \right)^{1/p}$$

$$\leq \frac{\lambda p + p - \gamma - 1}{\lambda p - \gamma - 1} \left(\int_0^\infty \int_0^\infty \frac{|g(x) - g(y)|^p}{|x - y|^{1 + \lambda p}} x^\gamma dx dy \right)^{1/p}$$

whenever $1 < p < \infty$, $\lambda \geq -1/p$ and $\gamma < \lambda p - 1$ (cf. Theorem 5.13).

The case $p \neq q$

Up to now, we have considered inequality (5.1) for the case $p = q$. The case $p \neq q$ is more complicated; in order to illustrate this fact we present a result for the case $1 < p \leq q < \infty$ where mixed norms appear in a natural way. For two weight functions $w_1(x)$, $w_2(x)$ on $(0, b)$, we introduce the notation

$$W_2(y) = \int_0^y w_2(x) dx$$

and

$$W_{p,q}(y) = \left(\frac{1}{y} \int_0^y \left(\frac{w_2^p(x)}{w_1(x)} \right)^{p'-1} dx \right)^{-q/p'} \left(\frac{1}{y} \int_0^y w_2(x)dx \right)^q w_2(y).$$

Theorem 5.23. *Let* $0 < b \leq \infty$, $1 < p \leq q < \infty$, $\lambda \geq -1/p$. *Denote*

$$u(x) = W_{p,q}(x)x^{-\lambda q}$$

and assume that

$$A_{p,q} := \sup_{0<y\leq b} \left(\int_y^b u(x)W_2^{-q}(x)dx \right)^{1/q}$$

$$\times \left(\int_0^y (u(x)w_2^{-q}(x))^{1-q'}dx \right)^{1/q'} < \infty \qquad (5.39)$$

and

$$K := \frac{A_{p,q}q}{(q-1)^{1/q'}} < 1. \qquad (5.40)$$

Then for $g \in L^q(u)$ *we have*

$$\left(\int_0^b |g(x)|^q u(x)dx \right)^{1/q}$$

$$\leq \frac{1}{1-K} \left(\int_0^b \left(\int_0^b \frac{|g(x)-g(y)|^p}{|x-y|^{1+\lambda p}} w_1(x)dx \right)^{q/p} w_2(y)dy \right)^{1/p}.$$

Proof. Hölder's inequality yields

$$\left(\int_0^y |g(x)-g(y)|w_2(x)dx \right)^q$$

$$\leq \left(\int_0^y |g(x)-g(y)|^p w_1(x)dx \right)^{q/p} \left(\int_0^y \left(\frac{w_2^p(x)}{w_1(x)} \right)^{p'-1} dx \right)^{q/p'}.$$

Consequently,

$$\int_0^b \left(\int_0^b \frac{|g(x) - g(y)|^p}{|x - y|^{1+\lambda p}} w_1(x)dx \right)^{q/p} w_2(y)dy$$

$$\geq \int_0^b \left(\int_0^y \frac{|g(x) - g(y)|^p}{y^{1+\lambda p}} w_1(x)dx \right)^{q/p} w_2(y)dy$$

$$\geq \int_0^b y^{-q/p-\lambda q} \left(\int_0^y \left(\frac{w_2^p(x)}{w_1(x)} \right)^{p'-1} dx \right)^{-q/p'}$$

$$\times \left| \int_0^y |g(x) - g(y)| w_2(x)dx \right|^q w_2(y)dy$$

$$\geq \int_0^b y^{-q/p-\lambda q} \left(\int_0^y \left(\frac{w_2^p(x)}{w_1(x)} \right)^{p'-1} dx \right)^{-q/p'}$$

$$\times \left| g(y) \int_0^y w_2(x)dx - \int_0^y g(x)w_2(x)dx \right|^q w_2(y)dy$$

$$= \int_0^b u(y) \left| g(y) - \frac{1}{W_2(y)} \int_0^y g(x)w_2(x)dx \right|^q dy.$$

Using this estimate together with the Minkowski and Hardy inequalities (the latter can be used due to (5.39)) we obtain that

$$\left(\int_0^b |g(y)|^q u(y)dy \right)^{1/q}$$

$$\leq \left(\int_0^b u(y) \left| g(y) - \frac{1}{W_2(y)} \int_0^y g(x)w_2(x)dx \right|^q dy \right)^{1/q}$$

$$+ \left(\int_0^b u(y)(W_2(y))^{-q} \left| \int_0^y g(x)w_2(x)dx \right|^q dy \right)^{1/q}$$

$$\leq \left(\int_0^b \left(\int_0^b \frac{|g(x) - g(y)|^p}{|x - y|^{1+\lambda p}} w_1(x)dx \right)^{q/p} w_2(y)dy \right)^{1/q}$$

$$+ K \left(\int_0^b |g(y)|^q u(y)dy \right)^{1/q}.$$

The result now follows by subtracting and using (5.40). $\qquad \square$

Notation. Now we will deal with inequalities of type (5.2). First, let us introduce some notation.

For two weight functions $w(x, y)$ and $w_1(x, y)$ defined on $(0, b) \times (0, b)$ with $0 < b \leq \infty$ and for parameters $p > 1$, $q > 0$, define a function $B_{q,p}(x)$ on $(0, b)$ by

$$B_{q,p}(x) := \sup_{x < y < b} \left(\int_y^b |x - t|^{-1 - \lambda q} w(x, t) dt \right)^{1/q}$$

$$\times \left(\int_x^y w_1^{1 - p'}(x, t) dt \right)^{1/p'}$$

for $1 < p \leq q < \infty$, and by

$$B_{q,p}(x) = \left(\int_x^b \left(\int_y^b |x - t|^{-1 - \lambda q} w(x, t) dt \right)^{r/q} \right.$$

$$\left. \times \left(\int_x^y w_1^{1 - p'}(x, t) dt \right)^{r/q'} w_1^{1 - p'}(x, y) dy \right)^{1/r}$$

for $0 < q < p < \infty$, $p > 1$, $\frac{1}{r} = \frac{1}{q} - \frac{1}{p}$.

Here we assume that $0 < \lambda < 1$.

Remark 5.24. Notice that for x fixed, $0 < x < b$, the condition $B_{q,p}(x) < \infty$ is necessary and sufficient for the Hardy inequality to hold on the interval (x, b) with weight functions $U_x(t) = w(x, t)|x - t|^{-1 - \lambda q}$ and $V_x(t) = w_1(x, t)$. This inequality will be used substantially in the proof of the following assertion.

Theorem 5.25. *Let $1 < p < \infty$, $0 < q < \infty$. Let $w(x, y)$ and $w_1(x, y)$ be weight functions on $(0, b) \times (0, b)$ and assume that w is symmetric: $w(x, y) = w(y, x)$. Assume that the function $B_{q,p}(x)$ defined above is finite a.e. in $(0, b)$, denote*

$$V(y) := \begin{cases} \left(\int_0^y B_{q,p}^q(x) w_1^{q/p}(x, y) dx \right)^{p/q} & \text{for } p \leq q, \\ \int_0^y w_1(x, y) dx & \text{for } p > q, \end{cases}$$

suppose that $V(y)$ is finite a.e. in $(0, b)$ and that for $p > q$ also

$$c_0 := \left(\int_0^b B_{q,p}^{qp/(p-q)}(x) dx \right)^{(p-q)/p} < \infty.$$

Then

$$\left(\int_0^b \int_0^b \frac{|g(x) - g(y)|^q}{|x - y|^{1+\lambda q}} w(x, y) dx dy \right)^{1/q}$$

$$\leq C \left(\int_0^b |g'(x)|^p V(x) dx \right)^{1/p}, \tag{5.41}$$

where

$$C = \begin{cases} 2^{1/q} k(q, p) & \text{for } p \leq q, \\ 2^{1/q} k(q, p) c_0^{1/q} & \text{for } p > q \end{cases}$$

with

$$k(q, p) = \begin{cases} \left(1 + \dfrac{p}{q'} \right)^{1/p} \left(1 + \dfrac{q'}{p} \right)^{1/q'} & \text{for } p \leq q, \\ p^{1/p} (q')^{1/p'} & \text{for } p > q. \end{cases}$$

Proof. Due to the symmetry of $w(x, y)$ we find — as in the proof of Theorem 5.3, see (5.6) — that

$$\int_0^b \int_0^b \frac{|g(x) - g(y)|^q}{|x - y|^{1+\lambda q}} w(x, y) dx dy$$

$$= 2 \int_0^b \left(\int_x^b \frac{w(x, y)}{|x - y|^{1+\lambda q}} \left| \int_x^y g'(t) dt \right|^q dy \right) dx.$$

Thus, using for the function g' the Hardy inequality on the interval (x, b) as mentioned in Remark 5.24 we find after integrating with respect to y over the interval $(0, b)$ that

$$\int_0^b \int_0^b \frac{|g(x) - g(y)|^q}{|x - y|^{1+\lambda q}} w(x, y) dx dy$$

$$\leq 2 k^q(q, p) \int_0^b B_{q,p}^p(x) \left(\int_x^b |g'(y)|^p w_1(x, y) dy \right)^{q/p} dx. \tag{5.42}$$

First, suppose that $1 < p \leq q$. Estimating the right-hand side in (5.42) by Minkowski's integral inequality if $p < q$ (and thus, $q/p > 1$) and by Fubini's theorem if $p = q$, we obtain

$$\int_0^b \int_0^b \frac{|g(x) - g(y)|^q}{|x - y|^{1+\lambda q}} w(x, y) dx dy$$

$$\leq 2k^q(q, p) \left(\int_0^b \left(\int_0^y B_{q,p}^q(x) w_1^{q/p}(x, y) |g'(y)|^q dx \right)^{p/q} dy \right)^{q/p}$$

$$= 2k^q(q, p) \left(\int_0^b |g'(y)|^p \left(\int_0^y B_{q,p}^q(x) w_1^{q/p}(x, y) dx \right)^{p/q} dy \right)^{q/p}$$

$$= 2k^q(p, q) \left(\int_0^b |g'(y)|^p V(y) dy \right)^{q/p},$$

i.e., we have inequality (5.41).

Now, suppose that $p > q$. Then we estimate the right-hand side in (5.42) by Hölder's inequality with parameters $s = p/q > 1$ and $s' = p/(p - q)$, and we find that, by Fubini's theorem,

$$\int_0^b \int_0^b \frac{|g(x) - g(y)|^q}{|x - y|^{1+\lambda p}} w(x, y) dx dy$$

$$\leq 2k^q(q, p) \left(\int_0^b B_{q,p}^{qp/(p-q)}(x) dx \right)^{(p-q)/p}$$

$$\times \left(\int_0^b \left(\int_x^b |g'(y)|^p w_1(x, y) dy \right) dx \right)^{q/p}$$

$$= 2k^q(q, p) c_0 \left(\int_0^b \left(\int_0^y w_1(x, y) dx \right) |g'(y)|^p dy \right)^{q/p}$$

$$= 2k^q(q, p) c_0 \left(\int_0^b |g'(y)|^p V(y) dy \right)^{q/p}.$$

This is again (5.41), and the proof is complete. □

Remark 5.26. (i) Notice that due to the symmetry of $w(x, y)$, we can also write

$$\int_0^b \int_0^b \frac{|g(x) - g(y)|^q}{|x - y|^{1+\lambda q}} w(x, y) dx dy$$

$$= 2 \int_0^b \left(\int_0^x \frac{w(x, y)}{|x - y|^{1+\lambda p}} \left| \int_y^x g'(t) dt \right|^q dy \right) dx$$

(see the first lines of the proof of Theorem 5.25). Consequently, we can modify the proof using now the Hardy inequality on the interval $(0, x)$ (with $x < b$ fixed) and modifying appropriately the definitions of $B_{q,p}(x)$ and $V(y)$. (Now we are dealing with the conjugate Hardy operator for $g' : \int_y^x g'(t) dt$.) The formulation of the corresponding theorem is left to the reader.

(ii) It is also possible to omit the assumption of the symmetry of w, but in that case, we have to replace $w(x, y)$ by $\widetilde{w}(x, y) := w(x, y) + w(y, x)$ in (5.41) and in the definition of the functions $B_{q,p}(x)$ and $V(y)$. Indeed: as in the proof of Theorem 5.3, see (5.6), we have that

$$\int_0^b \int_0^b \frac{|g(x) - g(y)|^q}{|x - y|^{1+\lambda q}} w(x, y) dx dy$$

$$= \int_0^b \left(\int_x^b \frac{w(x, y)}{|x - y|^{1+\lambda q}} \left| \int_x^y g'(t) dt \right|^q dy \right) dx$$

$$+ \int_0^b \left(\int_0^x \frac{w(x, y)}{|x - y|^{1+\lambda q}} \left| \int_y^x g'(t) dt \right|^q dy \right) dx$$

$$= \int_0^b \left(\int_x^b \frac{w(x, y) + w(y, x)}{|x - y|^{1+\lambda q}} \left| \int_x^y g'(t) dt \right|^q dy \right) dx$$

$$= \int_0^b \left(\int_x^b \frac{\widetilde{w}(x, y)}{|x - y|^{1+\lambda q}} \left| \int_x^y g'(t) dt \right|^q dy \right) dx$$

where $\widetilde{w}$ is symmetric. Now, we proceed as in the proof of Theorem 5.25.

Example 5.27. (i) If we apply Theorem 5.25 with $b = \infty$, $q = p$, $w(x, y) = 1$ and $w_1(x, y) = |x - y|^{-1-\lambda p + p}$, we obtain inequality (5.5) but with a different constant. Notice that in this case, $B_{p,q}(x)$ is a constant!

(ii) If we apply Theorem 5.25 with $q = p$, $w(x,y) = |x - y|^{1+\lambda p} w(|x - y|)$ and $w_1(x,y) = w_1(|x - y|)$ we obtain that

$$\int_0^b \int_0^b |g(x) - g(y)|^p w(|x - y|) dx dy \leq C^p \int_0^b |g'(x)|^p v(x) dx, \quad (5.43)$$

where

$$v(x) = \int_0^x B_{p,p}^p(t) w_1(|x - t|) dt$$

with

$$B_{p,p}^p(t) = \sup_{t < y < t+b} \int_{y-t}^{b-t} w(s) ds \left(\int_0^{y-t} w_1^{1-p'}(s) ds \right)^{p-1}.$$

Obviously (5.43) may be regarded as a counterpart to inequality (5.30) — see Theorem 5.19.

We close this section by stating an elementary result. The proof is left to the reader.

Theorem 5.28. *Let $1 < p, q < \infty$, $0 < b \leq \infty$, and let $v(x)$ and $w(x,y)$ be weight functions on $(0, b)$ and on $(0, b) \times (0, b)$, respectively. Moreover, denote*

$$V(x) = \int_0^x v^{1-p'}(t) dt$$

and suppose that

$$A := \int_0^b \int_0^b \frac{|V(x) - V(y)|^{q/p'}}{|x - y|^{1+\lambda q}} w(x,y) dx dy < \infty.$$

Then

$$\left(\int_0^b \int_0^b \frac{|g(x) - g(y)|^q}{|x - y|^{1+\lambda q}} w(x,y) dx dy \right)^{1/q} \leq C \left(\int_0^b |g'(x)|^p v(x) dx \right)^{1/p}$$

with $C = A^{1/q}$.

5.4. Hardy-type Inequalities and Interpolation Theory

Some basic facts

First we recall some concepts and results from the *real method of interpolation* (cf. e.g. J. Bergh and J. Löfström [BL1]).

Let (A_0, A_1) denote a compatible couple of Banach spaces (i.e., both are continuously imbedded into a Hausdorff topological vector space). A Banach space A is called an *intermediate space* between A_0 and A_1 if

$$A_0 \cap A_1 \subset A \subset A_0 + A_1.$$

Now, let (A_0, A_1) and (B_0, B_1) be two compatible couples. Then spaces A and B are said to be *interpolation spaces* with respect to (A_0, A_1) and (B_0, B_1) if A and B are intermediate spaces to the respective couples and if, for any bounded linear operator T such that $T : A_0 \to B_0$ and $T : A_1 \to B_1$, we have $T : A \to B$. In terms of inequalities this can be expressed as follows:

If $\quad \|Tf\|_{B_0} \le M_0 \|f\|_{A_0} \quad$ and $\quad \|Tf\|_{B_1} \le M_1 \|f\|_{A_1}$,

then $\quad \|Tf\|_B \le G(M_0, M_1) \|f\|_A$, $\qquad\qquad$ (5.44)

where $G(s, t)$, $s, t > 0$, is a positive bounded function.

Nowadays there exist many methods of constructing interpolation spaces but the most frequent methods are the *real method* $(\cdot, \cdot)_{\lambda, q}$, $0 < \lambda < 1$, $1 \le q \le \infty$, and the *complex method* $[\cdot, \cdot]_\lambda$, $0 \le \lambda \le 1$. In both cases we have

$$G(s, t) = s^{1-\lambda} t^\lambda.$$

For our purposes, it is sufficient to consider the real method defined in the following way: For a compatible Banach couple (A_0, A_1), the Peetre K-functional is defined for any $g \in A_0 + A_1$ and $t \ge 0$ as

$$K(t, g) = K(t, g; A_0, A_1)$$
$$= \inf\{\|g_0\|_{A_0} + t\|g_1\|_{A_1} : g = g_0 + g_1, g_0 \in A_0, g_1 \in A_1\}.$$

The interpolation space $(A_0, A_1)_{\theta,q}$, $0 < \theta < 1$, $1 \le q < \infty$, is normed by

$$\|g\|_{(A_0,A_1)_{\theta,q}} = \left(\int_0^\infty t^{-\theta q} K^q(t, g) \frac{dt}{t} \right)^{1/q};$$

for $q = \infty$, we apply the usual supremum norm.

Interpolation and the Hardy inequality

In Chap. 1 it was mentioned that the Hardy inequality (1.25) can be treated as a continuous imbedding

$$X \hookrightarrow L^q(u),$$

where X is either $W_L^{1,p}(v)$ or $W_R^{1,p}(v)$ (see (1.29) or (1.30), respectively), i.e., we have for every $g \in X$

$$\|g\|_{q,u} \le M_1 \|g\|_X,$$

where $\|g\|_X = \|g'\|_{p,v}$. Therefore, if we use the space X as A_1 and the space $L^q(u)$ as B_1 and if we can find Banach spaces A_0 and B_0 which are compatible with X and $L^q(u)$, respectively, and if $A_0 \hookrightarrow B_0$, i.e., for every $g \in A_0$,

$$\|g\|_{B_0} \le M_0 \|g\|_{A_0},$$

then we have the following *general fractional order Hardy inequality*:

$$\|g\|_{(B_0, L^q(u))_{\lambda,r}} \le M_0^{1-\lambda} M_1^\lambda \|g\|_{(A_0, X)_{\lambda,r}}.$$

Notice that in this case, the operator T from (5.44) is the imbedding operator, i.e., the identity.

The technique just described was first used by A. Kufner and H. Triebel [KT1]; some inequalities of the type

$$\int_0^\infty |g(x)|^p u(x)dx \le C \int_0^\infty \int_0^\infty \frac{|g(x) - g(y)|^p}{|x - y|^{1+\lambda p}} v(x)dxdy$$
$$+ \int_0^\infty |g(x)|^p w(x)dx$$

have been derived, with u, v and w weight functions on $(0, \infty)$.

In A. Kufner and L.E. Persson [KuP1] (see also [KuP2]), this technique was further investigated and applied. The key problem is to determine spaces A_0 and B_0 so that the norms of the interpolation spaces $(B_0, L^q(u))_{\lambda,r}$ and $(A_0, W_L^{1,p}(v))_{\lambda,r}$ or $(A_0, W_R^{1,p}(v))_{\lambda,r}$ can be identified with the norms of $g^{(\lambda)}$ required in (5.1) and (5.2). The results obtained there are not completely satisfactory and here we only present two examples where we consider $B_0 = W_L^{1,p}(u)$ with $u(x) \equiv 1$ and $A_0 = W_L^{1,p}(v)$ in the first case, while in the other case, we choose $B_0 = L^q(u)$ and $A_0 = L^p(v)$ with $v(x) \equiv 1$.

Example 5.29. Let $1 < p \leq q < \infty$, $0 < \lambda < 1$, $\frac{1}{r} = \frac{1-\lambda}{p} + \frac{\lambda}{q}$. Let g be a differentiable function on $(0, b)$, $0 < b < \infty$, such that $g(0) = 0$.

(i) If

$$\sup_{0<x<b} (b-x)^{1/q} \left(\int_0^x v^{p'/r(\lambda-1)}(t)dt \right)^{1/p'} < \infty,$$

then, for any δ, $0 < \delta < b$,

$$\int_0^\delta t^{-\lambda r - 1} \left(\int_0^b |g(x+t) - g(x)|^q dx \right)^{r/q} dt$$

$$\leq C \int_0^b |g'(x)|^r v(x)dx.$$

In particular, if $p = q$, then $r = p$ and we have that

$$\int_0^\delta \int_0^b \frac{|g(x+t) - g(x)|^p}{t^{1+\lambda p}} dx dt \leq C \int_0^b |g'(x)|^p v(x)dx.$$

Thus the last inequality is a modification of the fractional order inequality (5.2) with the special weight $u(x, t) \equiv 1$.

(ii) Assume additionally that $\lambda \neq 1/p$ and extend g by zero to $(-b, 0)$. If

$$\sup_{0<x<b} x^{1/p'} \left(\int_x^b u(t)dt \right)^{1/q} < \infty,$$

then, for any δ, $0 < \delta < b$,

$$\int_0^b |g(x)|^r u^{\lambda r/q}(x)dx$$

$$\leq C \int_0^\delta t^{-\lambda r-1} \left(\int_{-t}^b |g(x+t) - g(x)|^p dx \right)^{r/p} dt.$$

In particular, if $p = q$, then $r = p$ and we have

$$\int_0^b |g(x)|^p u^\lambda(x)dx \leq C \int_0^\delta \left(\int_{-t}^b \frac{|g(x+t) - g(x)|^p}{t^{\lambda p+1}} dx \right) dt.$$

This is a modified version of the fractional order Hardy inequality (5.1) with the special weight $v(x) \equiv 1$.

There are many examples of inequalities which hold except for one or more values of the parameters involved. Sometimes, this phenomenon can be explained via interpolation. Let us give two simple examples.

Example 5.30. The (classical) Hardy inequality

$$\left(\int_0^\infty \left| \frac{1}{x} g(x) \right|^p x^\beta dx \right)^{1/p} \leq C \left(\int_0^\infty |g'(x)|^p x^\beta dx \right)^{1/p} \qquad (5.45)$$

holds, with $p \geq 1$ and $g \in C_0^\infty(0, \infty)$, for every $\beta \neq p - 1$ but *does not hold* for $\beta = p - 1$.

In order to be able to understand this phenomenon we first note that inequality (5.45) can be rewritten as

$$\left(\int_0^\infty \left| \frac{1}{x} \int_0^x f(y)dy \right|^p x^\beta dx \right)^{1/p} \leq C \left(\int_0^\infty |f(x)|^p x^\beta dx \right)^{1/p} \qquad (5.46)$$

for $\beta < p - 1$ and as

$$\left(\int_0^\infty \left| \frac{1}{x} \int_x^\infty f(y)dy \right|^p x^\beta dx \right)^{1/p} \leq C \left(\int_0^\infty |f(x)|^p x^\beta dx \right)^{1/p} \qquad (5.47)$$

for $\beta > p - 1$. When (5.45) is written in this form, we see that it is impossible to interpolate between $\beta < p-1$ (see (5.46)) and $\beta > p-1$

(see (5.47)) to obtain the inequality for $\beta = p - 1$. The reason is that we have in fact *two different operators* involved:

$$(H_a f)(x) = \frac{1}{x} \int_0^x f(y) dy \qquad \text{for } \beta < p - 1,$$

$$(\tilde{H}_a f)(x) = -\frac{1}{x} \int_x^\infty f(y) dy \quad \text{for } \beta > p - 1. \qquad (5.48)$$

(We call H_a the *Hardy averaging operator*.) In order that these two operators coincide, it is necessary that

$$\frac{1}{x} \int_0^x f(y) dy = -\frac{1}{x} \int_x^\infty f(y) dy,$$

i.e.,

$$\int_0^\infty f(y) dy = 0.$$

Denote by N the set of locally integrable functions satisfying this condition. In order to interpolate the Hardy operator on weighted Lebesgue spaces, we have to consider not the spaces themselves but their intersection with N.

Remark 5.31. Notice that for weighted Lebesgue spaces, the following interpolation result holds:

$$(L^p(v_1), L^p(v_2))_{\theta,p} = L^p(v_1^{1-\theta} v_2^\theta),$$

i.e., the interpolation with two different weight functions v_1, v_2 "produces" a new weight function $v_\theta = v_1^{1-\theta} v_2^\theta$. In the case mentioned in Example 5.30, we need to interpolate between $L^p(v_1) \cap N$ and $L^p(v_2) \cap N$ with $v_1(x) = x^\gamma$, $\gamma < \beta - 1$, and $v_2(x) = x^\delta$, $\delta > \beta - 1$.

Example 5.32. Return to the fractional order Hardy inequality

$$\left(\int_0^\infty \left| \frac{g(x)}{x^\lambda} \right|^p dx \right)^{1/p} \leq C \left(\int_0^\infty \int_0^\infty \frac{|g(x) - g(y)|^p}{|x - y|^{1+\lambda p}} dx dy \right)^{1/p} \tag{5.49}$$

and recall that this inequality holds for $1 < p < \infty$ with $\lambda \in (0, 1)$ but $\lambda \neq 1/p$ (see Theorem 5.9 and Remark 5.10 (i)), and e.g. for every $g \in C_0^\infty(0, \infty)$.

As we have seen in the proof of Theorem 5.9, (5.49) is only an easy consequence of the inequality

$$\int_0^\infty \left| \frac{g(x)}{x^\alpha} \right|^p \frac{dx}{x} \le C \int_0^\infty \left| \frac{g(x) - \frac{1}{x}\int_0^x g(y)dy}{x^\alpha} \right|^p \frac{dx}{x} \qquad (5.50)$$

with $\alpha \ne 0$ (see Lemma 5.6). For $\alpha = 0$, it fails (see Remark 5.7 (ii)).

To become acquainted with this somewhat curious phenomenon, let us investigate the operator appearing in (5.50), namely

$$(Tg)(x) = \frac{g(x) - \frac{1}{x}\int_0^x g(y)dy}{x}.$$

In Example 5.30 we have introduced the set

$$N = \left\{ g \in L^1_{\text{loc}}(0, \infty) : \int_0^\infty g(x)dx = 0 \right\}.$$

We note that $Tg \in N$ e.g. for every locally integrable function with compact support. In terms of the operator T, inequality (5.50) reads

$$\|g\|_{p, x^{-\alpha p - 1}} \le C \|Tg\|_{p, x^{-\alpha p + p - 1}}, \qquad (5.51)$$

where $\alpha \ne 0$.

If T has an inverse T^{-1}, then (5.51) can be rewritten as

$$\|T^{-1}g\|_{p, x^{-\alpha p - 1}} \le C \|g\|_{p, x^{-\alpha p + p - 1}}, \quad \alpha \ne 0. \qquad (5.52)$$

Moreover, if the assumptions of Lemma 5.6 are satisfied — see (5.7) and (5.8), then T^{-1} exists and from the proof of the lemma mentioned we see that it can be given explicitly in the form (see (5.12) and (5.13))

$$(T^{-1}g)(x) = xg(x) + \int_0^x g(y)dy, \ \alpha < 0,$$

or

$$(T^{-1}g)(x) = xg(x) - \int_x^\infty g(y)dy, \ \alpha > 0.$$

Further, it follows from the Hardy inequality that

$$T^{-1} : L^p(x^{-\alpha p + p - 1}) \to L^p(x^{-\alpha p - 1}) \qquad (5.53)$$

boundedly for all $\alpha \neq 0$. Moreover, if we consider $g \in L^p(x^{-\alpha p - p + 1}) \cap N$, the two expressions for T^{-1} coincide and interpolation is possible between such special subspaces also across the critical point $\alpha = 0$.

These observations lead to the investigation of interpolation spaces of the type

$$(N \cap L^p(x^\beta), N \cap L^p(x^\gamma))_{\lambda, p}.$$

The following result was proved by N. Krugljak *et al.* [KrMP1] and is particularly relevant to the previous examples, in particular for the case from Example 5.30 where the exceptional value was $p - 1$.

Theorem 5.33. *Let* $1 \leq p < \infty$ *and* $\gamma < p - 1 < \beta$. *Then*

$$(N \cap L^p(x^\beta), N \cap L^p(x^\gamma))_{\lambda, p} = N \cap L^p(x^{(1-\lambda)\beta + \lambda\gamma}) \qquad (5.54)$$

if $\lambda \neq (\beta + 1 - p)/(\beta - \gamma)$, *and*

$$(N \cap L^p(x^\beta), N \cap L^p(x^\gamma))_{\lambda, p} = C^p(x^{p-1}) \cap L^p(x^{p-1})$$

if $\lambda = (\beta + 1 - p)/(\beta - \gamma)$.

Here $C^p(v)$ denotes the weighted Cesàro function space of non-absolute type:

$$C^p(v) = \left\{ g(x), x \in (0, \infty) : \right.$$

$$\left. \|g\|_{C^p(v)} := \left(\int_0^\infty \left| \frac{1}{x} \int_0^x g(y) dy \right|^p v(x) dx \right)^{1/p} < \infty \right\}.$$

Remark 5.34. If we use Theorem 5.33 for the Hardy averaging operator H_a considered in Example 5.30, we obtain for the exceptional value $p - 1$ the following (trivial) estimate:

$$\|H_a f\|_{L^p(x^{p-1}) \cap N} \leq C \|f\|_{L^p(x^{p-1}) \cap C^p(x^{p-1}) \cap N}, \qquad (5.55)$$

and for the operator T^{-1} from (5.53) in Example 5.32 we have in the case $\alpha = 0$ the inequality

$$\|T^{-1}g\|_{L^p(x^{-1}) \cap N} \leq C \|g\|_{L^p(x^{p-1}) \cap C^p(x^{p-1}) \cap N}. \qquad (5.56)$$

Sometimes, we can obtain estimates for the exceptional values *directly* without using interpolation. This is illustrated by the following proposition.

Proposition 5.35. *If $f \in N \cap L^1(|\log x|)$, then*

$$\int_0^\infty \left| \frac{1}{x} \int_0^x f(y)dy \right| dx \le C \int_0^\infty |f(x)| \, |\log x| dx. \qquad (5.57)$$

Proof. Using the assumption $f \in N$ and Fubini's theorem, we find that

$$\int_0^\infty \left| \frac{1}{x} \int_0^x f(y)dy \right| dx$$

$$= \int_0^1 \left| \frac{1}{x} \int_0^x f(y)dy \right| dx + \int_1^\infty \left| -\frac{1}{x} \int_x^\infty f(y)dy \right| dx$$

$$\le \int_0^1 \frac{1}{x} \int_0^x |f(y)|dydx + \int_1^\infty \frac{1}{x} \int_x^\infty |f(y)|dydx$$

$$= \int_0^1 \left(\int_y^1 \frac{1}{x}dx \right) |f(y)|dy + \int_1^\infty \left(\int_1^y \frac{1}{x}dx \right) |f(y)|dy$$

$$= \int_0^\infty |\log y| \, |f(y)|dy. \qquad \square$$

Remark 5.36. Inequality (5.57) means that the Hardy averaging operator $(H_a f)(x) = \frac{1}{x} \int_0^x f(y)dy$ is bounded from $N \cap L^1(|\log x|)$ into L^1. Let us recall that H_a is *not bounded* from $L^1(|\log x|)$ into L^1.

By using the estimates from [OK, Example 8.6 (v) and Remark 8.7], we can in the same way prove the following more general result:

If $1 \le p < \infty$ and $f \in N \cap L^p(x^{p-1}|\log x|^p)$, then

$$\left(\int_0^\infty \left| \frac{1}{x} \int_0^x f(y)dy \right|^p x^{p-1}dx \right)^{1/p}$$

$$\le C \left(\int_0^\infty |f(x)|^p |\log x|^p x^{p-1}dx \right)^{1/p}.$$

This estimate states that the Hardy averaging operator H_a is bounded from $N \cap L^p(x^{p-1}|\log x|^p)$ into $L^p(x^{p-1})$, i.e., that we have

$$\|H_a f\|_{L^p(x^{p-1})} \le C \|f\|_{N \cap L^p(x^{p-1}|\log x|^p)}.$$

Finally, let us note that Theorem 5.33 is only a special case of a more general result from [KrMP1] (with power weights replaced by general weights) which allows to describe situations where some Hardy inequality fails not just for one value of the parameter but even for an *interval* of parameters. For illustration, let us give an example.

Example 5.37. Let $v_0(x) = \max(x^{\alpha_0}, x^{\alpha_1})$ with $0 \le \alpha_0 \le \alpha_1$ and $v_1(x) = \min(x^{-\beta_0}, x^{-\beta_1})$ with $0 < \beta_0 \le \beta_1$. Assume that $\alpha_0/\alpha_1 \le \beta_0/\beta_1$. Then, for

$$\lambda \in (0,1) \setminus \left[\frac{\alpha_0}{\alpha_0 + \beta_0}, \frac{\alpha_1}{\alpha_1 + \beta_1} \right]$$

and $f \in N$, we have both the Hardy inequalities

$$\int_0^\infty \left| \frac{1}{x} \int_0^x f(y)dy \right| v_0^{1-\lambda}(x) v_1^\lambda(x) dx \le C \int_0^\infty |f(x)| v_0^{1-\lambda}(x) v_1^\lambda(x) dx$$

and

$$\int_0^\infty \left| \frac{1}{x} \int_x^\infty f(y)dy \right| v_0^{1-\lambda}(x) v_1^\lambda(x) dx \le C \int_0^\infty |f(x)| v_0^{1-\lambda}(x) v_1^\lambda(x) dx,$$

and therefore

$$(N \cap L^1(v_0),\ N \cap L^1(v_1))_{\lambda,1} = N \cap L^1(v_0^{1-\lambda} v_1^\lambda)$$

(see also Remark 5.31).

For $\lambda \in [\frac{\alpha_0}{\alpha_0 + \beta_0}, \frac{\alpha_1}{\alpha_1 + \beta_1}]$, none of these Hardy inequalities is true and we only have that (with $C^1(v)$ the Cesàro space, see p. 281)

$$(N \cap L^1(v_0),\ N \cap L^1(v_1))_{\lambda,1} = N \cap C^1(v_0^{1-\lambda} v_1^\lambda) \cap L^1(v_0^{1-\lambda} v_1^\lambda).$$

5.5. Further Results

A complement to Lemma 5.6

If $1 \leq p < \infty$ and $\alpha > -1$, then we easily obtain the following inequality which is reverse to inequality (5.9):

$$\left(\int_0^\infty \left| \frac{g(x) - \frac{1}{x} \int_0^x g(y)dy}{x^\alpha} \right|^p \frac{dx}{x} \right)^{1/p}$$

$$\leq \left(1 + \frac{1}{\alpha + 1} \right) \left(\int_0^\infty \left| \frac{g(x)}{x^\alpha} \right|^p \frac{dx}{x} \right)^{1/p}. \qquad (5.58)$$

Indeed, the Minkowski and Hardy inequalities (for $p > 1$) and Fubini's theorem (for $p = 1$) yield

$$\left(\int_0^\infty \left| \frac{g(x) - \frac{1}{x} \int_0^x g(y)dy}{x^\alpha} \right|^p \frac{dx}{x} \right)^{1/p}$$

$$\leq \left(\int_0^\infty \left| \frac{g(x)}{x^\alpha} \right|^p \frac{dx}{x} \right)^{1/p} + \left(\int_0^\infty \left| \frac{1}{x^{\alpha+1}} \int_0^x g(y)dy \right|^p \frac{dx}{x} \right)^{1/p}$$

$$\leq \left(1 + \frac{1}{1+\alpha} \right) \left(\int_0^\infty \left| \frac{g(x)}{x^\alpha} \right|^p \frac{dx}{x} \right)^{1/p}.$$

Hence we have the following result about equivalent norms in $L^p(x^{-\alpha p - 1})$:

Proposition 5.38. *Let* $g \in L^p(x^{-\alpha p - 1})$ *with* $p \geq 1$ *and* $\alpha > -1$, $\alpha \neq 0$. *Then*

$$\left(\int_0^\infty \left| \frac{g(x) - \frac{1}{x} \int_0^x g(y)dy}{x^\alpha} \right|^p \frac{dx}{x} \right)^{1/p} \approx \left(\int_0^\infty \left| \frac{g(x)}{x^\alpha} \right|^p \frac{dx}{x} \right)^{1/p}$$

with the equivalence constants $1 + 1/|\alpha|$ *and* $(\alpha + 1)/(\alpha + 2)$.

Proof. The upper estimate follows from (5.58), the lower from Lemma 5.6 since by Hölder's inequality we can show that $\frac{1}{x} \int_0^x g(t)dt$ tends to zero for $x \to 0$ (if $\alpha > 0$) and for $x \to \infty$ (if $\alpha < 0$) provided $g \in L^p(x^{-\alpha p - 1})$.

Remark 5.39. For the case that g is decreasing and $-1 < \alpha < 0$, this is essentially the result of C. Bennett *et al.* [BDS1]. They investigated the functional $f^{**} - f^*$ where f^* is the decreasing rearrangement of the measurable function f in the totally σ-finite measure space (Ω, μ), i.e.,

$$f^*(t) = \inf\{y : \mu\{x \in (0, \infty) : |f(x)| > y\} \leq t\}$$

and

$$f^{**}(t) := \frac{1}{t} \int_0^t f^*(y)dy.$$

More exactly, they proved that the norm in the Lorentz $L^{p,q}$-space, i.e.,

$$\|f\|_{L^{p,q}} := \left(\int_0^\infty (t^{1/p} f^*(t))^q \frac{dt}{t} \right)^{1/q}, \tag{5.59}$$

is equivalent to

$$\|f\|_{p,q} := \left(\int_0^\infty (t^{1/p}(f^{**}(t) - f^*(t)))^q \frac{dt}{t} \right)^{1/q}$$

provided that $1 < p < \infty$, $1 \leq q < \infty$ and $f^{**}(\infty) = \lim_{t \to \infty} f^{**}(t) = 0$.

Using the estimate (5.58) and Lemma 5.6 with $g = f^*$ and $\alpha = -1/p$ and p replaced by q, we obtain the following somewhat more precise result.

Corollary 5.40. *Let f^* and f^{**} be defined as above.*
 (i) *If $1 < p < \infty$, $1 \leq q < \infty$, then*

$$\|f\|_{p,q} \leq \left(1 + \frac{p}{p-1} \right) \|f\|_{L^{p,q}}.$$

 (ii) *If $0 < p < \infty$, $1 \leq q < \infty$ and $f^{**}(\infty) = 0$, then*

$$\|f\|_{L^{p,q}} \leq (p+1)\|f\|_{p,q}.$$

Remark 5.41. Let us denote

$$f_\#(t) = \frac{1}{t} \int_0^t f(y)dy - f(t). \qquad (5.60)$$

We have shown in Proposition 5.38 that the functions f and $f_\#$ are equivalent in the (norm of the) $L^p(x^{-\alpha p-1})$-spaces. M. Milman and Y. Sagher [MiS1] have shown that they are equivalent in the Lorentz $L^{p,q}$-spaces. Their proof is based on interpolation techniques; here, we will give a completely different proof based on the idea of the proof of Lemma 5.6.

Proposition 5.42. (i) *If* $1 < p < \infty$, $1 < q \leq \infty$, *then*

$$\|f_\#\|_{L^{p,q}} \leq C_{p,q}\|f\|_{L^{p,q}}.$$

(ii) *If* $1 < p < \infty$, $0 < q \leq \infty$ *and* $\lim_{t\to\infty} \frac{1}{t}\int_0^t f(y)dy = 0$, *then*

$$\|f\|_{L^{p,q}} \leq D_{p,q}\|f_\#\|_{L^{p,q}}.$$

Proof. (i) Since

$$|f_\#(t)| \leq \frac{1}{t}\int_0^t |f(y)|dy + |f(t)| \leq \frac{1}{t}\int_0^t f^*(y)dy + |f(t)|,$$

we obtain that

$$f_\#^*(t) \leq \frac{2}{t}\int_0^{t/2} f^*(y)dy + f^*\left(\frac{t}{2}\right)$$

(where $f_\#^* = (f_\#)^*$). The desired inequality now follows by using the definition of the norm in $L^{p,q}$, the Minkowski and the Hardy inequalities.

(ii) Similarly as in the proof of Lemma 5.6 (see formulas (5.10) and (5.13)) we obtain from (5.60) that

$$f(t) = -f_\#(t) + \int_t^\infty f_\#(y)\frac{dy}{y}.$$

Now, the result follows again by the Minkowski and Hardy inequalities, but we must consider separately the cases $1 \leq q \leq p$, $1 < p \leq q$ and $0 < q < 1 < p < \infty$. In the second case, we must apply duality, in the third case we must first do the following calculation:

$$
\begin{aligned}
\int_t^\infty f_\#(y) \frac{dy}{y} &= \int_0^\infty f_\#(y) \left(\frac{1}{y} \chi_{[t,\infty)}(y) \right) dy \\
&\leq \int_0^\infty f_\#^*(y) \left(\frac{1}{y} \chi_{[t,\infty)}(y) \right)^* dy = \int_0^\infty f_\#^*(y) \frac{1}{y+t} dy \\
&\leq \frac{1}{t} \int_0^t f_\#^*(y) dy + \int_t^\infty f_\#^*(y) \frac{dy}{y}.
\end{aligned}
$$

$\square$

Example 5.43. If we apply Proposition 5.38 with $p = 2$ and $\alpha = -1/2$ we obtain that

$$
\left(\int_0^\infty \left(g(x) - \frac{1}{x} \int_0^x g(y) dy \right)^2 dx \right)^{1/2} \approx \left(\int_0^\infty g^2(x) dx \right)^{1/2}
$$

for every $g \in L^2(0, \infty)$ with equivalence constants $1/3$ and 3. (Note that for such functions, the condition $\lim_{x \to \infty} \frac{1}{x} \int_0^x g(y) dy = 0$ is satisfied.)

Remark 5.44. The last equivalence relation can be rewritten as

$$
\frac{1}{3} \|g - H_a g\|_2 \leq \|g\|_2 \leq 3 \|g - H_a g\|_2
$$

with H_a the Hardy averaging operator (see (5.48)). Moreover, it is possible to replace the inequalities by equality signs:

$$
\|g\|_2 = \|g - H_a g\|_2 \tag{5.61}
$$

for every $g \in L^2(0, \infty)$, and in the next theorem, we will generalize this result to weighted spaces.

For this purpose, let us introduce the following *generalized Hardy averaging operator*:

$$
(H_w g)(x) := \frac{1}{W(x)} \int_0^x g(y) w(y) dy,
$$

where w is a weight function on $(0, \infty)$ and

$$W(x) = \int_0^x w(y)dy < \infty \quad \text{for every } x > 0.$$

Similarly,

$$(\widetilde{H}_w g)(x) := \frac{1}{\widetilde{W}(x)} \int_x^\infty g(y)w(y)dy$$

with

$$\widetilde{W}(x) = \int_x^\infty w(y)dy < \infty \quad \text{for every } x > 0.$$

The following theorem is due to N. Kaiblinger *et al.* [KaMP1], and their improvement of formula (5.61) reads as follows:

Theorem 5.45. *Let* $g \in L^2(w)$ *and suppose that* $W(\infty) = \widetilde{W}(0) = \infty$. *Then*

$$\|g\|_{2,w} = \|g - H_w g\|_{2,w} \tag{5.62}$$

and

$$\|g\|_{2,w} = \|g - \widetilde{H}_w g\|_{2,w}. \tag{5.63}$$

Proof. The function gw is integrable on $(0, x)$ for every $x < \infty$, since due to $g \in L^2(w)$ and $W(x) < \infty$, Hölder's inequality yields

$$\left(\int_0^x g(y)w(y)dy \right)^2 \leq \int_0^x g^2(y)w(y)dy \cdot \int_0^x w(y)dy$$

$$= W(x) \int_0^x g^2(y)w(y)dy$$

$$\leq W(x)\|g\|_{2,w}^2 < \infty. \tag{5.64}$$

For a.e. $x > 0$, we have

$$\frac{d}{dx}\left[\frac{1}{W(x)}\left(\int_0^x g(y)w(y)dy\right)^2\right]$$

$$= \frac{2g(x)w(x)}{W(x)}\int_0^x g(y)w(y)dy - \frac{w(x)}{W^2(x)}\left(\int_0^x g(y)w(y)dy\right)^2$$

$$= g^2(x)w(x) - \left(g(x) - \frac{1}{W(x)}\int_0^x g(y)w(y)dy\right)^2 w(x).$$

Integrating this identity from 0 to b with $b < \infty$ arbitrary but fixed, we obtain

$$\int_0^b g^2(x)w(x)dx - \int_0^b (g(x) - (H_w g)(x))^2 w(x)dx$$

$$= \frac{1}{W(b)}\left(\int_0^b g(y)w(y)dy\right)^2 - \lim_{x\to 0+}\left[\frac{1}{W(x)}\left(\int_0^x g(y)w(y)dy\right)^2\right].$$

The last limit is zero due to (5.64), and hence

$$\int_0^b g^2(x)w(x)dx - \int_0^b (g(x) - (H_w g)(x))^2 w(x)dx$$

$$= \frac{1}{W(b)}\left(\int_0^b g(y)w(y)dy\right)^2. \tag{5.65}$$

Choose $\varepsilon > 0$ arbitrary. Since $g \in L^2(w)$, there exists a constant $c = c(\varepsilon) > 0$ such that

$$\int_c^\infty g^2(y)w(y)dy < \frac{1}{4}\varepsilon.$$

Using Hölder's inequality, we obtain — similarly as in (5.64) — that for any $b > c$,

$$\left(\int_c^b g(y)w(y)dy\right)^2 \leq \int_c^b g^2(y)w(y)dy \cdot \int_c^b w(y)dy < \frac{1}{4}\varepsilon W(b).$$

Moreover, since $W(\infty) = \infty$, we have that for large b

$$\frac{1}{W(b)} \left(\int_0^c g(y)w(y)dy \right)^2 < \frac{1}{4}\varepsilon.$$

Consequently, for large b,

$$\frac{1}{W(b)} \left(\int_0^b g(y)w(y)dy \right)^2$$

$$= \frac{1}{W(b)} \left(\int_0^c g(y)w(y)dy + \int_c^b g(y)w(y)dy \right)^2$$

$$\leq \frac{2}{W(b)} \left[\left(\int_0^c g(y)w(y)dy \right)^2 + \left(\int_c^b g(y)w(y)dy \right)^2 \right]$$

$$< 2 \left(\frac{1}{4}\varepsilon + \frac{1}{4}\varepsilon \right) = \varepsilon,$$

and it follows from (5.65) that

$$\lim_{b \to \infty} \left[\int_0^b g^2(x)w(x)dx - \int_0^b (g(x) - (H_w g)(x))^2 w(x)dx \right] = 0,$$

i.e.,

$$\|g\|_{2,w}^2 = \|g - H_w g\|_{2,w}^2.$$

But this is (5.62), and the proof of (5.63) is similar.   □

Remark 5.46. (i) The assumption $W(\infty) = \int_0^\infty w(y)dy = \infty$ in Theorem 5.45 is essential. Indeed, if $0 < \int_0^\infty w(y)dy < \infty$, then for the constant function $g_0(x) \equiv c \neq 0$ we have $H_w g_0 = \tilde{H}_w g_0 = g_0$ and hence $\|g_0 - H_w g_0\|_{2,w} = \|g_0 - \tilde{H}_{w_0} g_0\|_{2,w} = 0$ while $\|g_0\|_{2,w} = |c|(\int_0^\infty w(y)dy)^{1/2} > 0$.

(ii) It follows from the proof of Theorem 5.45 that for the case $0 < \int_0^\infty w(y)dy < \infty$ we can replace formulas (5.62), (5.63) by

$$\|g\|_{2,w}^2 = \|g - H_w g\|_{2,w}^2 + \frac{(\int_0^\infty g(y)w(y)dy)^2}{\int_0^\infty w(y)dy}$$

and

$$\|g\|_{2,w}^2 = \|g - \widetilde{H}_w g\|_{2,w}^2 + \frac{(\int_0^\infty g(y)w(y)dy)^2}{\int_0^\infty w(y)dy},$$

respectively.

Theorem 5.45 has interesting consequences. Let us mention some of them.

Corollary 5.47. *Let g and w satisfy the assumptions of Theorem 5.45 and denote*

$$(S_w g)(x) := \int_x^\infty g(y)\frac{w(y)}{W(y)}dy,$$

$$(\widetilde{S}_w g)(x) := \int_0^x g(y)\frac{w(y)}{W(y)}dy.$$

Then

$$\|g\|_{2,w} = \|g - S_w g\|_{2,w} = \|g - \widetilde{S}_w g\|_{2,w}. \tag{5.66}$$

Proof. The weight function $u(x) = w(x)/W^2(x) = (-1/W(x))'$ satisfies the assumptions of Theorem 5.45 and

$$\widetilde{U}(x) = \int_x^\infty u(y)dy = -\int_x^\infty \left(\frac{1}{W(y)}\right)' dy = \frac{1}{W(x)}.$$

If we denote $h(x) := g(x)W(x)$, then

$$\|h\|_{2,u}^2 = \int_0^\infty g^2(x)W^2(x)u(x)dx = \int_0^\infty g^2(x)w(x)dx = \|g\|_{2,w}^2$$

and thus by Theorem 5.45,

$$\|g\|_{2,w}^2 = \|h\|_{2,u}^2 = \|h - \widetilde{H}_u h\|_{2,u}^2$$

$$= \int_0^\infty \left| g(x)W(x) - \frac{1}{\widetilde{U}(x)} \int_x^\infty g(y)W(y)u(y)dy \right|^2 u(x)dx$$

$$= \int_0^\infty \left| g(x)W(x) - W(x) \int_x^\infty g(y)\frac{w(y)}{W(y)}dy \right|^2 \frac{w(x)}{W^2(x)}dx$$

$$= \int_0^\infty |g(x) - (S_w g)(x)|^2 w(x)dx = \|g - S_w g\|_{2,w}^2,$$

which is the first part of (5.66). The second equality follows analogously. $\qquad \square$

Example 5.48. (i) If we choose $w(x) \equiv 1$ in Corollary 5.47, then we have for $g \in L^2$

$$\|g - H_a g\|_2 = \|g\|_2 = \|g - Sg\|_2,$$

where

$$(H_a g)(x) = \frac{1}{x}\int_0^x g(y)dy, \quad Sg(x) = \int_x^\infty \frac{g(y)}{y}dy.$$

(ii) If we choose $w(x) = x^{-2}$ in Corollary 5.47, then we have for $g \in L^2(x^{-2})$

$$\|g - \widehat{H}_w g\|_{2,x^{-2}} = \|g\|_{2,x^{-2}} = \|g - \widehat{S}_w g\|_{2,x^{-2}},$$

where

$$(\widehat{H}_w g)(x) = x \int_x^\infty \frac{g(y)}{y^2}dy \quad \text{and} \quad (\widehat{S}_w g)(x) = \int_0^x \frac{g(y)}{y}dy.$$

For the case $p = 2$, we also have the following sharp version of Proposition 5.38.

Proposition 5.49. *If* $g \in L^2(x^\beta)$, $\beta < 1$ *and* $\beta \neq -1$, *then*

$$\min\left(1, \frac{|1+\beta|}{1-\beta}\right)\|g\|_{2,x^\beta} \leq \|g - H_a g\|_{2,x^\beta}$$

$$\leq \max\left(1, \frac{|1+\beta|}{1-\beta}\right)\|g\|_{2,x^\beta}, \quad (5.67)$$

where $(H_a g)(x) = \frac{1}{x}\int_0^x g(y)dy$. *Both inequalities are sharp. In particular, for* $\beta = 0$, (5.67) *reduces to identity* (5.61).

Proof. For $\beta < 1$ and $\beta \neq 0$ we have

$$\|g\|_{2,x^\beta} = \|g - (1-\beta)H_a g\|_{2,x^\beta} \tag{5.68}$$

since, using Theorem 5.45 with $w(x) = x^{-\beta}$ and $f(x) = g(x)x^\beta$, we obtain that

$$\|g\|_{2,x^\beta}^2 = \|f\|_{2,x^{-\beta}}^2 = \|f - H_w f\|_{2,x^{-\beta}}^2$$

$$= \int_0^\infty \left| g(x)x^\beta - x^{\beta-1}(1-\beta)\int_0^x g(y)dy \right|^2 x^{-\beta}dx$$

$$= \|g - (1-\beta)H_a g\|_{2,x^\beta}^2 .$$

Using (5.68), the Minkowski and the reversed Minkowski inequalities, we obtain that

$$(1-\beta)\|g - H_a g\|_{2,x^\beta} = \|-\beta g + g - (1-\beta)H_a g\|_{2,x^\beta}$$

$$\leq |\beta|\|g\|_{2,x^\beta} + \|g - (1-\beta)H_a g\|_{2,x^\beta}$$

$$= (|\beta| + 1)\|g\|_{2,x^\beta}$$

and

$$(1-\beta)\|g - H_a g\|_{2,x^\beta} = \|-\beta g + g - (1-\beta)H_a g\|_{2,x^\beta}$$

$$\geq \left| \, |\beta| \, \|g\|_{2,x^\beta} - \|g - (1-\beta)H_a g\|_{2,x^\beta} \right|$$

$$= |(|\beta| - 1)| \, \|g\|_{2,x^\beta} .$$

Hence

$$\left| \frac{|\beta| - 1}{1 - \beta} \right| \|g\|_{2,x^\beta} \leq \|g - H_a g\|_{2,x^\beta} \leq \frac{|\beta| + 1}{1 - \beta}\|g\|_{2,x^\beta}$$

and (5.67) follows.

For $r > (-\beta - 1)/2$, the functions

$$g_r(x) := x^r \chi_{[0,1]}(x)$$

belong to $L^2(x^\beta)$ and

$$Q_r := \|g_r - H_a g_r\|_{2,x^\beta} / \|g_r\|_{2,x^\beta}$$

$$= \left(\left(\frac{r}{r+1}\right)^2 + \frac{2r + \beta + 1}{(r+1)^2(1-\beta)} \right)^{1/2} .$$

Since $Q_r \to \frac{|1+\beta|}{1-\beta}$ as $r \to [\frac{-\beta-1}{2}]^+$ and $Q_r \to 1$ as $r \to \infty$, we see that both the inequalities in (5.67) are sharp. $\square$

Remark 5.50. (i) The crucial formula (5.68) holds trivially for $\beta = 1$ but it can never hold for any $\beta > 1$.

(ii) Proposition 5.49 implies that the following sharper version of Corollary 5.40 holds for $q = 2$:

If $f \in L^{p,2}$, $1 < p < \infty$, and $f^{**}(\infty) = 0$, then

$$\min\left(1, \frac{1}{p-1}\right) \le \left(\frac{\int_0^\infty [t^{1/p}(f^{**}(t) - f^*(t))]^2 \frac{dt}{t}}{\int_0^\infty [t^{1/p} f^*(t)]^2 \frac{dt}{t}}\right)^{1/2}$$

$$\le \max\left(1, \frac{1}{p-1}\right).$$

The lower estimate is the best one for $1 < p \le 2$ and the upper estimate for $2 \le p < \infty$.

Remark 5.51. Some new information concerning fractional order Hardy's inequalities can be found in the papers [GO1] and [GO2] by P. Gurka and B. Opic. In particular, in [GO1] it was proved that the power weights in inequality (5.20) in Theorem 5.13 are in a sense optimally chosen. Moreover, in [GO2] some results of this type involving more general weights were proved. Hence, this paper further motivates "Open Problem 1" below.

5.6. Comments and Remarks

5.6.1. Inequality (5.4) in Proposition 5.1 was established independently by P. Grisvard [Gr1] and G.N. Jakovlev [J1] (see also [J2]). For the case $p = 2$, see also N. Aronszajn and K.T. Smith [AS1] (cf. also R. Adams *et al.* [AAS1]). Similar ideas can be found in the book by J.L. Lions and E. Magenes [LM1].

5.6.2. The approach in Sec. 5.2 is taken over from N. Krugljak *et al.* [KrMP2]. One motivation for this approach is to supplement the theory so that the "disturbing" condition $\lambda \ne 1/p$ in Proposition 5.1 could be understood also from an interpolation point of view. This theory was further developed in N. Krugljak *et al.* [KrMP1].

5.6.3. The results and proofs in Sec. 5.3 are mostly taken from H.P. Heinig *et al.* [HKP1], [HKP2] (cf. also A. Kufner and L.E. Persson [KuP2] and [PeK1]); Theorem 5.19 is due to V.I Burenkov and W.D. Evans [BE1] (cf. also V.I. Burenkov *et al.* [BEG1]).

5.6.4. The results mentioned in Secs. 5.2 and 5.3 contain only *sufficient* conditions for the validity of the corresponding fractional order Hardy inequalities (5.1) and (5.2). The problem to find *necessary and sufficient* characterizations of the weights seems to be still not solved in all cases (cf. Remark 5.52 and Proposition 5.53 below).

Open Problem 1. *Find necessary and sufficient conditions on the weights $u = u(x)$, $0 \leq x \leq b$, and $v = v(x,y)$, $0 \leq x,y \leq b$, so that* (5.1) *holds (for the case $p = q$), i.e., characterize u and v so that for $p > 1$, $0 < \lambda < 1$, $\lambda \neq 1/p$,*

$$\left(\int_0^b |g(x)|^p u(x) dx \right)^{1/p}$$
$$\leq K \left(\int_0^b \int_0^b \frac{|g(x) - g(y)|^p}{|x - y|^{1 + \lambda p}} v(x,y) dx dy \right)^{1/p} \tag{5.69}$$

holds for some finite $K > 0$ (cf. Theorem 5.20).

Problem 2. *Find necessary and sufficient conditions on the weights $v = v(x)$, $0 \leq x \leq b$, and $u = u(x,y)$, $0 \leq x,y \leq b$, so that* (5.2) *holds, i.e., characterize u and v so that for $1 < p < \infty$, $0 < q < \infty$, $0 < \lambda < 1$,*

$$\left(\int_0^b \int_0^b \frac{|g(x) - g(y)|^q}{|x - y|^{1 + \lambda p}} u(x,y) dx dy \right)^{1/q}$$
$$\leq K \left(\int_0^b |g'(x)|^p v(x) dx \right)^{1/p} \tag{5.70}$$

holds for some finite $K > 0$ (cf. Theorem 5.28).

Remark 5.52. In the previous version of the book "Problem 2" was posed as "Open Problem 2." The reason for this change is that in 2015 M. Nasyrova and E. Ushakova [NU1] gave a more or less satisfactory solution of old "Open Problem 2." It is interesting to note that this (fairly technical) solution is heavily depending on some new results concerning the Hardy-Steklov operator discussed in Chap. 3.

5.6.5. Let us formulate a proposition which contains some necessary and sufficient conditions for the validity of inequality (5.70) (or, more precisely, of its modification (5.72) below) provided the weight function u on the left hand side satisfies an *additional* assumption.

Proposition 5.53. *Let* $1 < p \leq q < \infty$. *Let* $u = u(x,y)$ *and* $v = v(x)$ *be weight functions on* $(a,b) \times (a,b)$ *and on* (a,b), *respectively, and denote* $\tilde{u}(x,y) = u(x,y) + u(y,x)$. *Suppose that the weight function* u *satisfies*

$$\sup_{a<t<b} \int_a^t \int_t^b \tilde{u}(x,y) \left(\int_a^x \int_t^b \tilde{u}(\sigma,\tau)d\tau d\sigma \right)^{-1/q}$$

$$\times \left(\int_a^t \int_y^b \tilde{u}(\sigma,\tau)d\tau d\sigma \right)^{-1/q'} dy dx < \infty. \qquad (5.71)$$

Then the inequality

$$\left(\int_a^b \int_a^b |g(x) - g(y)|^q u(x,y) dy dx \right)^{1/q}$$

$$\leq K \left(\int_a^b |g'(x)|^p v(x) dx \right)^{1/p} \qquad (5.72)$$

holds if and only if

$$\sup_{(\alpha,\beta)\subset(a,b)} \left(\int_a^\alpha \int_\beta^b \tilde{u}(x,y) dy dx \right)^{1/q} \left(\int_\alpha^\beta v^{1-p'}(x) dx \right)^{1/p'} < \infty. \qquad (5.73)$$

This result is due to H.P. Heinig (for $p = q$) and A. Kufner (for $p < q$). Condition (5.73) is *necessary* for any choice of p, q, i.e., also for $1 < q < p < \infty$, and for the proof of its necessity (which follows from (5.72) by the special choice $g'(x) = v^{1-p'}(x)\chi_{(\alpha,\beta)}(x)$,

$a < \alpha < \beta < b$) we *do not need* the additional condition (5.71). For details, see A. Kufner [Ku5].

Let us also emphasize that condition (5.73) is a certain modification of the condition $A = A(a, b; u, v) < \infty$ with A from (0.20) (cf. also Remark 1.5).

5.6.6. As far as it concerns inequality (5.1) (i.e., the analogue of inequality (5.69), but with a weighted L^q-norm on the left-hand side), the case $p \neq q$ seems to be still *not solved* in general. As mentioned on pp. 267 and 268 only for the case $p \leq q$ there are some results, but with a *mixed norm* on the right-hand side (cf. Theorem 5.23) while the case $p > q$ seems to be open.

5.6.7. Interpolation theory is widely described e.g. in the books by J. Bergh and J. Löfström [BL1]; H. Triebel [Tr1]; C. Bennett and R. Sharpley [BeS1]; Yu.A. Brudnyĭ and N. Krugljak [BrK1]. The idea of using interpolation theory to prove fractional order Hardy inequalities was initiated in A. Kufner and H. Triebel [KT1] (but, of course, this idea had appeared much earlier in the paper by P. Grisvard [Gr1] mentioned in **5.6.1** in connection with inequality (5.4)) and further developed in A. Kufner and L.E. Persson [KuP1] (see also [PeK1]). Attempts to understand in full the exception $\lambda \neq 1/p$ in (5.4) has led to some problems, but by using interpolation between special subspaces also this phenomenon can be completely understood, see N. Krugljak *et al.* [KrMP1]. In particular, (a generalized form of) Theorem 5.33 and its consequences (e.g., the fairly surprising Hardy-type inequality in Example 5.37) can be found in this paper.

5.6.8. The equivalence formula for weighted L^p-spaces pointed out in Proposition 5.38 was proved in N. Krugljak *et al.* [KrMP2]. For decreasing functions, this was discovered by C. Bennett *et al.* [BDS1] in connection with the definition and investigation of weak L^∞-spaces (see also C. Bennett and R. Sharpley [BeS1]). Proposition 5.42 is due to M. Milman and Y. Sagher [MiS1] but the proof here is taken from N. Krugljak *et al.* [KrMP2]. Our proof of (a weighted version of) the remarkable identity (5.61) (see Theorem 5.45), i.e.,

$$\|g\|_2 = \|(I - H_a)g\|_2, \quad g \in L^2 = L^2(0, \infty),$$

is taken from N. Kaiblinger *et al.* [KaMP1] where also another proof, based on isometries in Hilbert spaces, can be found.

In this connection (cf. Example 5.48 (i)) we also propose another open problem:

Open Problem 3. *Find necessary and sufficient conditions on the* (*averaging*) *operator T such that*

$$\|g\|_2 = \|(I - T)g\|_2$$

holds for all $g \in L^2 = L^2(0, \infty)$.

5.6.9. Some results mentioned in this chapter can be generalized also in other directions. For example, in H.P. Heinig *et al.* [HKP2] some multidimensional fractional order Hardy inequalities and also some extensions to Orlicz norms are proved. Another related question was investigated by H. Triebel [Tr2].

6

Integral Operators on the Cone
of Monotone Functions

6.1. Introduction

In the previous chapters, we discussed the mapping properties of the Hardy, Hardy-type, Hardy-Steklov and other integral operators in weighted Lebesgue spaces. It is of interest — and sometimes even inevitable — to consider weight characterization of such operators on some subsets of the Lebesgue spaces, e.g. in the case when the operator is defined on the cone of decreasing functions.

Motivating example

If we want to obtain a weight characterization of operators in *weighted Lorentz spaces* $\Lambda^p(w)$, $1 < p < \infty$, then it is necessary to consider operators defined on decreasing functions.

Recall that, for $0 < p < \infty$,

$$\Lambda^p(w) := \left\{ f : \|f^*\|_{p,w} = \left(\int_0^\infty (f^*(t))^p w(t) dt \right)^{1/p} < \infty \right\}, \quad (6.1)$$

where f is a measurable function on a measure space X, f^* being the equimeasurable decreasing rearrangement of $|f|$ defined by

$$f^*(t) := \inf\{y > 0 : \lambda_f(y) \le t\}. \tag{6.2}$$

Here λ_f is the *distribution function* defined by

$$\lambda_f(y) := \text{meas}\{x \in X : |f(x)| > y\}.$$

Note that $\|f^*\|_{p,w}$ is a norm on $\Lambda^p(w)$ if and only if w is decreasing. But the expression $\|f^{**}\|_{p,w}$ with

$$f^{**}(x) = \frac{1}{x} \int_0^x f^*(t)dt \tag{6.3}$$

is a norm which is equivalent to $\|f^*\|_{p,w}$ (see, e.g., C. Bennett and R. Sharpley [BeS1]).

In what follows, we take the measure in $\lambda_f(y)$ to be the *Lebesgue measure*.

Recall that the rearrangement of the Hardy-Littlewood maximal function Mf is equivalent to the Hardy (averaging) operator of the rearrangement of $|f|$. To be more precise, if

$$(Mf)(x) := \sup_{x \in Q} \frac{1}{|Q|} \int_Q |f(z)|dz, \quad x \in \mathbb{R}^n, \tag{6.4}$$

where Q is a cube in $\mathbb{R}^n$ with sides parallel to the coordinate axes and $|Q|$ is its Lebesgue measure, then

$$(Mf)^*(t) \approx \frac{1}{t} \int_0^t f^*(s)ds, \quad t > 0.$$

Hence, to prove that

$$M : \Lambda^p(v) \to \Lambda^q(u), \quad 1 < p, q < \infty,$$

is a *bounded* mapping, or, in other words, to characterize the weight functions u and v for which M is bounded between Lorentz spaces, it is equivalent to prove that the Hardy (averaging) operator H_a, defined now by

$$(H_a f)(t) := \frac{1}{t} \int_0^t f(s) ds, \quad t \geq 0, \tag{6.5}$$

and considered on non-negative decreasing functions (notation: $0 \leq f \downarrow$), is bounded from $L^p(v)$ to $L^q(u)$, $1 < p, q < \infty$. This means that one desires to characterize the weight functions u and v for which the (Hardy) inequality

$$\left(\int_0^\infty \left(\frac{1}{t} \int_0^t f(s) ds \right)^q u(t) dt \right)^{1/q} \leq C \left(\int_0^\infty f^p(t) v(t) dt \right)^{1/p} \tag{6.6}$$

holds whenever $f \geq 0$ is *decreasing*.

Remark 6.1. (i) Although problems of the type just mentioned were studied by B. Steckin in the early '60s for the discrete case, the first characterization of weights for which (6.6) is satisfied (with $0 \leq f \downarrow$!) was given by M. Ariño and B. Muckenhoupt [ArM1] in 1990 when $p = q$ and $u = v$. Shortly afterwards E. Sawyer [Sa3] proved the general case in the index (exponent) range $1 < p, q < \infty$ thereby characterizing the weight functions for which the operator M from (6.4) is bounded as a mapping from $\Lambda^p(v)$ into $\Lambda^q(u)$. Similarly he proved such results for the Hilbert transform

$$p.v. \int_{\mathbb{R}} \frac{f(y)}{x - y} dy$$

and other mappings arising in harmonic analysis (as usual p.v. means principal value).

(ii) The main purpose of this chapter is to provide the duality principle of Sawyer, given in Sec. 6.2, together with a corresponding duality principle in the case $0 < p \leq 1$. In Sec. 6.3, applications involving Hardy-type operators are given, while Sec. 6.4 is devoted to the case of integral operators with more general (positive) kernels.

6.2. The Duality Principle of Sawyer

The duality principle in $L^p(\mu)$-spaces

For (X, μ) a measure space with positive Borel measure μ, $1 < p < \infty$, and f a measurable function on X, the *standard duality principle* in $L^p(\mu)$-spaces, defined by the norm

$$\|f\|_{p,\mu} := \left(\int_X |f(x)|^p d\mu \right)^{1/p},$$

is expressed by the formula

$$\|g\|_{p',\mu} = \sup \left| \int_X f(x)g(x)d\mu(x) \right|, \qquad (6.7)$$

where the supremum is taken over all $f \in L^p(\mu)$ such that $\|f\|_{p,\mu} = 1$.

Here as usual $p' = \frac{p}{p-1}$ and $L^{p'}(\mu)$ is the space dual to $L^p(\mu)$.

Recall that duality was considered already on pp. 17 and 223. The duality in the latter case coincides with the concept given here in the special case $d\mu(x) = v(x)dx$ and $X = (0, 1)$. Here, we will be interested in the cases of (6.7), where X is $\mathbb{R}^+$, $\mathbb{R}$ or $\mathbb{R}^n$ and μ is absolutely continuous, i.e., $d\mu(x) = v(x)dx$ with some weight function v.

The other form of the duality principle (already introduced in Sec. 1.1) is given by

$$\sup_{f \geq 0} \frac{\int_0^\infty f(x)g(x)dx}{(\int_0^\infty f^p(x)v(x)dx)^{1/p}} = \|g\|_{p',v^{1-p'}}, \qquad (6.8)$$

where v is a measurable locally integrable weight function.

The aim of this section is to obtain an expression for the left hand side of (6.8) when the supremum is taken over all non-negative decreasing functions.

Before stating the duality principle, we recall the following well-known result.

Lemma 6.2. *If $f \geq 0$ is decreasing, then for any measurable weight function w we have*

$$\int_0^\infty f^p(x)w(x)dx = p \int_0^\infty y^{p-1} \left(\int_0^{\lambda_f(y)} w(x)dx \right) dy, \quad 0 < p < \infty,$$

$$(6.9)$$

where λ_f is the distribution function defined above.

Proof. Denote

$$E(f, t) = \{x \in (0, \infty) : f(x) > t\}.$$

Then $\lambda_f(t)$ is the Lebesgue measure of $E(f, t)$. Now, by Fubini's theorem,

$$
\begin{aligned}
\int_0^\infty f^p(x) w(x) dx &= \int_0^\infty \left(\int_0^{f(x)} pt^{p-1} dt \right) w(x) dx \\
&= \int_0^\infty \left(\int_0^\infty pt^{p-1} \chi_{E(f,t)}(x) dt \right) w(x) dx \\
&= \int_0^\infty pt^{p-1} \left(\int_0^\infty \chi_{E(f,t)}(x) w(x) dx \right) dt \\
&= p \int_0^\infty t^{p-1} \left(\int_{E(f,t)} w(x) dx \right) dt.
\end{aligned}
$$

But, since $f \geq 0$ is decreasing, the set $E(f, t)$ is the interval $(0, \lambda_f(t))$. Hence (6.9) holds. $\qquad \square$

The main result of this chapter is the following (Sawyer's) duality principle for decreasing functions f:

Theorem 6.3. *Suppose* $1 < p < \infty$. *Let* g, v *be non-negative measurable functions on* $(0, \infty)$ *with* v *locally integrable. Then*

$$
\sup_{0 \leq f \downarrow} \frac{\int_0^\infty f(x) g(x) dx}{\left(\int_0^\infty f^p(x) v(x) dx \right)^{1/p}}
$$

$$
\approx \left(\int_0^\infty \left(\int_0^x g(t) dt \right)^{p'} \left(\int_0^x v(t) dt \right)^{-p'} v(x) dx \right)^{1/p'}
$$

$$
+ \frac{\int_0^\infty g(x) dx}{\left(\int_0^\infty v(x) dx \right)^{1/p}}. \tag{6.10}
$$

Remark 6.4. (i) Note that integration by parts shows that the first term on the right hand side of (6.10) can be replaced by

$$\left(\int_0^\infty \left(\int_0^x g(t)dt \right)^{p'-1} \left(\int_0^x v(t)dt \right)^{1-p'} g(x)dx \right)^{1/p'}.$$

(ii) If $\int_0^\infty v(x)dx = \infty$, then by convention the second term on the right hand side of (6.10) is taken to be zero.

Proof of Theorem 6.3

First choose $f(x) = C$ a positive constant, and denote the left-hand side of (6.10) by $L(g)$. Then f is decreasing and

$$L(g) \geq \sup_{f=C} \frac{\int_0^\infty f(x)g(x)dx}{\left(\int_0^\infty f^p(x)v(x)dx \right)^{1/p}}$$

$$= \frac{\int_0^\infty g(x)dx}{\left(\int_0^\infty v(x)dx \right)^{1/p}}. \tag{6.11}$$

Next, let

$$f(x) = \int_x^\infty h(t)dt$$

where $h \geq 0$ is arbitrary. Then f is again decreasing and

$$L(g) \geq \sup_{f(x)=\int_x^\infty h(t)dt} \frac{\int_0^\infty f(x)g(x)dx}{\left(\int_0^\infty f^p(x)v(x)dx \right)^{1/p}}$$

$$= \sup_{h \geq 0} \frac{\int_0^\infty g(x) \int_x^\infty h(t)dtdx}{\left(\int_0^\infty v(x)(\int_x^\infty h(t)dt)^p dx \right)^{1/p}}.$$

Denote

$$G(t) = \int_0^t g(s)ds, \quad V(t) = \int_0^t v(s)ds.$$

The conjugate Hardy inequality (cf. Introduction, formula (0.18)) shows that

$$\left(\int_0^\infty v(x) \left(\int_x^\infty h(t)dt \right)^p dx \right)^{1/p}$$

$$\leq C \left(\int_0^\infty h^p(x) V^p(x) v^{1-p}(x) dx \right)^{1/p}$$

holds if and only if (see (0.20))

$$\sup_{r>0} \left(\int_0^r v(x)dx \right)^{1/p} \left(\int_r^\infty V^{p(1-p')}(x) v^{(1-p)(1-p')}(x)dx \right)^{1/p'} < \infty.$$

But, since $(1-p)(1-p') = 1$ and $p(1-p') = -p'$, integration by parts shows that the last supremum is finite. Consequently, if we apply (6.7) with $f = \frac{hV}{v}$, $g = \frac{G}{V}$ and $d\mu(t) = v(t)dt$, then

$$L(g) \geq \frac{1}{C} \sup_{h \geq 0} \frac{\int_0^\infty g(x) \int_x^\infty h(t)dtdx}{(\int_0^\infty h^p(x)[V(x)/v(x)]^p v(x)dx)^{1/p}}$$

$$= \frac{1}{C} \sup_{h \geq 0} \frac{\int_0^\infty h(t)G(t)dt}{(\int_0^\infty h^p(x)[V(x)/v(x)]^p v(x)dx)^{1/p}}$$

$$= \frac{1}{C} \sup_{h \geq 0} \frac{\int_0^\infty h(t)\frac{V(t)}{v(t)}\frac{G(t)}{V(t)}v(t)dt}{(\int_0^\infty [\frac{h(t)V(t)}{v(t)}]^p v(t)dt)^{1/p}}$$

$$= \frac{1}{C} \left(\int_0^\infty \left[\frac{G(t)}{V(t)} \right]^{p'} v(t)dt \right)^{1/p'}$$

$$= \frac{1}{C} \left(\int_0^\infty \left(\int_0^t g(s)ds \right)^{p'} \left(\int_0^t v(s)ds \right)^{-p'} v(t)dt \right)^{1/p'}. \quad (6.12)$$

The lower bound for (6.10) follows now from (6.11) and (6.12).

To prove the upper bound of $L(g)$ in (6.10), let

$$h(t) = \left[\int_t^\infty \left(\int_0^x g(s)ds \right)^{p'-1} \left(\int_0^x v(s)ds \right)^{-p'} v(x)dx \right.$$

$$\left. + C \left[\frac{\int_0^\infty g(s)ds}{\int_0^\infty v(s)ds} \right]^{p'-1} \right]^{1/p'}$$

with $C \geq \frac{1}{p'-1}$.

By Hölder's inequality we have

$$\int_0^\infty f(t)g(t)dt = \int_0^\infty f(t)g(t)h(t)h^{-1}(t)dt$$

$$\leq \left(\int_0^\infty f^p(t)h^{-p}(t)g(t)dt \right)^{1/p} \left(\int_0^\infty h^{p'}(t)g(t)dt \right)^{1/p'}.$$

$$(6.13)$$

Now, by Fubini's theorem,

$$\left(\int_0^\infty h^{p'}(t)g(t)dt \right)^{1/p'}$$

$$= \left(\int_0^\infty g(t) \left[\int_t^\infty \left(\int_0^x g(s)ds \right)^{p'-1} \left(\int_0^x v(s)ds \right)^{-p'} v(x)dx \right.\right.$$

$$\left.\left. + C \left[\frac{\int_0^\infty g(s)ds}{\int_0^\infty v(s)ds} \right]^{p'-1} \right] dt \right)^{1/p'}$$

$$= \left(\int_0^\infty \left(\int_0^x g(t)dt \right)^{p'-1} \left(\int_0^x v(s)ds \right)^{-p'} v(x) \left(\int_0^x g(t)dt \right) dx \right.$$

$$\left. + C \left[\frac{(\int_0^\infty g(s)ds)^{p'}}{(\int_0^\infty v(s)ds)^{p'-1}} \right] \right)^{1/p'}$$

$$\approx \left(\int_0^\infty \left(\int_0^x g(t)dt \right)^{p'} \left(\int_0^x v(t)dt \right)^{-p'} v(x)dx \right)^{1/p'}$$

$$+ \frac{\int_0^\infty g(s)ds}{(\int_0^\infty v(s)ds)^{1/p}}.$$

But this expression is the right-hand side of (6.10). Hence if we show that

$$\left(\int_0^\infty f^p(x)h^{-p}(x)g(x)dx \right)^{1/p} \le C \left(\int_0^\infty f^p(x)v(x)dx \right)^{1/p} \quad (6.14)$$

for $0 \le f \downarrow$, then the upper bound for $L(g)$ follows from (6.13) by dividing by $(\int_0^\infty f^p(x)v(x)dx)^{1/p}$.

To prove (6.14), observe first that integration yields

$$h^{-p}(t) = \left[\int_t^\infty \left(\int_0^x g(s)ds \right)^{p'-1} \left(\int_0^x v(s)ds \right)^{-p'} v(x)dx \right.$$

$$\left. + C \left(\int_0^\infty g(s)ds \right)^{p'-1} \left(\int_0^\infty v(s)ds \right)^{1-p'} \right]^{1-p}$$

$$\le \left\{ \left(\int_0^t g(s)ds \right)^{p'-1} \left[\frac{1}{p'-1} \left(\left(\int_0^t v(s)ds \right)^{1-p'} \right. \right. \right.$$

$$\left. \left. - \left(\int_0^\infty v(s)ds \right)^{1-p'} \right) \right] + C \left(\int_0^t g(s)ds \right)^{p'-1}$$

$$\left. \times \left(\int_0^\infty v(s)ds \right)^{1-p'} \right\}^{1-p}$$

$$= \left\{ \left(\int_0^t g(s)ds \right)^{p'-1} \left[\frac{1}{p'-1} \left(\int_0^t v(s)ds \right)^{1-p'} \right. \right.$$

$$-\frac{1}{p'-1}\left(\int_0^\infty v(s)ds\right)^{1-p'}$$

$$+C\left(\int_0^\infty v(s)ds\right)^{1-p'}\Bigg]\Bigg\}^{1-p}$$

$$\leq (p'-1)^{p-1}\left(\int_0^t g(s)ds\right)^{-1}\left(\int_0^t v(s)ds\right) \tag{6.15}$$

since $C - \frac{1}{p'-1} \geq 0$. Now Lemma 6.2 with $w = h^{-p}g$ yields

$$\int_0^\infty f^p(x)h^{-p}(x)g(x)dx = p\int_0^\infty y^{p-1}\left(\int_0^{\lambda_f(y)} h^{-p}(x)g(x)dx\right)dy. \tag{6.16}$$

Integrating by parts, taking into account that h^{-p} is increasing and applying (6.15) we obtain

$$\int_0^{\lambda_f(y)} h^{-p}(x)g(x)dx = h^{-p}(x)\int_0^x g(s)ds\Bigg|_0^{\lambda_f(y)}$$

$$-\int_0^{\lambda_f(y)}\left(\int_0^x g(s)ds\right)dh^{-p}(x)$$

$$\leq h^{-p}(\lambda_f(y))\int_0^{\lambda_f(y)} g(s)ds$$

$$\leq (p'-1)^{p-1}\left(\int_0^{\lambda_f(y)} g(s)ds\right)^{-1}$$

$$\times\left(\int_0^{\lambda_f(y)} v(s)ds\right)\left(\int_0^{\lambda_f(y)} g(s)ds\right)$$

$$= (p'-1)^{p-1}\int_0^{\lambda_f(y)} v(s)ds.$$

Substituting into (6.16), we conclude by Lemma 6.2 with $w = v$ that

$$\int_0^\infty f^p(x)h^{-p}(x)g(x)dx \le p(p'-1)^{p-1} \int_0^\infty y^{p-1} \left(\int_0^{\lambda_f(y)} v(x)dx \right) dy$$

$$= (p'-1)^{p-1} \int_0^\infty f^p(x)v(x)dx.$$

Hence (6.14) is established and so is the upper bound for $L(g)$. $\square$

The case $0 < p \le 1$

In order to prove an analogue of Theorem 6.3 in the case $0 < p \le 1$ the following result is required:

If $h \ge 0$ is decreasing, then for $0 < p \le 1$ and $0 < b \le \infty$,

$$\left(\int_0^b h(x)dx \right)^p \le p \int_0^b x^{p-1}h^p(x)dx \tag{6.17}$$

is satisfied.

The *proof* is easy: If we define

$$F(t) := p \int_0^t x^{p-1}h^p(x)dx - \left(\int_0^t h(x)dx \right)^p,$$

we have that

$$F'(t) = pt^{p-1}h^p(t) - p \left(\int_0^t h(x)dx \right)^{p-1} h(t) \ge 0$$

since $\int_0^t h(x)dx \ge th(t)$ and $p - 1 \le 0$. Hence $F(t) \ge F(0) = 0$ and (6.17) follows for $t = b$. $\square$

Theorem 6.5. *Suppose $0 < p \le 1$. Let g, v be non-negative measurable functions on $(0, \infty)$ with v locally integrable. Then*

$$\sup_{0 \le f \downarrow} \frac{\int_0^\infty f(x)g(x)dx}{\left(\int_0^\infty f^p(x)v(x)dx \right)^{1/p}} = \sup_{r>0} \left(\int_0^r g(x)dx \right) \left(\int_0^r v(x)dx \right)^{-1/p}. \tag{6.18}$$

Proof.   If $f(x) = \chi_{(0,r)}(x)$, $r > 0$, then

$$\sup_{0 \le f \downarrow} \frac{\int_0^\infty f(x)g(x)dx}{(\int_0^\infty f^p(x)v(x)dx)^{1/p}} \ge \frac{\int_0^r g(x)dx}{(\int_0^r v(x)dx)^{1/p}}$$

holds for any $r > 0$, hence the lower bound for (6.18) follows.

To establish the upper bound, we apply (6.9) with $w = g$ and $p = 1$. Then

$$\int_0^\infty f(x)g(x)dx = \int_0^\infty \left(\int_0^{\lambda_f(y)} g(x)dx \right) dy$$

$$= \int_0^\infty \frac{\int_0^{\lambda_f(y)} g(x)dx}{(\int_0^{\lambda_f(y)} v(x)dx)^{1/p}} \left(\int_0^{\lambda_f(y)} v(x)dx \right)^{1/p} dy$$

$$\le \sup_{r>0} \frac{\int_0^r g(x)dx}{(\int_0^r v(x)dx)^{1/p}} \left[\int_0^\infty \left(\int_0^{\lambda_f(y)} v(x)dx \right)^{1/p} dy \right].$$

Applying inequality (6.17) with $h(y) = (\int_0^{\lambda_f(y)} v(x)dx)^{1/p}$ and with $b = \infty$, and then by Lemma 6.2 with $w = v$, we obtain that

$$\int_0^\infty f(x)g(x)dx \le \sup_{r>0} \frac{\int_0^r g(x)dx}{(\int_0^r v(x)dx)^{1/p}}$$

$$\times \left[p \int_0^\infty y^{p-1} \left(\int_0^{\lambda_f(y)} v(x)dx \right) dy \right]^{1/p}$$

$$= \sup_{r>0} \frac{\int_0^r g(x)dx}{(\int_0^r v(x)dx)^{1/p}} \left[\int_0^\infty f^p(x)v(x)dx \right]^{1/p}.$$

The upper bound follows after dividing by $(\int_0^\infty f^p(x)v(x)dx)^{1/p}$. $\square$

Remark 6.6. If $f(x) = h^q(x)$, $q > 0$, then $h \downarrow$ if and only if $f \downarrow$. Hence, from Theorem 6.5 with $g(x) = u(x)$ we obtain that

$$\int_0^\infty h^q(x)u(x)dx \le \sup_{r>0} \frac{\int_0^r u(x)dx}{(\int_0^r v(x)dx)^{1/p}} \left(\int_0^\infty h^{pq}(x)v(x)dx \right)^{q/(pq)}.$$

With $s = qp < q$ it follows that

$$\left(\int_0^\infty h^q(x)u(x)dx \right)^{1/q} \le \sup_{r>0} \left(\int_0^r u(x)dx \right)^{1/q} \left(\int_0^r v(x)dx \right)^{-1/s}$$

$$\times \left(\int_0^\infty h^s(x)v(x)dx \right)^{1/s}$$

or

$$\|h\|_{q,u} \le C\|h\|_{s,v}, \qquad 0 < s < q, \qquad 0 \le h \downarrow, \tag{6.19}$$

with

$$C = \sup_{r>0} \left(\int_0^r u(x)dx \right)^{1/q} \left(\int_0^r v(x)dx \right)^{-1/s}.$$

Taking $h(x) = \chi_{(0,r)}(x)$, $r > 0$, in (6.19), we see that the condition $C < \infty$ is also necessary. Let us point out that inequality (6.19) is a weighted norm inequality for the identity operator defined on monotone functions. Such characterizations appear quite frequently in literature. See also Corollary 6.15 (iv) below.

6.3. Applications of the Duality Principle

An integral operator

Let us consider an operator T defined by

$$(Tf)(x) = \int_0^\infty k(x,y)f(y)dy, \tag{6.20}$$

where k is a non-negative kernel. In order to characterize the weight functions u and v for which the inequality

$$\left(\int_0^\infty (Tf)^q(x)u(x)dx \right)^{1/q} \leq C \left(\int_0^\infty f^p(x)v(x)dx \right)^{1/p} \tag{6.21}$$

with $1 < p, q < \infty$ holds for all f, $0 \leq f \downarrow$, we can use the duality principles (6.8) and (6.10). They show that (6.21) is equivalent to the inequality

$$\left(\int_0^\infty \left(\int_0^x (\widetilde{T}g)(t)dt \right)^{p'} \left(\int_0^x v(t)dt \right)^{-p'} v(x)dx \right)^{1/p'}$$
$$\leq C \left(\int_0^\infty g^{q'}(x)u^{1-q'}(x)dx \right)^{1/q'}, \tag{6.22}$$

where $\widetilde{T}$ is the conjugate of T and g is an *arbitrary* non-negative measurable function. Indeed: If we assume for simplicity that

$$\int_0^\infty v(x)dx = \infty \tag{6.23}$$

and use (6.10) with g replaced by $\widetilde{T}g$, we get from (6.22) that

$$\sup_{0 \leq f \downarrow} \frac{\int_0^\infty f(x)(\widetilde{T}g)(x)dx}{(\int_0^\infty f^p(x)v(x)dx)^{1/p}} \leq C\|g\|_{q',u^{1-q'}},$$

i.e.,

$$\sup_{0 \leq f \downarrow} \frac{\int_0^\infty (Tf)(x)g(x)dx}{(\int_0^\infty f^p(x)v(x)dx)^{1/p}} \leq C\|g\|_{q',u^{1-q'}}.$$

But then for $0 \leq f \downarrow$ we have

$$\|Tf\|_{q,u} = \sup_{\|g\|_{q',u^{1-q'}}=1} \int_0^\infty (Tf)(x)g(x)dx \leq C\|f\|_{p,v},$$

which is (6.21). Similarly we can show that (6.21) implies (6.22).

In some special cases, weight characterizations for which (6.22) is satisfied are known. Let us consider the following example:

Example 6.7. If $T = H_a$ is the Hardy averaging operator,

$$(H_a f)(x) = \frac{1}{x} \int_0^x f(t)dt, \qquad (6.24)$$

then its conjugate is given by

$$(\widetilde{H}_a g)(y) = \int_y^\infty \frac{g(t)}{t} dt.$$

Now

$$\int_0^x (\widetilde{H}_a g)(y) dy = \int_0^x \left(\int_y^\infty \frac{g(t)}{t} dt \right) dy$$

$$= \int_0^x \left(\int_y^x \frac{g(t)}{t} dt + \int_x^\infty \frac{g(t)}{t} dt \right) dy$$

$$= \int_0^x \frac{g(t)}{t} \left(\int_0^t dy \right) dt + x \int_x^\infty \frac{g(t)}{t} dt$$

$$= \int_0^x g(t) dt + x \int_x^\infty \frac{g(t)}{t} dt,$$

so that $\int_0^x (\widetilde{H}_a g)(y) dy$ is essentially the sum of the Hardy operator and the adjoint of the Hardy averaging operator. Therefore, if we assume that condition (6.23) is satisfied, then (6.22) is equivalent – for our special operator H from (6.24) – to proving weight characterizations for which both

$$\left(\int_0^\infty \left(\int_0^x g(t) dt \right)^{p'} V^{-p'}(x) v(x) dx \right)^{1/p'}$$

$$\leq C \left(\int_0^\infty g^{q'}(x) u^{1-q'}(x) dx \right)^{1/q'} \qquad . \qquad (6.25)$$

and

$$\left(\int_0^\infty \left(\int_x^\infty \frac{g(t)}{t} dt \right)^{p'} x^{p'} V^{-p'}(x) v(x) dx \right)^{1/p'}$$

$$\leq C \left(\int_0^\infty g^{q'}(x) u^{1-q'}(x) dx \right)^{1/q'} \qquad (6.26)$$

are satisfied with

$$V(x) = \int_0^x v(t)dt. \tag{6.27}$$

Now suppose

$$1 < p \le q < \infty.$$

Then $1 < q' \le p' < \infty$ and hence by formula (0.20), (6.25) is satisfied if and only if

$$\sup_{r>0} \left(\int_r^\infty V^{-p'}(x)v(x)dx \right)^{1/p'} \left(\int_0^r u^{(1-q')(1-q)}(x)dx \right)^{1/q}$$

is finite. But since $(1 - q')(1 - q) = 1$, we obtain — using (6.27) with $V(\infty) = \infty$ and integrating — that (6.25) is equivalent to

$$A_0 := \sup_{r>0} \left(\int_0^r v(x)dx \right)^{1/p} \left(\int_0^r u(x)dx \right)^{1/q} < \infty. \tag{6.28}$$

Similarly, formula (0.24) shows that (6.26) is satisfied if and only if

$$A_1 := \sup_{r>0} \left(\int_0^r x^{p'} V^{-p'}(x)v(x)dx \right)^{1/p'} \left(\int_r^\infty x^{-q}u(x)dx \right)^{1/q} < \infty. \tag{6.29}$$

If we suppose

$$1 < q < p < \infty,$$

then $1 < p' < q' < \infty$, and hence, by formula (0.21), (6.25) is satisfied if and only if

$$\left(\int_0^\infty \left(\int_x^\infty V^{-p'}(t)v(t)dt \right)^{r/p'} \left(\int_0^x u(t)dt \right)^{r/p} u(x)dx \right)^{1/r} < \infty$$

where $\frac{1}{r} = \frac{1}{p'} - \frac{1}{q'} = \frac{1}{q} - \frac{1}{p}$. But again, calculation of the first inner integral shows that (6.25) is equivalent to

$$B_0 := \left(\int_0^\infty \left(\int_0^x v(t)dt \right)^{-r/p} \left(\int_0^x u(t)dt \right)^{r/p} u(x)dx \right)^{1/r} < \infty. \tag{6.30}$$

In the same way, by formula (0.25), (6.26) is equivalent to

$$B_1 := \left(\int_0^\infty \left(\int_0^x t^{p'} V^{-p'}(t) v(t) dt \right)^{r/p'} \right.$$

$$\left. \times \left(\int_x^\infty t^{-q} u(t) dt \right)^{r/p} x^{-q} u(x) dx \right)^{1/r} < \infty. \quad (6.31)$$

Hence, we have proved the following assertion.

Theorem 6.8. *Let* $1 < p, q < \infty$ *and let* u, v *be weight functions such that* $V(x) = \int_0^x v(t) dt$ *satisfies* $V(\infty) = \infty$. *Then the inequality*

$$\left(\int_0^\infty u(x) \left(\frac{1}{x} \int_0^x f(t) dt \right)^q dx \right)^{1/q} \le C \left(\int_0^\infty v(x) f^p(x) dx \right)^{1/p}$$

is satisfied for all non-negative decreasing functions if and only if
(i) *for* $1 < p \le q < \infty$, $\max(A_0, A_1) < \infty$, *where* A_0, A_1 *are given by* (6.28), (6.29), *respectively;*
(ii) *for* $1 < q < p < \infty$, $\max(B_0, B_1) < \infty$, *where* B_0, B_1 *are given by* (6.30), (6.31), *respectively.*

The Hardy-Littlewood maximal function M

Let M be defined by (6.4). Since, as mentioned on p. 300, $(Mf)^*(t) \approx \frac{1}{t} \int_0^t f^*(s) ds$, we obtain from Theorem 6.8 immediately the following result.

Corollary 6.9. *Let* $1 < p, q < \infty$ *and let* u, v *be weight functions such that* $V(x) = \int_0^x v(t) dt$ *satisfies* $V(\infty) = \infty$. *Then the mapping*

$$M : \Lambda^p(u) \to \Lambda^q(v)$$

is bounded if and only if (i) *or* (ii) *of Theorem 6.8 is satisfied.*

Analogously as in Example 6.7, we can derive the following result.

Theorem 6.10. *Let u, v be weight functions such that $V(\infty) = \infty$. If $1 < p \le q < \infty$, then the inequality*

$$\left(\int_0^\infty u(x) \left[\int_x^\infty \frac{f(t)}{t} dt \right]^q dx \right)^{1/q} \le C \left(\int_0^\infty f^p(x) v(x) dx \right)^{1/p} \tag{6.32}$$

holds for all $f \ge 0$, $f \downarrow$, if and only if the expressions

$$\sup_{r>0} \left\{ \int_r^\infty u(t) \left[\int_t^\infty v^{q/p'}(y) V^{-q}(y) \ln \frac{y}{t} dy \right]^q dt \right\}^{1/q} \left\{ \int_0^r v(t) dt \right\}^{1/p}$$

and

$$\sup_{r>0} \left\{ \int_r^\infty u(t) \left[\int_t^\infty v^{q/p'}(y) V^{-q}(y) \ln^{p'} \frac{y}{t} dy \right]^q dt \right\}^{1/q} \cdot$$

$$\times \left\{ \int_r^\infty V^{-p'}(t) v(t) \ln^{p'} \frac{t}{r} dt \right\}^{1/p}$$

are finite.

Proof. If $(Tf)(x) = \int_x^\infty \frac{f(t)}{t} dt$, then the adjoint of T has the form

$$(\tilde{T}g)(y) = \frac{1}{y} \int_0^y g(t) dt.$$

But then

$$\int_0^x (\tilde{T}g)(t) dt = \int_0^x g(t) \ln \frac{x}{t} dt$$

and by Theorem 6.3 (cf. (6.22)), (6.32) is equivalent to

$$\left(\int_0^\infty \left[\int_0^x g(t) \ln \frac{x}{t} dt \right]^{p'} V^{-p'}(x) v(x) dx \right)^{1/p'}$$

$$\le C \left(\int_0^\infty g^{q'}(x) u^{1-q'}(x) dx \right)^{1/q'}. \tag{6.33}$$

However, we observed in Chap. 2 (see Example 2.7 (iii)) that $\ln \frac{x}{t}$ is an Oinarov kernel, i.e., a kernel satisfying conditions (2.25), (2.26). Hence, applying Theorem 2.10 with p replaced by q', q by p', u by

$V^{-p'}v$ and v by $u^{1-q'}$, it follows that (6.33) holds if and only if the conditions of Theorem 6.10 are satisfied. □

Remark 6.11. (i) An analogous result holds also in the case $1 < q < p < \infty$. The argument is as above only now we apply Theorem 2.15 to obtain equivalent weight conditions for which (6.33) is satisfied.

(ii) In Theorems 6.8 and 6.10 we gave weight characterizations for certain Hardy-type operators defined on non-negative decreasing functions. This indicates that also more general Hardy-type operators can be considered, which is indeed the case provided we impose additional conditions on the corresponding kernel. Thus if

$$(Tf)(x) = \int_0^x k(x,y)f(y)dy, \quad 0 \le f \downarrow,$$

then

$$(\tilde{T}g)(y) = \int_y^\infty k(x,y)g(x)dx, \quad 0 \le g$$

and

$$\int_0^x (\tilde{T}g)(y)dy = \int_0^x \int_y^\infty k(t,y)g(t)dtdy$$

$$= \int_0^x \int_y^x k(t,y)g(t)dtdy + \int_0^x \int_x^\infty k(t,y)g(t)dtdy$$

$$= \int_0^x g(t)\left(\int_0^t k(t,y)dy\right)dt$$

$$+ \int_x^\infty g(t)\left(\int_0^x k(t,y)dy\right)dt.$$

If we denote

$$K(t) = \int_0^t k(t,y)dy,$$

then, by (6.21), the inequality

$$\left(\int_0^\infty u(x)(Tf(x))^q dx\right)^{1/q} \le C \left(\int_0^\infty v(x)f^p(x)dx\right)^{1/p}$$

is satisfied for $0 \le f \downarrow$ if and only if (cf. (6.22)) the inequalities

$$\left(\int_0^\infty \left(\int_0^x K(t)g(t)dt \right)^{p'} \left(\int_0^x v(t)dt \right)^{-p'} v(x)dx \right)^{1/p'}$$

$$\le C \left(\int_0^\infty g^{q'}(x)u^{1-q'}(x)dx \right)^{1/q'}$$

and

$$\left(\int_0^\infty \left(\int_x^\infty K(t)g(t)dt \right)^{p'} \left(\int_0^x v(t)dt \right)^{-p'} v(x)dx \right)^{1/p'}$$

$$\le C \left(\int_0^\infty g^{q'}(x)u^{1-q'}(x)dx \right)^{1/q'}$$

are satisfied for *any g*, $g \ge 0$. But weight characterizations for which these estimates are satisfied follow from the Hardy and conjugate Hardy inequalities given in the Introduction.

(iii) Up to now we have considered operators on the cone of *decreasing* functions. However, the main duality theorem — Theorem 6.3 — can be applied to obtain an analogous result for non-negative *increasing* functions (notation: $0 \le f \uparrow$). For simplicity, let us consider only the case $V(\infty) = \infty$.

A change of variables shows that

$$\sup_{0<f\uparrow} \frac{\int_0^\infty f(x)g(x)dx}{\left(\int_0^\infty f^p(x)v(x)dx \right)^{1/p}} = \sup_{0<f\uparrow} \frac{\int_0^\infty f(\frac{1}{x})g(\frac{1}{x})\frac{dx}{x^2}}{\left(\int_0^\infty f^p(\frac{1}{x})v(\frac{1}{x})\frac{dx}{x^2} \right)^{1/p}}.$$

Now write $\widehat{f}(x) = f(\frac{1}{x})$, $\widehat{g}(x) = g(\frac{1}{x})x^{-2}$, $\widehat{v}(x) = v(\frac{1}{x})x^{-2}$. Then $\widehat{f} \downarrow$ and by Theorem 6.3 we arrive at

$$\sup_{0<f\uparrow} \frac{\int_0^\infty f(x)g(x)dx}{\left(\int_0^\infty f^p(x)v(x)dx \right)^{1/p}}$$

$$= \sup_{0<\widehat{f}\downarrow} \frac{\int_0^\infty \widehat{f}(x)\widehat{g}(x)dx}{\left(\int_0^\infty \widehat{f}^p(x)\widehat{v}(x)dx \right)^{1/p}}$$

$$\approx \left(\int_0^\infty \left(\int_0^x \widehat{g}(t)dt \right)^{p'} \left(\int_0^x \widehat{v}(t)dt \right)^{-p'} \widehat{v}(x)dx \right)^{1/p'}$$

$$= \left(\int_0^\infty \left(\int_{1/x}^\infty g(t)dt \right)^{p'} \left(\int_{1/x}^\infty v(t)dt \right)^{-p'} \frac{v(\frac{1}{x})}{x^2}dx \right)^{1/p'}$$

$$= \left(\int_0^\infty \left(\int_x^\infty g(t)dt \right)^{p'} \left(\int_x^\infty v(t)dt \right)^{-p'} v(x)dx \right)^{1/p'}.$$

Consequently, the analogue of Theorem 6.3 for increasing functions consists in changing the integrals on the right-hand side of (6.10) from $(0, x)$ to (x, ∞).

The corresponding analogue of Theorem 6.5 holds too.

6.4. More General Integral Operators

Introduction

The results mentioned in the foregoing section indicate that it is of interest to investigate more general integral inequalities of the type

$$\left(\int_0^\infty (Tf)^q(x)u(x)dx \right)^{1/q} \leq C \left(\int_0^\infty f^p(x)v(x)dx \right)^{1/p} \tag{6.34}$$

where now

$$(Tf)(x) = \int_0^\infty k(x, t)f(t)dt$$

with $k(x, y) \geq 0$. Moreover, it can also be interesting to study inequalities reversed to (6.34), in particular on the cone of non-negative decreasing (or increasing) functions. The aim of this section is to provide such an investigation in the following more general frame.

Let $(\mathcal{M}_j, \mu_j)$, $j = 1, 2$, denote two σ-finite measure spaces. Further, for every $x \in \mathcal{M}_j$ let $d\sigma_j^x(y)$ denote a positive measure on $(0, \infty)$ and define T_j by

$$(T_j f)(x) := \int_0^\infty f(y) d\sigma_j^x(y), \quad j = 1, 2.$$

We investigate inequalities of the type

$$\left(\int_{\mathcal{M}_1} (T_1 f)^q(x) d\mu_1(x) \right)^{1/q} \leq C \left(\int_{\mathcal{M}_2} (T_2 f)^p(x) d\mu_2(x) \right)^{1/p}. \quad (6.35)$$

Example 6.12. Let $\mathcal{M}_1 = \mathcal{M}_2 = (0, \infty)$ and $d\mu_1(x) = u(x)dx$, $d\mu_2(x) = v(x)dx$, where u, v are weight functions. Then inequality (6.35) takes the form

$$\left(\int_0^\infty (T_1 f)^q(x) u(x) dx \right)^{1/q} \leq C \left(\int_0^\infty (T_2 f)^p(x) v(x) dx \right)^{1/p},$$

i.e., we have a general (two operators) weight inequality.

If $d\sigma_j^x(y)$ is the Dirac delta function $\delta_x(y)$ then T_j is the identity operator I. If $d\sigma_j^x(y) = k_j(x, y) dy$, then

$$(T_j f)(x) = \int_0^\infty k_j(x, y) f(y) dy. \quad (6.36)$$

If T_1 is of the form (6.36) and T_2 is the identity operator, then we obtain (6.34). If T_2 has the form (6.36) and $T_1 = I$, we obtain the reverse to (6.34).

An important constant

A crucial role will be played by the following constant connected with the framework mentioned above:

$$C_1 = C_1(p, q, T_1, T_2) := \sup_{r > 0} \frac{(\int_{M_1} (T_1 \chi_{(0,r)})^q(x) d\mu_1(x))^{1/q}}{(\int_{M_2} (T_2 \chi_{(0,r)})^p(x) d\mu_2(x))^{1/p}}. \quad (6.37)$$

Before we formulate the main result of this section, we shall prove a useful lemma. The proof given here is elementary and the lemma is in fact a special case of some more general results from J. Bergh *et al.* [BBP2] (see also [BBP1]).

Lemma 6.13. *Let* $0 < p \le q < \infty$, $\alpha > 0$. *Then every decreasing function* f *on* $(0, \infty)$ *satisfies*

$$\left(\int_0^\infty (x^\alpha f(x))^q \frac{dx}{x} \right)^{1/q} \le p^{1/p} q^{-1/q} \alpha^{1/p - 1/q} \left(\int_0^\infty (x^\alpha f(x))^p \frac{dx}{x} \right)^{1/p}$$
(6.38)

and the constant is sharp.

Proof. Applying inequality (6.17) with $b = \infty$ and with p replaced by p/q we obtain that

$$\left(\int_0^\infty h(x) dx \right)^{p/q} \le \frac{p}{q} \int_0^\infty x^{p/q - 1} h^{p/q}(x) dx$$

holds for any function h, $0 \le h \downarrow$. Hence also

$$\left(\int_0^\infty h(x) dx \right)^{p/q} \le \int_0^\infty h^{p/q}(x) dx^{p/q}$$

and

$$\left(\int_0^\infty h(x^{\alpha q}) dx^{\alpha q} \right)^{p/q} \le \int_0^\infty h^{p/q}(x^{\alpha q}) dx^{\alpha p}.$$

The choice $h(t) = f^q(t^{1/(\alpha q)})$ yields

$$\left(\int_0^\infty f^q(x) dx^{\alpha q} \right)^{1/q} \le \left(\int_0^\infty f^p(x) dx^{\alpha p} \right)^{1/p},$$

which is (6.38). The sharpness of the estimate follows by inserting $f(x) = \chi_{(0,a)}(x)$, $0 < a < \infty$. $\quad\square$

Now we are ready to formulate the main result.

Theorem 6.14. *Let* C_1 *be defined by* (6.37) *and let* f *be non-negative and decreasing.*

(i) *Let* $0 < p \le 1 < q < \infty$. *Then the inequality*

$$\left(\int_{\mathcal{M}_1} (T_1 f)^q(x) d\mu_1(x) \right)^{1/q} \le C \left(\int_{M_2} (T_2 f)^p(x) d\mu_2(x) \right)^{1/p}$$

holds with $C > 0$ *independent of* f *if and only if* $C_1 < \infty$.

(ii) *Let* $0 < \max(1, p) \leq q < \infty$ *and* $T_1 = I$. *Then the inequality*

$$\left(\int_0^\infty f^q(x) d\mu_1(x) \right)^{1/q} \leq C \left(\int_{M_2} (T_2 f)^p(x) d\mu_2(x) \right)^{1/p}$$

holds with $C > 0$ *independent of* f *if and only if* $C_1 < \infty$.

(iii) *Let* $0 < p \leq \min(1, q) < \infty$ *and* $T_2 = I$. *Then the inequality*

$$\left(\int_{M_1} (T_1 f)^q(x) d\mu_1(x) \right)^{1/q} \leq C \left(\int_0^\infty f^p(x) d\mu_2(x) \right)^{1/p}$$

holds with $C > 0$ *independent of* f *if and only if* $C_1 < \infty$.

(iv) *Let* $0 < p \leq q < \infty$ *and* $T_1 = T_2 = I$. *Then the inequality*

$$\left(\int_0^\infty f^q(x) d\mu_1(x) \right)^{1/q} \leq C \left(\int_0^\infty f^p(x) d\mu_2(x) \right)^{1/p}$$

holds with $C > 0$ *independent of* f *if and only if* $C_1 < \infty$.

(v) *In all cases,* $C = C_1$ *is the sharp constant.*

Proof. Our main tool will be the following identity for the operator T_j which can be easily proved by using Fubini's theorem:

$$(T_j f)(x) = \int_0^\infty f(y) d\sigma_j^x(y) = \int_0^\infty \sigma_j^x(\{y : f(y) > t\}) dt$$

$$= \int_0^\infty \left(\int_{\{y : f(y) > t\}} d\sigma_j^x(y) \right) dt = \int_0^\infty (T_j \chi_{E(t)})(x) dt,$$

$$(6.39)$$

where

$$E(t) := E(f, t) = \{y : f(y) > t\}.$$

(i) In this case we just use (6.39) and twice Minkowski's integral inequality for $q \geq 1$ and $1/p \geq 1$:

$$\left(\int_{\mathcal{M}_1} (T_1 f)^q (x) d\mu_1(x) \right)^{1/q}$$

$$= \left(\int_{\mathcal{M}_1} \left(\int_0^\infty (T_1 \chi_{E(t)})(x) dt \right)^q d\mu_1(x) \right)^{1/q}$$

$$\leq \int_0^\infty \left(\int_{\mathcal{M}_1} (T_1 \chi_{E(t)})^q (x) d\mu_1(x) \right)^{1/q} dt$$

$$\leq C_1 \int_0^\infty \left(\int_{\mathcal{M}_2} (T_2 \chi_{E(t)})^p (x) d\mu_2(x) \right)^{1/p} dt$$

$$\leq C_1 \left(\int_{\mathcal{M}_2} \left(\int_0^\infty (T_2 \chi_{E(t)})(x) dt \right)^p d\mu_2(x) \right)^{1/p}$$

$$= C_1 \left(\int_{\mathcal{M}_2} (T_2 f)^p (x) d\mu_2(x) \right)^{1/p}. \qquad \square$$

(ii) In view of (i), we only need to prove the case $1 \leq p \leq q < \infty$. As in the proof of Lemma 6.2 we see that

$$\left(\int_0^\infty f^q(x) d\mu_1(x) \right)^{1/q} = q^{1/q} \left(\int_0^\infty t^q \mu_1(E(t)) \frac{dt}{t} \right)^{1/q}.$$

Therefore, using Lemma 6.13 for the decreasing function $\mu_1(E(t))$ and with $q = 1$, $\alpha = q$ and p replaced by p/q, we find that

$$\left(\int_0^\infty f^q(x) d\mu_1(x) \right)^{1/q}$$

$$\leq q^{1/q} \left[\left(\frac{p}{q} \right)^{q/p} q^{q/p-1} \left(\int_0^\infty (t^q \mu_1(E(t)))^{p/q} \frac{dt}{t} \right)^{q/p} \right]^{1/q}$$

$$= p^{1/p} \left(\int_0^\infty t^{p-1} \mu_1^{p/q}(E(t)) dt \right)^{1/p}$$

$$\leq C_1 p^{1/p} \left(\int_0^\infty t^{p-1} \int_{\mathcal{M}_2} (T_2 \chi_{E(t)})^p(x) d\mu_2(x) dt \right)^{1/p}$$

$$= C_1 p^{1/p} \left(\int_{\mathcal{M}_2} \left(\int_0^\infty t^{p-1} (T_2 \chi_{E(t)})^p(x) dt \right) d\mu_2(x) \right)^{1/p}.$$

Therefore, using Lemma 6.13 again, this time for the decreasing function $(T_2 \chi_{E(t)})(x)$, q replaced by p, $\alpha = 1$ and $p = 1$, we obtain that

$$\left(\int_0^\infty f^q(x) d\mu_1(x) \right)^{1/q}$$

$$\leq C_1 p^{1/p} \left(\int_{\mathcal{M}_2} \left(p^{-1/p} \int_0^\infty t (T_2 \chi_{E(t)})(x) \frac{dt}{t} \right)^p d\mu_2(x) \right)^{1/p}$$

$$= C_1 \left(\int_{\mathcal{M}_2} \left(\int_0^\infty (T_2 \chi_{E(t)})(x) dt \right)^p d\mu_2(x) \right)^{1/p}$$

$$= C_1 \left(\int_{\mathcal{M}_2} (T_2 f)^p(x) d\mu_2(x) \right)^{1/p}.$$

(iii) In this case, again according to (i), it suffices to prove the assertion for $0 < p \leq q < 1$. We use again Lemma 6.13 twice and obtain that

$$\left(\int_{\mathcal{M}_1} (T_1 f)^q(x) d\mu_1(x) \right)^{1/q}$$

$$= \left(\int_{\mathcal{M}_1} \left(\int_0^\infty (T_1 \chi_{E(t)})(x) dt \right)^q d\mu_1(x) \right)^{1/q}$$

$$\leq \int_{\mathcal{M}_1} \left(q \int_0^\infty t^{q-1} (T_1 \chi_{E(t)})^q(x) dt d\mu_1(x) \right)^{1/q}$$

$$= \left(q \int_0^\infty t^{q-1} \left(\int_{\mathcal{M}_1} (T_1 \chi_{E(t)})^q(x) d\mu_1(x) \right) dt \right)^{1/q}$$

$$\leq q^{1/q} \left(\int_0^\infty t^{q-1} C_1^q \mu_2^{q/p}(E(t)) dt \right)^{1/q}$$

$$\leq C_1 q^{1/q} \left(\left(\frac{q}{p} \right)^{-p/q} p^{1-p/q} \int_0^\infty t^p \mu_2(E(t)) \frac{dt}{t} \right)^{1/p}$$

$$= C_1 \left(p \int_0^\infty t^{p-1} \mu_2(E(t)) dt \right)^{1/p} = C_1 \left(\int_0^\infty f^p(x) d\mu_2(x) \right)^{1/p}.$$

(iv) We use again Lemma 6.13 and similar arguments as above and obtain that

$$\left(\int_0^\infty f^q(x) d\mu_1(x) \right)^{1/q}$$

$$= \left(q \int_0^\infty t^{q-1} \mu_1(E(t)) dt \right)^{1/q}$$

$$\leq \left(q C_1^q \int_0^\infty t^{q-1} \mu_2^{q/p}(E(t)) dt \right)^{1/q}$$

$$\leq C_1 q^{1/q} \left(\left(\frac{q}{p} \right)^{-p/q} p^{1-p/q} \int_0^\infty t^{p-1} \mu_2(E(t)) dt \right)^{1/p}$$

$$= C_1 \left(p \int_0^\infty t^{p-1} \mu_2(E(t)) dt \right)^{1/p} = C_1 \left(\int_0^\infty f^p(x) d\mu_2(x) \right)^{1/p}.$$

(v) We observe that for the decreasing function $f(x) = \chi_{(0,r)}(x)$ with any $r > 0$, we have equality in each of the cases above. Thus, (i) – (iv) are proved and $C = C_1$ is the sharp constant in all cases. $\qquad\square$

For the special cases of $\mathcal{M}_i$, T_i and μ_i ($i = 1, 2$) mentioned in Example 6.12, we obtain the following result:

Corollary 6.15. *Let f be a non-negative decreasing function and u, v weight functions on $(0, \infty)$.*
(i) *Let $0 < p \leq 1 \leq q < \infty$. Then the inequality*

$$\left(\int_0^\infty (T_1 f)^q(x) u(x) dx \right)^{1/q} \leq C \left(\int_0^\infty (T_2 f)^p(x) v(x) dx \right)^{1/p}$$

holds if and only if

$$C_0 := \sup_{r>0} \frac{(\int_0^\infty (\int_0^r k_1(x,y)dy)^q u(x)dx)^{1/q}}{(\int_0^\infty (\int_0^r k_2(x,y)dy)^p v(x)dx)^{1/p}} < \infty.$$

(ii) *Let* $0 < p \le 1$ *and* $p \le q < \infty$. *Then the inequality*

$$\left(\int_0^\infty (T_1 f)^q(x)u(x)dx \right)^{1/q} \le C \left(\int_0^\infty f^p(x)v(x)dx \right)^{1/p}$$

holds if and only if

$$C_0 := \sup_{r>0} \left(\int_0^\infty \left(\int_0^r k_1(x,y)dy \right)^q u(x)dx \right)^{1/q}$$
$$\times \left(\int_0^r v(x)dx \right)^{-1/p} < \infty.$$

(iii) *Let* $0 < p \le q < \infty$ *and* $1 \le q < \infty$. *Then the inequality*

$$\left(\int_0^\infty f^q(x)u(x)dx \right)^{1/q} \le C \left(\int_0^\infty (T_2 f)^p(x)v(x)dx \right)^{1/p}$$

holds if and only if

$$C_0 := \sup_{r>0} \left(\int_0^r u(x)dx \right)^{1/q}$$
$$\times \left(\int_0^\infty \left(\int_0^r k_2(x,y)dy \right)^p v(x)dx \right)^{-1/p} < \infty.$$

(iv) *Let* $0 < p \le q < \infty$. *Then the inequality*

$$\left(\int_0^\infty f^q(x)u(x)dx \right)^{1/q} \le C \left(\int_0^\infty f^p(x)v(x)dx \right)^{1/p}$$

holds if and only if

$$C_0 := \sup_{r>0} \left(\int_0^r u(x)dx \right)^{1/q} \left(\int_0^r v(x)dx \right)^{-1/p} < \infty.$$

The constant $C = C_0$ *is in all cases the best possible.*

Remark 6.16. (i) If we choose in Corollary 6.15 (i) $(T_i f)(x) = \int_0^x f(t)dt = g(x)$, $i = 1, 2$, then g is concave decreasing, and we obtain an imbedding of concave decreasing functions in weighted Lebesgue spaces.

(ii) The result of Corollary 6.15 (iv) was derived by a different method in Remark 6.11.

Example 6.17. If we choose in Corollary 6.15 the kernels $k_j(x, y)$ as

$$k_j(x, y) = \frac{1}{x} \quad \text{for} \quad 0 \le y \le x,$$
$$k_j(x, y) = 0 \quad \text{for} \quad y > x,$$

then T_j is the Hardy (averaging) operator,

$$(H_a f)(x) := \frac{1}{x} \int_0^x f(y)dy,$$

and we obtain for functions f, $0 \le f \downarrow$, the following results:

(i) Let $0 < p \le 1 \le q < \infty$. Then the inequality

$$\left(\int_0^\infty \left(\frac{1}{x} \int_0^x f(y)dy \right)^q u(x)dx \right)^{1/q}$$
$$\le C \left(\int_0^\infty \left(\frac{1}{x} \int_0^x f(y)dy \right)^p v(x)dx \right)^{1/p}$$

holds if and only if

$$C_0 := \sup_{r>0} \frac{(\int_0^\infty (\min(1, \frac{r}{x}))^q u(x)dx)^{1/q}}{(\int_0^\infty (\min(1, \frac{r}{x}))^p v(x)dx)^{1/p}} < \infty.$$

(ii) Let $0 < p \le 1$ and $p \le q < \infty$. Then the inequality

$$\left(\int_0^\infty \left(\frac{1}{x} \int_0^x f(y)dy \right)^q u(x)dx \right)^{1/q} \le C \left(\int_0^\infty f^p(x)v(x)dx \right)^{1/p}$$

holds if and only if

$$C_0 := \sup_{r>0} \left(\int_0^\infty \left(\min\left(1, \frac{r}{x} \right) \right)^q u(x)dx \right)^{1/q} \left(\int_0^r v(x)dx \right)^{-1/p} < \infty.$$

(iii) Let $0 < p \le q < \infty$, $1 \le q < \infty$. Then the inequality

$$\left(\int_0^\infty f^q(x)u(x)dx \right)^{1/q} \le C \left(\int_0^\infty \left(\frac{1}{x} \int_0^x f(y)dy \right)^p v(x)dx \right)^{1/p}$$

holds if and only if

$$C_0 := \sup_{r>0} \left(\int_0^r u(x)dx \right)^{1/q} \left(\int_0^\infty \left(\min\left(1, \frac{r}{x}\right) \right)^p v(x)dx \right)^{-1/p} < \infty.$$

The constant $C = C_0$ is in all cases the best possible. Note that for the case (ii) a more general result was stated in Theorem 6.8 but without the sharp constant.

Remark 6.18. (On the Ariño-Muckenhoupt B_p condition) The condition mentioned in Remark 6.1 is called the B_p condition and it reads: the weight $v = v(s) \in B_p$ if there exists a constant $C > 0$ so that

$$\int_t^\infty s^{-p} v(s)ds \le C t^{-p} \int_0^t v(s)ds \text{ for all } t > 0,$$

where $p \ge 1$. It was later on proved that in fact the B_p condition can be replaced by eight other different but equivalent conditions, see [Sol], [CS3], [Sa3] and [CM1]. In fact, in 2009 A. Gogatishvili *et al.* [GKP1] proved that the B_p condition can be replaced by infinite many conditions, namely six scales of conditions. Each scale depends on a parameter $t, 0 < t < \infty$, and all nine mentioned conditions above are just points on some of these scales. In particular, this means that the Hardy inequality (6.6) holds for all decreasing functions if and only if one of these infinite many conditions holds. Concerning scales of conditions cf. also Sec. 7.3.

Remark 6.19. (Functions satisfying two monotonicity conditions, quasi-concave functions) A function $f(t), 0 < t < \infty$, is said to be quasi-concave if $f(t)$ is non-decreasing and $f(t)/t$ is non-increasing. Many crucial objects in analysis are quasi-concave e.g. the Peetre K-functional $K(t, x; A_0, A_1)$ in interpolation theory, the integral moduli of continuity $\omega_p(f, x)$ in approximation theory, the function

$f^{**}(x) := H(f^*)(x) = \frac{1}{x} \int_0^x f^*(t)dt$ in the theory of Lorentz spaces. Some new information about quasi-concave or even more general quasi-monotone functions and their relations to indices, can be found in the paper [PSW1] by L.E. Persson *et al.* Moreover, some new results concerning the mapping properties of Hardy-type operators on the cone of more general quasi-concave functions were proved in 2014 by L.E. Persson *et al.* [PPS1]. See also the Ph.D. thesis [Po1] of O. Popova.

Remark 6.20. (Two-sided Hardy-type inequalities for monotone functions) It is well known that for the case when f is monotone we can get even two-sided Hardy-type inequalities. We refer to the papers [BBP1] and [BBP2] by J. Bergh *et al.* (see also the references therein). Some new results can be found in the paper [PPS1] by L.E. Persson *et al.* See also the Ph.D. thesis [Po1] of O. Popova.

Remark 6.21. Concerning further new results in this connection we refer to the Ph.D. theses [Ar1], [Jo1] and [Po1] by L. Arendarenko, M. Johansson and O. Popova, respectively. See also [PPS1], [PSU1], [GJOP1], [JSU1], [AOP1] and [KPT1].

6.5. Comments and Remarks

6.5.1. The boundedness of the Hardy-Littlewood maximal operator M (see (6.4)) on L^p, $p > 1$, have had a considerable significance for the development of analysis, specifically harmonic analysis. One of the reasons is that a large class of convolution operators are dominated by the Hardy-Littlewood maximal operator and hence the boundedness of M on L^p, $p > 1$, implies the boundedness of the class of convolution operators on L^p. As usual, the convolution operator T_g is defined by $T_g : T_g(f) = \int_0^x g(x - t)f(t)dt$ for fixed measurable function g.

However, the mapping properties of M were also studied in *weighted* L^p-spaces, and in fact it was shown by B. Muckenhoupt [Mu2] (see also [Mu3]) that $M : L^p(w) \to L^p(w)$, $p > 1$ (where we now have in mind functions defined on $\mathbb{R}^N$ so that $f \in L^p(w)$

means that

$$\int_{\mathbb{R}^N} |f(x)|^p w(x)dx = \|f\|_{p,w}^p < \infty),$$

is bounded if and only if the function w belongs to the so-called Muckenhoupt class A_p (i.e., that

$$C = \sup_Q \left(\frac{1}{|Q|}\int_Q w(x)dx\right)^{1/p}\left(\frac{1}{|Q|}\int_Q w^{1-p'}(x)dx\right)^{1/p'} < \infty,$$

where Q are cubes in $\mathbb{R}^N$ with edges parallel to the coordinate axes). It follows from this result that convolution operators which are majorized by M are also bounded on $L^p(w)$.

6.5.2. With the introduction, popularity and use of weighted Lorentz spaces $\Lambda^p(w)$ (see (6.1)) in the '50s and '60s the question arose whether the weight characterization of the maximal operator M on $L^p(w)$ would carry over to a corresponding result for the weighted Lorentz spaces $\Lambda^p(w)$. (For $w \equiv 1$ this is trivial!) Since the spaces $\Lambda^p(w)$ are defined in terms of the equimeasurable decreasing rearrangement f^* of f, it became necessary to relate this quantity with Mf (namely, with $(Mf)^*$). In fact (see p. 300),

$$(Mf)^*(x) \approx \frac{1}{x}\int_0^x f^*(t)dt. \tag{6.40}$$

The estimate $\lesssim$ in (6.40) was established by F. Riesz [R1] as a consequence of his nice "sunrise" lemma. Moreover, C. Herz [Her1] (under the influence of a paper of E.M. Stein) proved the inequality $\gtrsim$ in (6.40). Some further information about these inequalities (and about the corresponding inequalities in terms of the distribution function of Mf) can be found in the paper I. Asekritova *et al.* [AKMP1].

In view of (6.40), it is clear that $M : \Lambda^p(v) \to \Lambda^q(u)$ is bounded if and only if the Hardy averaging operator H_a defined on decreasing functions is bounded from $L^p(v)$ into $L^p(u)$. In 1990, M. Ariño and B. Muckenhoupt [ArM1] characterized the weights u, v for which $M : \Lambda^p(u) \to \Lambda^p(v)$ is bounded ($p > 1$), or equivalently, for which the operator H_a defined on decreasing functions is bounded on $L^p(u)$. Shortly after the result mentioned, E. Sawyer [Sa3] established an explicit duality theorem for weighted L^p spaces on decreasing functions. This duality theorem was used to show that the mapping properties of an operator defined on *decreasing* functions are equivalent to the mapping properties of an operator defined on *arbitrary* functions but with different weights. Hence he was able to characterize weights u, v for which $T : \Lambda^p(v) \to \Lambda^q(u)$ with $1 < p, q < \infty$ is bounded, in particular, if T is the Hardy-Littlewood maximal operator M.

In this chapter, we have proved this Sawyer's duality theorem, but our proof (see Theorem 6.3) follows closely that of V.D. Stepanov [St3], which seems to be more elementary. A multidimensional version of the duality principle was proved by S. Barza *et al.* [BHP1].

6.5.3. The case $0 < q < p$, $p > 1$, of Theorem 6.8 is also known and is due to G. Sinnamon [Si1], [Si3], V.D. Stepanov [St3] and others.

6.5.4. The duality theorem extends in a natural way to *increasing* functions so that weight characterizations for operators defined on increasing functions can also be given, see Remark 6.11 (iii).

6.5.5. Among spaces related to the weighted Lorentz spaces $\Lambda^p(w)$, let us mention the spaces

$$\Gamma^p(w) = \left\{ f : \left(\int_0^\infty \left(\frac{1}{x} \int_0^x f^*(t)dt \right)^p w(x)dx \right)^{1/p} < \infty \right\}.$$

Since $\frac{1}{x} \int_0^x f^*(t)dt \geq f^*(x)$, it is clear that $\Gamma^p(w) \subset \Lambda^p(w)$. Moreover, if $w \in A_p$, then $\Lambda^p(w) \subset \Gamma^p(w)$, $p > 1$, so that for such weights and $p > 1$, these spaces are equivalent. A duality theorem for functions in such spaces has been obtained by M.L. Goldman [G1] and by M.L. Goldman *et al.* [GHS1] with subsequent weight characterizations for the maximal operator and Hilbert transform in these spaces.

6.5.6. Inequality (6.17) (cf. also (6.38)) was probably first discovered by G.G. Lorentz [Lor1], p. 39. Various other proofs can be found in literature, see e.g., J. Bergh *et al.* [BBP2], H.P. Heinig and

L. Maligranda [HM2], V.G. Maz'ja [Maz1] and E.M. Stein and G. Weiss [StW1].

6.5.7. The proof of Theorem 6.14 is a simplified form of that presented in S. Barza *et al.* [BPSo1]. Let us note that

(i) this proof has the advantage that it can be carried over to the multidimensional case where, in comparison to the one-dimensional case, some new problems appear, see the paper just mentioned and the Ph.D. thesis of S. Barza [B1];

(ii) special cases of Theorem 6.14 have been proved (in other ways) by several authors; see M.J. Carro and J. Soria [CS1], [CS2]; H.P. Heinig and L. Maligranda [HM2]; Q. Lai [L2], L. Maligranda [M1] and E.A. Myasnikov *et al.* [MPS1];

(iii) for the case $q > p$ fairly little is known; however, for the onedimensional case, V.D. Stepanov [St4] proved a result corresponding to Theorem 6.14 (iv), and this result was generalized to a multidimensional setting in S. Barza *et al.* [BPSt1].

6.5.8. As far as it concerns multidimensional generalizations, we refer to the Ph.D. thesis by S. Barza [B1], where also a fairly complete list of references and some open questions can be found.

6.5.9. Extensions of the results mentioned, e.g. to modular inequalities and weighted Orlicz function and sequence spaces, were given by H.P. Heinig and A. Kufner [HK2]; H.P. Heinig and L. Maligranda [HM1] and Q. Lai [L1], [L3].

7

New and Complementary Results

7.1. On the Prehistory and History of the Hardy Inequality

We again consider the following original versions of the Hardy inequality: the discrete inequality asserts that if $\{a_n\}_1^\infty$ is a sequence of non-negative real numbers, then

$$\sum_{n=1}^\infty \left(\frac{1}{n}\sum_{i=1}^n a_i\right)^p \leq \left(\frac{p}{p-1}\right)^p \sum_{i=1}^\infty a_n^p, \quad p > 1, \qquad (7.1)$$

the continuous inequality informs us that if f is a non-negative p-integrable function on $(0, \infty)$, then f is integrable over the interval $(0, x)$ for each positive x and

$$\int_0^\infty \left(\frac{1}{x}\int_0^x f(y)dy\right)^p dx \leq \left(\frac{p}{p-1}\right)^p \int_0^\infty f^p(x)dx), \quad p > 1. \quad (7.2)$$

The development of the famous Hardy inequality in both discrete and continuous forms during the period 1906 to 1928 has its own history or, as we call it, prehistory. Contributions of mathematicians other than G.H. Hardy, such as E. Landau, G. Pólya, E. Schur and

M. Riesz, are important here. This prehistory was recently described in detail in [KMPe1].

In particular, the following is clear:

(a) Inequalities (7.1) and (7.2) are the standard forms of the Hardy inequalities that can be found in many textbooks on Analysis and were highlighted first in the famous book *Inequalities* [HLP] by Hardy, Littlewood and Pólya.

(b) By restricting (7.2) to the class of step functions one proves easily that (7.2) implies (7.1).

(c) The constant $(p/(p-1))^p$ in both (7.2) and (7.1) is *sharp*: it cannot be replaced with a smaller number such that (7.2) and (7.1) remain true for all relevant sequences and functions, respectively.

(d) The main motivation for Hardy to begin this dramatic history in 1915 was to find a simpler proof of the Hilbert inequality from 1906:

$$\sum_{n=1}^{\infty}\sum_{m=1}^{\infty}\frac{a_m b_n}{m+n} \le \pi \left(\sum_{m=1}^{\infty}a_m^2\right)^{1/2}\left(\sum_{n=1}^{\infty}b_n^2\right)^{1/2}. \qquad (7.3)$$

(In Hilbert's version of (7.3) the constant 2π appears instead of the sharp one π.) In fact, concerning the case $a_n = b_n$, Hardy wrote the following in the introduction of his 1920 paper [Ha2]: *It was proved by D. Hilbert, in the course of his theory of integral equations, that the double series $\sum_{n=1}^{\infty}\sum_{m=1}^{\infty}\frac{a_m a_n}{m+n}$ is convergent whenever $\sum_{n=1}^{\infty}a_n^2$ is convergent. Of this theorem, which is one of the simplest and most beautiful in the theory of double series of positive terms at least five essentially different proofs have been published. Hilbert's own proof, which depends on the theory of Fourier's series is outlined by H. Weyl in his Inaugural-Dissertation [Wey1]. Another proof was given by F. Wiener [Wi1] and two more by I. Schur [Sh1]; but none of these proofs is as simple and elementary as might be desired. To these four proofs I added recently a fifth, which seemed to me to lack nothing of this simplicity. I observed first that Hilbert's theorem is an immediate corollary of another theorem which seems of some interest in itself.*

We also remark that nowadays the following more general form of (7.3) is also sometimes referred in the literature as Hilbert's inequality

$$\sum_{n=1}^{\infty}\sum_{m=1}^{\infty}\frac{a_m b_n}{m+n} \le \frac{\pi}{\sin\frac{\pi}{p}}\left(\sum_{m=1}^{\infty}a_m^p\right)^{1/p}\left(\sum_{n=1}^{\infty}b_n^q\right)^{1/q}, \qquad (7.4)$$

where $p > 1$ and $q = p/(p-1)$. However, Hilbert was not even close to consider this case (the l_p-spaces appeared only around 1910).

The continuous version of (7.3) reads:

$$\int_0^{\infty}\int_0^{\infty}\frac{f(x)g(y)}{x+y}dxdy$$

$$\le \frac{\pi}{\sin\frac{\pi}{p}}\left(\int_0^{\infty}f^p(x)dx\right)^{1/p}\left(\int_0^{\infty}g^q(y)dy\right)^{1/q}, \qquad (7.5)$$

where $p > 1$ and $q = p/(p-1)$.

The constant $\frac{\pi}{\sin\frac{\pi}{p}}$ is sharp in both (7.4) and (7.5).

Remark 7.1. There is nowadays a huge number of papers generalizing (7.4)–(7.5) to more general situations. Such inequalities are usually referred to as Hilbert-type or Hardy-Hilbert-type inequalities in the literature. We note that a great number of these generalizations can be unified by presenting them as a unique result concerning operators containing kernels, which are homogeneous of degree -1. We refer to the recent papers [LPS1] and [LPSW1] (and references therein), where this is described in detail. See also Sec. 7.4(e)–(f).

(e) Concerning the history we note that the first weighted version of (7.2) was proved by Hardy himself in 1928, see [Ha5]:

$$\int_0^{\infty}\left(\frac{1}{x}\int_0^x f(y)dy\right)^p x^a dx \le \left(\frac{p}{p-1-a}\right)^p \int_0^{\infty} f^p(x)x^a dx, \quad (7.6)$$

where f is a measurable and non-negative function on $(0,\infty)$ whenever $a < p-1, p > 1$.

The continued almost unbelievable development of Hardy-type inequalities has been partly described not only in this book but also in other books, e.g. [OK], [KMP] and [KoMP]. Especially the book [KMP] has one main focus to describe this history up to 2007. In the remaining part of this chapter we mainly give some striking examples of newer result and developments which cannot be found in these books.

7.2. A Convexity Approach to Prove Hardy Inequalities

First we present the following basic

Observation 7.2. *We note that for $p > 1$*

$$\int_0^\infty \left(\frac{1}{x} \int_0^x f(y)dy\right)^p dx \leq \left(\frac{p}{p-1}\right)^p \int_0^\infty f^p(x)dx,$$

$$\Leftrightarrow$$

$$\int_0^\infty \left(\frac{1}{x} \int_0^x g(y)dy\right)^p \frac{dx}{x} \leq 1 \cdot \int_0^\infty g^p(x)\frac{dx}{x}, \qquad (7.7)$$

where $f(x) = g(x^{1-1/p})x^{-1/p}$.

This means that Hardy's inequality (7.2) is equivalent to (7.7) for $p > 1$ and, thus, that Hardy's inequality can be proved in the following simple way: By Jensen's inequality and Fubini's theorem we have that

$$\int_0^\infty \left(\frac{1}{x} \int_0^x g(y)dy\right)^p \frac{dx}{x} \leq \int_0^\infty \left(\frac{1}{x} \int_0^x g^p(y)dy\right) \frac{dx}{x}$$

$$= \int_0^\infty g^p(y) \int_y^\infty \frac{dx}{x^2}dy$$

$$= \int_0^\infty g^p(y)\frac{dy}{y}.$$

Instead by making the substitution $f(t) = g(t^{\frac{p-1-a}{p}})t^{-\frac{1+a}{p}}$ in (7.6) we also see that this inequality is equivalent to (7.7). These facts imply especially the following:

(a) Hardy's inequalities (7.2) and (7.6) hold also for $p < 0$ (because the function $\varphi(u) = u^p$ is convex also for $p < 0$) and (7.6) holds in the reverse direction for $0 < p < 1$ (with sharp constant $\left(\frac{p}{p-a-1}\right)^p, a < p - 1$). This fact follows by just making the same variable substitution and noting that Jensen's inequality holds in the reverse direction in this case.

(b) The inequalities (7.2) and (7.6) are equivalent for any $a < p - 1$, $p > 1$ or $p < 0$. This fact was not observed by Hardy himself since he proved (7.6) separately later on in his 1928 paper [Ha5].

(c) The inequality (7.7) holds also for $p = 1$, which gives us a possibility to interpolate and get more information about the mapping properties of the Hardy operator in other function spaces than weighted L^p spaces (in fact the mapping $L^\infty \to L^\infty$ at the other endpoint is trivial).

Remark 7.3. For the simplest case $\alpha = 0$, $p > 1$, this convexity approach to prove Hardy's inequality seems first to have been presented in 1965 by E.K. Godunova in [G1], but this wonderful discovery seems not to have been observed very much until it was rediscovered and complemented in 2002 by S. Kaijser *et al.* [KPO1]. After that a great number of results based on this idea have been presented and applied. See e.g. the Ph.D. thesis by K. Krulic [Kru1] from 2010 and the recent review articles [PeO1] and [PeS3] and the references given there.

A further development of this idea was presented by L.E. Persson and N. Samko in 2012 in [PeS3], where the case with finite interval was also covered. First we make the following basic

Observation 7.4. *It yields that*

$$\int_0^\infty \exp\left(\frac{1}{x}\int_0^x \ln f(y)dy\right)dx \leq e\int_0^\infty f(x)dx \qquad (7.8)$$

$$\Leftrightarrow$$

$$\int_0^\infty \exp\left(\frac{1}{x}\int_0^x \ln g(y)dy\right)\frac{dx}{x} \leq 1\cdot\int_0^\infty g(x)\frac{dx}{x}, \qquad (7.9)$$

where $f(x) = \frac{g(x)}{x}$.

Remark 7.5. According to Observation 7.4, Jensen's inequality and Fubini's theorem we see that the limit Pólya-Knopp inequality (7.8) also follows directly from Jensen's inequality via (7.9). Concerning (7.8) see also Remark 1.17.

In the same way as above, we can also prove the following more general statement:

Proposition 7.6. *Let f be a measurable function on $\mathbb{R}_+$ and let Φ be a convex function on $D_0 = \{f(x)\}$. Then*

$$\int_0^\infty \Phi\left(\frac{1}{x}\int_0^x f(y)dy\right)\frac{dx}{x} \leq \int_0^\infty \Phi\big(f(x)\big)\frac{dx}{x}. \qquad (7.10)$$

If Φ instead is positive and concave, then the reversed inequality holds.

Example 7.7. Consider the convex function $\Phi(u) = u^p, p \geq 1$ or $p < 0$. Then (7.10) reads

$$\int_0^\infty \left(\frac{1}{x}\int_0^x f(y)dy\right)^p\frac{dx}{x} \leq \int_0^\infty f^p(x)\frac{dx}{x}, \qquad (7.11)$$

i.e., according to basic Observation 7.2, we obtain Hardy's original inequality (7.1) for $p > 1$, but also that it holds for $p < 0$ (if we assume that $f(x) > 0$ a.e.). For $0 < p < 1$, (7.11) holds in the reversed direction.

Moreover, by applying (7.10) with $\Phi(u) = \exp u$ and $f(y)$ replaced by $\ln f(y)$ we get directly Pólya-Knopp's inequality (7.8) without going via some limit argument.

The case including finite intervals

It is also known that the Hardy inequality (7.2) holds for finite intervals, e.g. that

$$\int_0^\ell \left(\frac{1}{x}\int_0^x f(y)dy\right)^p dx \le \left(\frac{p}{p-1}\right)^p \int_0^\ell f^p(x)dx, \quad p > 1 \quad (7.12)$$

holds for any $\ell, 0 < \ell \le \infty$, and still the constant $\left(\frac{p}{p-1}\right)^p$ is sharp. Moreover, we shall point out an improved variant of (7.12) with the same sharp constant.

We begin by giving the following auxiliary result of independent interest.

Lemma 7.8. *Let g be a non-negative and measurable function on $(0, \ell), 0 < \ell \le \infty$.*
  a) *If $p < 0$ or $p \ge 1$, then*

$$\int_0^\ell \left(\frac{1}{x}\int_0^x g(y)dy\right)^p \frac{dx}{x} \le 1 \cdot \int_0^\ell g^p(x)\left(1 - \frac{x}{\ell}\right)\frac{dx}{x} \quad (7.13)$$

(in the case $p < 0$ we assume, that $g(x) > 0, 0 < x \le \ell$).
  b) *If $0 < p \le 1$, then (7.13) holds in the reversed direction.*
  c) *The constant $C = 1$ is sharp in both a) and b).*

Proof. By using Jensen's inequality with the convex function $\Phi(u) = u^p, p \ge 1, p < 0$, and reversing the order of integration, we find that

$$\int_0^\ell \left(\frac{1}{x}\int_0^x g(y)dy\right)^p \frac{dx}{x} \le \int_0^\ell \frac{1}{x}\int_0^x g^p(y)dy\frac{dx}{x}$$

$$= \int_0^\ell g^p(y)\left(\int_y^\ell \frac{1}{x^2}dx\right)dy$$

$$= \int_0^\ell g^p(y)\left(\frac{1}{y} - \frac{1}{\ell}\right)dy$$

$$= \int_0^\ell g^p(y)\left(1 - \frac{y}{\ell}\right)\frac{dy}{y}.$$

The only inequality in this proof holds in the reversed direction when $0 < p \leq 1$ so the proof of b) follows in the same way.

Concerning the sharpness of the inequality (7.33) we first let $\ell < \infty$ and assume that

$$\int_0^\ell \left(\frac{1}{x} \int_0^x g(y) dy \right)^p \frac{dx}{x} \leq C \cdot \int_0^\ell g^p(x) \left(1 - \frac{x}{\ell} \right) \frac{dx}{x} \qquad (7.14)$$

for all non-negative and measurable functions g on $(0, \ell)$ with some constant $C, 0 < C < 1$. Let $p \geq 1$ and $\varepsilon > 0$ and consider $g_\varepsilon(x) = x^\varepsilon$ (for the case $p < 0$ we assume that $\varepsilon < 0$). By inserting this function into (7.14) we obtain that

$$C \geq (\varepsilon p + 1)^{1-p},$$

so that, by letting $\varepsilon \to 0_+$ we have that $C \geq 1$. This contradiction shows that the best constant in (7.13) is $C = 1$. In the same way we can prove that the constant $C = 1$ is sharp also in the case b). For the case $\ell = \infty$ the sharpness follows by just making a limit procedure with the result above in mind. The proof is complete. $\qquad \square$

We note that for the case $\ell = \infty$ (7.13) coincides with the inequality (7.11) and, thus, the constant $C = 1$ is sharp, which in its turn, implies the well-known fact that the constant $C = \left(\frac{p}{p-1} \right)^p$ in (7.2) is sharp for $p > 1$ and as we see above also for $p < 0$.

A generalization of (7.6) for the interval $(0, \ell), 0 < \ell \leq \infty, p \geq 1$, reads:

$$\int_0^\ell \left(\frac{1}{x} \int_0^x f(y) dy \right)^p x^a dx$$

$$\leq \left(\frac{p}{p - 1 - a} \right)^p \int_0^\ell f^p(x) x^a \left[1 - \left(\frac{x}{\ell} \right)^{\frac{p-a-1}{p}} \right] dx, \qquad (7.15)$$

where $a < p - 1, p \geq 1$.

Remark 7.9. An inequality of type (7.15) was proved in [YZD1] (see also [YD2]). Inequality (7.15) with sharp constant was proved in [CP3] (see also [CP1], [CP2]). However, in these papers it was not observed that (7.15) holds also for $p < 0$ and that the inequality is in fact equivalent to (7.13) (see our Theorem 7.10 below).

In our next theorem we in particular give another proof of (7.15) based on the fact that (7.15) is equivalent to (7.13) and it directly follows that the constant $\left(\frac{p}{p-1-a}\right)^p$ in (7.15) is sharp. Moreover, the constant is sharp also for the case $p < 0, a < p-1$. More generally, it turns out that all the inequalities in our next Theorem are equivalent to the basic inequality (7.13):

Theorem 7.10. *Let* $0 \leq \ell \leq \infty$, *let* $p \in \mathbb{R} \setminus \{0\}$ *and let* f *be a non-negative function. Then*
  a) *the inequality* (7.15) *holds for all measurable functions* f, *each* $\ell, 0 < \ell \leq \infty$ *and all* a *in the following cases:*

$$(\text{a}_1)\ p \geq 1,\ a < p - 1,$$

$$(\text{a}_2)\ p < 0,\ a > p - 1.$$

  b) *for the case* $0 < p < 1, a < p - 1$, *inequality* (7.15) *holds in the reversed direction under the conditions considered in* a).
  c) *the inequality*

$$\int_\ell^\infty \left(\frac{1}{x}\int_x^\infty f(y)dy\right)^p x^{a_0}dx$$
$$\leq \left(\frac{p}{a_0+1-p}\right)^p \int_\ell^\infty f^p(x)x^{a_0}\left[1-\left(\frac{\ell}{x}\right)^{\frac{a_0+1-p}{p}}\right]dx \quad (7.16)$$

holds for all measurable functions f, *each* $\ell, 0 \leq \ell < \infty$ *and all* a_0 *in the following cases:*

$$(\text{c}_1)\qquad p \geq 1,\ a_0 > p - 1,$$

$$(\text{c}_2)\qquad p < 0,\ a_0 < p - 1.$$

  d) *for the case* $0 < p \leq 1$, $a > p - 1$, *inequality* (7.16) *holds in the reversed direction under the conditions considered in* c).
  e) *All inequalities above are sharp.*

f) *Let $p \geq 1$ or $p < 0$. Then, the statements in* a) *and* c) *are equivalent (for fixed p) for all permitted a and a_0 because they are in all cases equivalent to* (7.13) *via substitutions.*

g) *Let $0 < p < 1$. Then, the reversed inequalities in the statements in* b) *and* d) *are equivalent for all permitted a and a_0.*

Proof.   First we prove that (7.15) in the case (a_1) is equivalent to (7.13) via the relation

$$f(x) = g\left(x^{\frac{p-a-1}{p}}\right) x^{-\frac{a+1}{p}}.$$

In fact, with $f(x) = g\left(x^{\frac{p-a-1}{p}}\right)x^{-\frac{a+1}{p}}$ and $\ell_0 = \ell^{\frac{p}{p-a-1}}$ the right-hand side (RHS) in (7.15) is equal to

$$\text{RHS} = \left(\frac{p}{p-1-a}\right)^p \int_0^{\ell_0} g^p \left(x^{\frac{p-a-1}{p}}\right) \left[1 - \left(\frac{x}{\ell_0}\right)^{\frac{p-1-a}{p}}\right] \frac{dx}{x}$$

$$= \left(\frac{p}{p-1-a}\right)^{p+1} \int_0^{\ell^{\frac{p-a-1}{p}}} g^p(y) \left[1 - \frac{y}{\ell_0^{\frac{p-1-a}{p}}}\right] \frac{dy}{y}$$

$$= \left(\frac{p}{p-1-a}\right)^{p+1} \int_0^{\ell} g^p(y) \left[1 - \frac{y}{\ell}\right] \frac{dy}{y},$$

where $y = x^{\frac{p-a-1}{p}}$, $dy = x^{-\frac{a+1}{p}}\left(\frac{p-1-a}{p}\right) dx$, and for the left-hand side (LHS) in (7.15) we have

$$\text{LHS} = \int_0^{\ell_0} \left(\frac{1}{x} \int_0^x g\left(y^{\frac{p-a-1}{p}}\right) y^{-\frac{a+1}{p}} dy\right)^p x^a dx$$

$$= \left(\frac{p}{p-1-a}\right)^p \int_0^{\ell_0} \left(\frac{1}{x^{\frac{p-a-1}{p}}} \int_0^{x^{\frac{p-a-1}{p}}} g(s)ds\right)^p \frac{dx}{x}$$

$$= \left(\frac{p}{p-1-a}\right)^{p+1} \int_0^{\ell} \left(\frac{1}{y} \int_0^y g(s)ds\right)^p \frac{dy}{y}.$$

Since we have only equalities in the calculations above we conclude that (7.13) and (7.15) are equivalent and, thus, by Lemma 7.8, a) is proved for the case (a_1).

For the case (a_2) all calculations above are still valid and, according to Lemma 7.8, (7.13) holds also in this case and a) is proved.

For the case $0 < p \leq 1, a < p - 1$, all calculations above are still true and both (7.13) and (7.15) hold in the reversed direction according to Lemma 7.8. Hence also b) is proved.

For the proof of c) we consider (7.15) with $f(x)$ replaced by $f(1/x)$, with a replaced by a_0 and with ℓ replaced by $\ell_0 = 1/\ell$:

$$\int_0^{\ell_0} \left(\frac{1}{x} \int_0^x f(1/y) dy \right)^p x^{a_0} dx$$

$$\leq \left(\frac{p}{p-1-a_0} \right)^p \int_0^{\ell_0} f^p(1/x) x^{a_0} \left[1 - \left(\frac{x}{\ell_0} \right)^{\frac{p-a_0-1}{p}} \right] dx.$$

Moreover, by making the variable substitution $y = 1/s$, we find that with $g(s) := \frac{f(s)}{s^2}$

$$\text{LHS} = \int_0^{\ell_0} \left(\frac{1}{x} \int_{1/x}^\infty \frac{f(s)}{s^2} ds \right)^p x^{a_0} dx$$

$$= \int_\ell^\infty \left(y \int_y^\infty \frac{f(s)}{s^2} ds \right)^p y^{-a_0-2} dy$$

$$= \int_\ell^\infty \left(\frac{1}{y} \int_y^\infty \frac{f(s)}{s^2} ds \right)^p y^{-a_0-2+2p} dy$$

$$= \int_\ell^\infty \left(\frac{1}{y} \int_y^\infty g(y) \right)^p y^{2p-a_0-2} dy,$$

and

$$
\text{RHS} = \left(\frac{p}{p-1-a_0}\right)^p \int_\ell^\infty f^p(y) y^{-a_0} \left[1 - \left(\frac{\ell}{y}\right)^{\frac{p-a_0-1}{p}}\right] y^{-2} dy
$$

$$
= \left(\frac{p}{p-1-a_0}\right)^p \int_\ell^\infty g^p(y) y^{2p-a_0-2} \left[1 - \left(\frac{\ell}{y}\right)^{\frac{p-a_0-1}{p}}\right] dy.
$$

Now replace $2p-a_0-2$ by a and g by f and we have $a_0 = 2p-a-2$, so that $p - 1 - a_0 = a + 1 - p$. Hence, it yields that

$$
\int_0^\ell \left(\frac{1}{x} \int_x^\infty f(s) ds\right)^p x^a dx
$$

$$
\leq \left(\frac{p}{a+1-p}\right)^p \int_\ell^\infty f^p(x) x^a \left[1 - \left(\frac{\ell}{x}\right)^{\frac{a+1-p}{p}}\right] dx
$$

and, moreover,

$$
a_0 < p - 1 \Leftrightarrow 2p - a - 2 < p - 1 \Leftrightarrow a > p - 1.
$$

We conclude that c) with the conditions (c_1) and (c_2) are in fact equivalent to a) with the conditions (a_1) and (a_2), respectively, and also c) is proved.

The calculations above hold also in the case d) and the only used inequality (7.13) holds in the reverse direction in this case. Hence, d) is also proved.

Next we note that the proof above only consists of suitable substitutions and equalities to reduce all inequalities to the sharp inequality (7.13) and we obtain a proof also of the statements e), f) and g) according to Lemma 7.8. The proof is complete. □

The case with piecewise constant $p = p(x)$

By using similar arguments as above, we can derive the following result:

Theorem 7.11. *Let $a > 0$ and*

$$
p(x) = \begin{cases} p_0, & 0 \leq x \leq a, \\ p_1, & x > a, \end{cases}
$$

where $p_0, p_1 \in \mathbb{R} \setminus \{0\}$. Moreover, let $\alpha < 1, 0 < a \leq \ell \leq \infty$. Then

$$\int_0^\ell \left(\frac{1}{x} \int_0^x f(t) dt \right)^{p(x)} x^\alpha \frac{dx}{x}$$

$$\leq \frac{1}{1-\alpha} \int_0^\ell (f(x))^{p(x)} x^\alpha \left(1 - \left(\frac{x}{\ell} \right)^{1-\alpha} \right) \frac{dx}{x}$$

$$+ \max \left\{ 0, \frac{a^{\alpha-1} - \ell^{\alpha-1}}{1-\alpha} \right\} \int_0^\ell \left[(f(x))^{p_1} - (f(x))^{p_0} \right] dx, \quad (7.17)$$

whenever $p(x) \geq 1$ or $p(x) < 0$.

For the case $0 < p(x) < 1$ (7.17) holds in the reverse direction. The inequality (7.17) is sharp in the sense that the constant $C = \frac{1}{1-\alpha}$ in front of the first integral on the RHS of (7.17) is sharp.

Remark 7.12. By using Theorem 7.11 with $p_0 = p_1 = p$ we obtain the following weighted generalization of our basic inequality (7.13):

$$\int_0^\ell \left(\frac{1}{x} \int_0^x g(t) dt \right)^p x^\alpha \frac{dx}{x} \leq \frac{1}{1-\alpha} \int_0^\ell (f(x))^p x^\alpha \left(1 - \left(\frac{x}{\ell} \right)^{1-\alpha} \right) \frac{dx}{x}$$

$$(7.18)$$

for any $\alpha < 1$. For the case $0 < p \leq 1$ (7.18) holds in the reverse direction. The inequality is sharp in both cases (for the case when $p < 0$ we also assume that $f(x) > 0$ a.e.). This, in particular, means that all power weighted Hardy inequalities presented in this section can be derived from Theorem 7.11.

Remark 7.13. It is obvious from the proof above that Theorem 7.11 can be generalized to the situation when $p(x) = p_i$, $a_i \leq x \leq a_{i+1}$, $a_0 = 0$, $a_{N+1} \leq \infty$, $i = 0, 1, \ldots, N$, $N \geq 2$. The only difference is that the second term on the RHS in (7.17) will be more complicated.

Some complementary information can be found in [PeS3], where the ideas used in this section were introduced.

Another limit case

As pointed out before the Pólya-Knopp inequality (7.8) is a limit case of the Hardy inequality (7.2). Another limit case is achieved by trying to prove an inequality of the type (7.15) with $a = p - 1$ (and $\left(\frac{p}{p-1-a}\right)^p$ replaced by some finite constant C).

It is obvious that such type of inequality can never hold without modification. A classical result with a modification in this direction is the following assertion:

Theorem 7.14. *Let $\alpha > 0, 1 \leq p \leq \infty$ and f be a non-negative and measurable function on $[0, 1]$. Then*

$$\left(\int_0^1 [\log(e/x)]^{\alpha p-1} \left(\int_0^x f(y)dy\right)^p \frac{dx}{x}\right)^{1/p}$$
$$\leq \alpha^{-1} \left(\int_0^1 x^p [\log(e/x)]^{(1+\alpha)p-1} f^p(x) \frac{dx}{x}\right)^{1/p} \qquad (7.19)$$

and

$$\left(\int_0^1 [\log(e/x)]^{-\alpha p-1} \left(\int_x^1 f(y)dy\right)^p \frac{dx}{x}\right)^{1/p}$$
$$\leq \alpha^{-1} \left(\int_0^1 x^p [\log(e/x)]^{(1-\alpha)p-1} f^p(x) \frac{dx}{x}\right)^{1/p} \qquad (7.20)$$

with the usual modification if $p = \infty$.

Proof.   The proof is given in [Ben1] only for $1 \leq p < \infty$. For completeness we present a short proof for $p = \infty$. We consider first (7.19).

We have that

$$\int_0^x f(t)dt \leq \left(\sup_{0<t<x} t\log^{1+\alpha}\left(\frac{e}{t}\right)f(t)\right) \int_0^x \log^{-(1+\alpha)}\left(\frac{e}{t}\right)\frac{dt}{t}.$$

After a change of variable and easy calculations we get that

$$\left(\log \frac{e}{x}\right)^\alpha \int_0^x f(t)dt \leq \alpha^{-1} \sup_{0<t<1} t\log^{1+\alpha}\left(\frac{e}{t}\right)f(t)$$

and, hence,

$$\sup_{0<x<1} \left(\log \frac{e}{x}\right)^{\alpha} \int_0^x f(t)dt \le \alpha^{-1} \sup_{0<x<1} x\log^{1+\alpha}\left(\frac{e}{x}\right) f(x). \quad (7.21)$$

In the same way we can prove the inequality corresponding to $p = \infty$ in (7.20),

$$\sup_{0<x<1} \left(\log \frac{e}{x}\right)^{-\alpha} \int_x^1 f(t)dt \le \alpha^{-1} \sup_{0<x<1} x\log^{1-\alpha}\left(\frac{e}{x}\right) f(x). \quad (7.22)$$
$$\square$$

Remark 7.15. This result is due to C. Bennett [Ben1]. He derived it as an important tool when he described the intermediate spaces between L and $L\log^+ L$ with the Peetre real $(K-)$ method. This result was later on completed and applied in various ways in e.g. [Pe1], [BW1] and [NT1]. In fact, the constant α^{-1} in both (7.19) and (7.20) is sharp. This was not pointed out in [Ben1], but it is a consequence of our next Theorem.

The following generalization, refinement and more precise version of Bennett's result was proved in 2014 by S. Barza *et al.* in [BPS1]:

Theorem 7.16. *Let* $\alpha, p > 0$ *and* f *be a non-negative and measurable function on* $[0,1]$.
(a) *If* $p > 1$, *then*

$$\alpha^{p-1} \left(\int_0^1 f(x)dx\right)^p + \alpha^p \int_0^1 [\log(e/x)]^{\alpha p-1} \left(\int_0^x f(y)dy\right)^p \frac{dx}{x}$$

$$\le \int_0^1 x^p [\log(e/x)]^{(1+\alpha)p-1} f^p(x) \frac{dx}{x} \quad (7.23)$$

and

$$\alpha^{p-1} \left(\int_0^1 f(x)dx\right)^p + \alpha^p \int_0^1 [\log(e/x)]^{-\alpha p-1} \left(\int_x^1 f(y)dy\right)^p \frac{dx}{x}$$

$$\le \int_0^1 x^p [\log(e/x)]^{(1-\alpha)p-1} f^p(x) \frac{dx}{x}. \quad (7.24)$$

Both constants α^{p-1} and α^p in (7.23) and (7.24) are sharp. Equality is never attained unless f is identically zero.

(b) *If $0 < p < 1$, then both (7.23) and (7.24) hold in the reverse direction and the constants in both inequalities are sharp. Equality is never attained unless f is identically zero.*

(c) *If $p = 1$ we have equality in (7.23) and (7.24) for any measurable function f and any $\alpha > 0$.*

The proof presented in [BPS1] is completely different from that in [B1] and mainly dependent on a convexity argument, namely the information in the following Lemma:

Lemma 7.17. *It yields that*

$$
h^p - ph + p - 1 \begin{cases} \geq 0 & \text{if } p \geq 1 \\ \leq 0 & \text{if } 0 < p < 1, \end{cases} \tag{7.25}
$$

for all $h > 0$. Equality holds if and only if $h = 1$.

Remark 7.18. The crucial inequality (7.25) is called "a fundamental relationship" in the book [BB1], p. 12. The proof of (7.25) is just simple calculus e.g. by observing that $y(x) = x^p$ is convex for $p \geq 1$ or $p < 0$ and concave for $0 < p \leq 1$, and that the equation of the tangent at the point $x = 1$ is $y = px - p + 1$.

Proof (of Theorem 7.16). (a) Let $p > 1$. Suppose that f is a continuous function and define for $x \in (0, 1]$ the function

$$
F(x; \alpha, p) := \int_0^x y^p \left(\log(e/y) \right)^{(1+\alpha)p-1} f^p(y) \frac{dy}{y}
$$

$$
- \alpha^p \int_0^x \left(\log(e/y) \right)^{\alpha p - 1} \left(\int_0^y f(s) \right)^p \frac{dy}{y}
$$

$$
- \alpha^{p-1} \left(\log(e/x) \right)^{\alpha p} \left(\int_0^x f(y) dy \right)^p .
$$

Differentiation gives that

$$\frac{d}{dx}F(x;\alpha,p) = (\log(e/x))^{\alpha p - 1} \cdot \frac{1}{x}\left(\alpha \int_0^x f(s)\right)^p$$
$$\times [h^p(x;\alpha) - ph(x;\alpha) + p - 1]$$

with

$$h(x;\alpha) := \frac{x(\log(e/x))f(x)}{\alpha \int_0^x f(y)dy}.$$

Thus, according to Lemma 7.17, we have that $\frac{d}{dx}F(x;\alpha,p) > 0$, i.e., $F(x;\alpha,p)$ is strictly increasing. Hence, in particular,

$$F(1;\alpha,p) \geq \lim_{x \to 0+} F(x;\alpha,p).$$

By applying Hölder's inequality we find that

$$\int_0^x f(y)dy = \int_0^x \left[y^{1-1/p}\,(\log(e/y))^{\alpha+1-1/p}\,f(y) \right]$$
$$\times y^{-1+1/p}\,(\log(e/y))^{-\alpha-1+1/p}\,dy$$
$$\leq \left(\frac{p-1}{\alpha p}\right)^{(p-1)/p}\,(\log(e/x))^{-\alpha}\,I(x)$$

with

$$(I(x))^p := \int_0^x y^{p-1}\,(\log(e/y))^{(1+\alpha)p-1}\,f^p(y)dy.$$

Hence, we get that

$$0 < (\log(e/x))^{\alpha p}\left(\int_0^x f(y)dy\right)^p \leq \left(\frac{p-1}{\alpha p}\right)^{p-1}\,I^p(x).$$

Since $I(x) \to 0$ as $x \to 0_+$, we have that $\lim_{x \to 0_+} F(x;\alpha,p) = 0$ and, in particular

$$F(1;\alpha,p) \geq \lim_{x \to 0_+} F(x;\alpha,p) = 0.$$

Hence, we have proved that (7.23) holds for all continuous functions. By standard approximation arguments (7.23) holds for all measurable functions.

Now we prove the sharpness of the inequality (7.23). We consider the inequality

$$K_1 \left(\int_0^1 f(x)dx \right)^p + K_2 \int_0^1 (\log(e/x))^{\alpha p-1} \left(\int_0^x f(y)dy \right)^p \frac{dx}{x}$$

$$\leq \int_0^1 x^p \left(\log(e/x) \right)^{(1+\alpha)p-1} f^p(x) \frac{dx}{x} \tag{7.26}$$

for $\alpha > 0, p > 1$ and some constants $0 < K_1, K_2 < \infty$. Assume that (7.26) holds for some constant $K_2 > \alpha^p$ and consider the test function

$$f_\varepsilon(x) = \frac{1}{x} \left(\log(e/x) \right)^{-(1+\varepsilon+\alpha)}, \ \varepsilon > 0. \tag{7.27}$$

Then, we have that

$$\int_0^1 f_\varepsilon(x)dx = \frac{1}{\varepsilon + \alpha},$$

$$\int_0^1 (\log(e/x))^{\alpha p-1} \left(\int_0^x f_\varepsilon(y)dy \right)^p \frac{dx}{x} = \frac{1}{\varepsilon p(\varepsilon + \alpha)^p}$$

and

$$\int_0^1 x^p \left(\log(e/x) \right)^{(1+\alpha)p-1} f_\varepsilon^p(x) \frac{dx}{x} = \frac{1}{\varepsilon p}.$$

Hence, by (7.26), we have that

$$\frac{K_1}{(\varepsilon + \alpha)^p} + \frac{K_2}{\varepsilon p(\varepsilon + \alpha)^p} \leq \frac{1}{\varepsilon p},$$

i.e., that

$$\varepsilon p K_1 + K_2 \leq (\varepsilon + \alpha)^p.$$

By letting $\varepsilon \to 0_+$ we find that $K_2 \leq \alpha^p$. This contradiction shows that the sharp constant K_2 in (7.26) is $K_2 = \alpha^p$. We consider now (7.26) with $K_2 = \alpha^p$ and assume that it holds with some $K_1 > \alpha^{p-1}$. By using the same test function f_ε defined in (7.27) we get from

(7.26) that

$$\frac{K_1}{(\varepsilon + \alpha)^p} + \frac{\alpha^p}{\varepsilon p (\varepsilon + \alpha)^p} \leq \frac{1}{\varepsilon p},$$

i.e.,

$$K_1 \leq \frac{(\varepsilon + \alpha)^p - \alpha^p}{\varepsilon p}.$$

Hence, by letting $\varepsilon \to 0_+$ we obtain that $K_1 \leq \alpha^{p-1}$. This shows that $K_1 = \alpha^{p-1}$ is also the sharp constant in (7.26) and consequently in (7.23).

The proof of (7.24) is similar. For this case we define

$$G(x; \alpha, p) := \int_0^x y^{p-1} \left(\log(e/y)\right)^{(1-\alpha)p-1} f^p(y) dy$$

$$- \alpha^p \int_0^x y^{-1} \left(\log(e/y)\right)^{-\alpha p - 1} \left(\int_y^1 f(s) ds\right)^p dy$$

$$- \alpha^{p-1} \left(\log(e/x)\right)^{-\alpha p} \left(\int_0^x f(s) ds\right)^p$$

and argue similarly as before. The proof of the sharpness of (7.24) consists only of small modifications of the proof above. By Lemma 7.17 it is clear that we cannot have equality in (7.23) or (7.24) unless f is identically zero. The proof is complete.

(b) Let $0 < p < 1$. In this case the crucial convexity inequality (7.25) from Lemma 7.17 holds in the reversed direction. Hence, the proofs of the reverse of (7.23) and (7.24) consist only of small modifications of the proofs of (7.23) and (7.24), respectively.

(c) The equality for $p = 1$ in both (7.23) and (7.24) is just a consequence of integration by parts and limiting arguments or by straightforward modifications of the proof above. $\square$

Remark 7.19. Easy calculations show that if $p = \infty$ we get equality in the inequality (7.21) for $f(x) = \frac{1}{x}(\log \frac{e}{x})^{-(1+\alpha)}$. Hence, in this case, the inequality (7.21) cannot be improved in the same manner as above to a refined inequality of the type (7.23). In the same way we find that for the case $p = \infty$, inequality (7.22) cannot be improved to some refined inequality of the type (7.24).

Remark 7.20. Some complementary information concerning this limit case is given in the paper [BPS1]. Here we just mention the following sharp limit case of a Hardy inequality for $p > 1$:

$$\int_0^1 \left(\int_0^x f(y)dy \right)^p \frac{dx}{x} \leq \min \left(p^p \int_0^1 x^{p-1} \left(\log \frac{e}{x} \right)^p f^p(x)dx, \right.$$

$$\left. \frac{1}{p-1} \int_0^1 (1 - x^{p-1}) f^p(x)dx \right).$$

7.3. Scales of Conditions to Characterize Hardy-type Inequalities

Already in Chap. 1 we mentioned some alternative criteria to characterise the Hardy inequality for the cases $1 < p \leq q < \infty$ (see Theorem 1.1) and $0 < q < p < \infty$, $p > 1$ (see Theorem 1.2). In this section we shall complement this information by also mentioning a number of other conditions to characterize the Hardy inequality (see Subsec. 7.3.1). The most important new information is given in Subsec. 7.3.2, namely that all these characterizations can be unified and extended to infinite many conditions, even to scales of conditions. The known conditions (e.g. those in Theorems 1.1 and 1.2 and those in Subsec. 7.3.1) then just appear as points on some of these scales. The first proofs of this type were given via a direct approach but nowadays it is known that for the case $1 < p \leq q < \infty$ it can also be proved and extended via a general equivalence theorem of independent interest. This general equivalence theorem is presented and discussed in Subsec. 7.3.3 and also how the proof of Theorem 7.23 follows from this Theorem and the known Muckenhoupt-Bradley characterization.

7.3.1. *Some alternative conditions to describe the Hardy inequality*

We consider again the general Hardy inequality

$$\left(\int_0^b \left(\int_0^x f(t)dt \right)^q u(x)dx \right)^{1/q} \leq C \left(\int_0^b f(x)^p v(x)dx \right)^{1/p} \quad (7.28)$$

with a fixed $b, 0 < b \le \infty$, for measurable function $f \ge 0$, weights u and v and for parameters p, q satisfying $0 < q < \infty$ and $p \ge 1$.

The case $p = q > \infty$

As mentioned in Chap. 1 a classical result here reads: Let $1 \le p < \infty$. Then the inequality (7.28) holds for all measurable functions $f \ge 0$ on $(0, b), 0 < b \le \infty$, if and only if

$$A := \sup_{r \in (0,b)} \left(\int_r^b u(x)dx \right)^{1/p} \left(\int_0^r v^{1-p'}(x)dx \right)^{1/p'} < \infty, \quad (7.29)$$

where as usual $p' = p/(p-1)$ when $p > 1$ and $p' = \infty$ when $p = 1$ (so the second integral must be interpreted as a supremum).

Remark 7.21. The condition (7.29) is frequently called the Muckenhoupt condition since B. Muckenhoupt [Mu1] presented in 1972 a nice and direct proof, which has influenced the further development in a crucial way. However, Muckenhoupt mentioned that G. Talenti [T1] and G. Tomaselli [To1] had already proved this result in 1969, but in these papers the result was not so explicitly stated as in [Mu1].

Remark 7.22. In [To1] Tomaselli also derived two other conditions for characterizing the Hardy inequality, namely the following:

$$A^* := \sup_{r \in (0,b)} \left(\int_0^r u(x) \left(\int_0^x v^{1-p'}(t)dt \right)^p dx \right)$$

$$\times \left(\int_0^r v^{1-p'}(x)dx \right)^{-1} < \infty$$

and

$$A^{**} := \inf_{h>0} \sup_{x \in (0,b)} \frac{1}{h(x)} \int_0^x u(t) \left[h(t) + \int_0^t v^{1-p'}(s)ds \right]^p dt < \infty.$$

Also this result has influenced the further development. Moreover, for the best constant C in (7.28) it yields that

$$C \approx A \approx A^* \approx A^{**}.$$

The case $1 < p \leq q < \infty$

Inequality (7.28) is usually characterized by the (so-called Muckenhoupt-Bradley) condition

$$A_{MB} := \sup_{0<x<b} A_{MB}(x) < \infty, \qquad (7.30)$$

where

$$A_{MB}(x) := \left(\int_x^b u(t)dt \right)^{1/q} \left(\int_0^x v^{1-p'}(t)dt \right)^{1/p'}. \qquad (7.31)$$

Further, let us denote

$$U(x) := \int_x^b u(t)dt, \qquad V(x) := \int_0^x v^{1-p'}(t)dt, \qquad (7.32)$$

and assume that $U(x) < \infty$, $V(x) < \infty$ for every $x \in (0, b)$.

The index MB in $A_{MB} := \sup_{0<x<b} U^{1/q}(x)V^{1/p'}(x)$ indicates the efforts of B. Muckenhoupt and J.S. Bradley. As mentioned before in 1972 B. Muckenhoupt [Mu1] gave a nice proof of the fact that $A_{MB} < \infty$ is necessary and sufficient for (7.28) to hold for the case $p = q$ and in 1978 J.S. Bradley [Br1] extended the Muckenhoupt result to the case $p \leq q$ and gave a complete and simple proof of Muckenhoupt type of this result. However, this result was also independently derived in 1979 by V.G. Maz'ja and L. Rozin (see [Maz1]) and by V.M. Kokilashvili (see [Ko1]).

Besides the condition $A_{MB} < \infty$, some other equivalent conditions have been derived during the next decades, e.g. the conditions $A_{TG} < \infty$ or $A_{TG}^* < \infty$, where

$$A_{TG} := \inf_{h>0} \sup_{0<x<b} \left(\frac{1}{h(x)} \int_0^x u(t)(h(t) + V(t))^{\frac{q}{p'}+1} dt \right)^{1/q};$$

$$\qquad (7.33)$$

$$A_{TG}^* := \inf_{h>0} \sup_{0<x<b} \left(\frac{1}{h(x)} \int_x^b v^{1-p'}(t)(h(t) + U(t))^{\frac{p'}{q}+1} dt \right)^{1/p'}.$$

This result was proved by P. Gurka [Gu1] in 1984; he extended to the case $p \leq q$ the result proved for $p = q$ in 1969 by G. Tomaselli [To1].

Some other alternative characterizations read $A_{PS} < \infty$ or $A_{PS}^* < \infty$ (see Theorem 1.1), where

$$A_{PS} := \sup_{0<x<b} \left(\int_0^x u(t)V^q(t)dt \right)^{1/q} V^{-1/p}(x);$$

$$A_{PS}^* := \sup_{0<x<b} \left(\int_x^b v^{1-p'}(t)U^{p'}(t)dt \right)^{1/p'} U^{-1/q'}(x).$$

(7.34)

This result was proved in 2002 by L.E. Persson and V.D. Stepanov [PeSt1] but as we have seen it was proved for the case $p = q$ already in 1969 in [To1].

Moreover, for the best constant C in (7.28) it yields that

$$C \approx A_{MB} \approx A_{TG} \approx A_{TG}^* \approx A_{PS} \approx A_{PS}^*.$$

The case $1 \leq q < p < \infty$

A necessary and sufficient condition for (7.28) to hold in this case was derived by V. Maz'ja and L. Rozin in the late '70s (see [Maz1]) and it reads (see the Introduction):

$$B_{MR} := \left(\int_0^\infty U^{r/p}(x)V^{r/p'}(x)u(x)dx \right)^{1/r} < \infty, \qquad (7.35)$$

where $1/r := 1/q - 1/p$. An alternative condition was found by L.E. Persson and V. Stepanov in 2002 (see [PeSt1]) and it reads (see Theorem 1.2):

$$B_{PS} := \left(\int_0^\infty \left[\int_0^x u(t)V^q(t)dt \right]^{r/p} u(x)V^{q-r/p}(x)\,dx \right)^{1/r} < \infty.$$

(7.36)

Moreover, for the best constant C in (7.28) it yields that

$$C \approx B_{MR} \approx B_{PS}.$$

7.3.2. *Unification and extensions: Scales of conditions*

The case $1 < p \leq q < \infty$

The first scale of conditions for general $b > 0$ was derived in 2004 by A. Kufner *et al.* (see [KPW1]). It reads $A(r) < \infty$ (with $1 < r < p$) or $A^*(r) < \infty$ (with $1 < r < q'$), where

$$A(r) := \sup_{0<x<b} \left(\int_x^b u(t) V^{q(p-r)/p}(t) dt \right)^{1/q} V^{(r-1)/p}(x), \quad 1 < r < p;$$

$$(7.37)$$

$$A^*(r) := \sup_{0<x<b} \left(\int_0^x v^{1-p'}(t) U^{p'(q'-r)/q'}(t) dt \right)^{1/p'} U^{(r-1)/q'}(x),$$

$$1 < r < q'.$$

For the case $b = \infty$ see the Ph.D. thesis [We1] of A. Wedestig and also [We2]. Note that the endpoint condition $A(p) < \infty$ is just the Muckenhoupt-Bradley condition $A_{MB} < \infty$ mentioned above.

Moreover, A. Gogatishvili *et al.* in [GKPW1] derived four new scales of equivalent integral conditions. This result was used to characterize the inequality (7.28) by four scales of conditions, namely the scales including the Muckenhoupt-Bradley condition, the Persson-Stepanov condition and the dual of these scales. This result reads:

Theorem 7.23. *Let* $1 < p \leq q < \infty$ *,$0 < s < \infty$, and define, for the weight functions u, v, the functions U and V by (7.32), and the functions $A_i(s)$, $i = 1, 2, 3, 4$, as follows:*

$$A_1(s) := \sup_{0<x<b} \left(\int_x^b u(t) V^{q\left(\frac{1}{p'}-s\right)}(t) dt \right)^{1/q} V^s(x),$$

$$A_2(s) := \sup_{0<x<b} \left(\int_0^x v^{1-p'}(t) U^{p'\left(\frac{1}{q}-s\right)}(t) dt \right)^{1/p'} U^s(x),$$

$$(7.38)$$

$$A_3(s) := \sup_{0<x<b} \left(\int_0^x u(t) V^{q(\frac{1}{p'}+s)}(t) dt \right)^{1/q} V^{-s}(x),$$

$$A_4(s) := \sup_{0<x<b} \left(\int_x^b v^{1-p'}(t) U^{p'(\frac{1}{q}+s)}(t) dt \right)^{1/p'} U^{-s}(x). \qquad (7.38)$$

Then the Hardy inequality (7.28) holds for all measurable functions $f \geq 0$ if and only if any of the quantities $A_i(s)$, $i = 1,2,3,4$, in (7.38) is finite for any $s, 0 < s < \infty$. Moreover, for the best constant C in (7.28) we have $C \approx A_i(s)$, $i = 1,2,3,4$. The constants in the equivalence relations depend only on s.

Remark 7.24. The previous mentioned conditions can be described in the following way:

$$A_{MB} = A_1\left(\frac{1}{p'}\right) = A_2\left(\frac{1}{q}\right), \quad A_{PS} = A_3\left(\frac{1}{p}\right),$$

$$A(r) = A_1\left(\frac{r-1}{p}\right) \quad \text{with } 1 < r < p,$$

$$A_{PS}^* = A_4\left(\frac{1}{q'}\right), \quad A^*(r) = A_2\left(\frac{r-1}{q'}\right) \quad \text{with } 1 < r < q'.$$

Hence, Theorem 7.23 contains all previous conditions except those by Tomaselli and Gurka: $A_{TG} < \infty$ and $A_{TG}^* < \infty$. In a later paper [GKP2] it was proved that Theorem 7.23 can in fact be generalized by adding ten further equivalent scales of conditions to characterize the Hardy inequality (7.28). Two of these scales read:

$$A_5(s) := \inf_{h>0} \sup_{0<x<b} \left(\int_0^x u(t)(h(t) + V(t))^{q(\frac{1}{p'}+s)} dt \right)^{1/q} h^{-s}(x);$$

$$\qquad (7.39)$$

$$A_6(s) := \inf_{h>0} \sup_{0<x<b} \left(\int_x^b v^{1-p'}(t)(h(t) + U(t))^{p'(\frac{1}{q}+s)}(t) \right)^{1/p'} h^{-s}(x).$$

In particular, we see that

$$A_{TG} = A_5 \left(\frac{1}{q}\right) \text{ and } A_{TG}^* = A_6 \left(\frac{1}{p'}\right).$$

The case $0 < q < p < \infty,\ p > 1, q \neq 1$

For simplicity in this case we only consider the case $b = \infty$. We suppose the following concerning the involved weight functions:

$$0 < U(x) := \int_x^\infty u(t)dt < \infty,$$

$$0 < V(x) := \int_0^x v^{1-p'}(t)dt < \infty \quad \text{for all } x > 0. \qquad (7.40)$$

Cf. also (7.32). Let $1/r := 1/q - 1/p$.

We now introduce the following scales of constants related to previous constants and their dual ones:

For $s > 0$ we define the following functionals:

$$B_{MR}^{(1)}(s) := \left(\int_0^\infty \left[\int_t^\infty uV^{q(1/p'-s)}dx\right]^{r/p} V^{q(1/p'-s)+rs}(t)u(t)\,dt\right)^{1/r},$$

$$B_{PS}^{(1)}(s) := \left(\int_0^\infty \left[\int_0^t uV^{q(1/p'+s)}dx\right]^{r/p} u(t)V^{q(1/p'+s)-sr}(t)\,dt\right)^{1/r},$$

$$B_{MR}^{(2)}(s) := \left(\int_0^\infty \left[\int_0^t U^{p'(1/q-s)}dV\right]^{r/p'} U^{rs-1}(t)u(t)\,dt\right)^{1/r},$$

$$B_{PS}^{(2)}(s) := \left(\int_0^\infty \left[\int_t^\infty U^{q(1/p'+s)}dV\right]^{r/p} U^{q(1/p'+s)-rs}(t)\,dV(t)\right)^{1/r}.$$

The main theorem in this case was derived in 2007 by L.E. Persson *et al.* (see [PStW1]):

Theorem 7.25. a) *Let* $0 < q < p < \infty,\ 1 < p < \infty$ *and* $q \neq 1$. *Then the Hardy inequality* (7.28) *with* $b = \infty$ *holds for some finite constant* $C > 0$ *if and only if any of the constants* $B_{MR}^{(1)}(s)$ *or* $B_{PS}^{(1)}(s)$

is finite for some $s > 0$. Moreover, for the best constant C in (7.28)
we have

$$C \approx B_{MR}^{(1)}(s) \approx B_{PS}^{(1)}(s). \qquad (7.41)$$

b) *Let $1 < q < p < \infty$. Then the Hardy inequality (7.28) with*
$b = \infty$ *holds for some finite constant $C > 0$ if and only if any of the*
constants $B_{MR}^{(2)}(s)$ or $B_{PS}^{(2)}(s)$ is finite for some $s > 0$. Moreover, for
the best constant C in (7.28) we have

$$C \approx B_{MR}^{(2)}(s) \approx B_{PS}^{(2)}(s).$$

Remark 7.26. Theorem 7.25 is a generalization of the original
results by Maz'ja-Rosin and Persson-Stepanov since

$$B_{MR}^{(1)}\left(\frac{1}{p'}\right) = B_{MR}, \quad \text{and} \quad B_{PS}^{(1)}\left(\frac{1}{p}\right) = B_{PS}.$$

The four scales in Theorem 7.25 correspond to the four scales in
Theorem 7.23. The difference depends on the fact that in the case
$0 < q < 1 < p < \infty$ no duality exists. In fact, the statement in
b) is just a corollary of a statement similar to that in a) in a dual
situation, namely when $\int_0^x$ in inequality (7.28) is replaced by $\int_x^\infty$ (see
Theorem 2 in [PStW1]).

We finish this section by giving some examples on how alternative
conditions to characterize Hardy-type inequalities can be used and
illustrated. The most obvious advantage is that in a concrete
situation one criterion can be easier to verify than another. Another
is that we get more possibilities to estimate the best constant in the
Hardy inequality.

Example 7.27. Consider the classical unweighted Hardy inequality

$$\int_0^\infty \left(\frac{1}{x}\int_0^x f(y)dy\right)^p dx \leq \left(\frac{p}{p-1}\right)^p \int_0^\infty f^p(x)dx, \ p > 1, \quad (7.42)$$

and its limit (Pólya-Knopp) inequality

$$\int_0^\infty \exp\left(\frac{1}{x}\int_0^x \ln f(y)dy\right) dx \leq e \int_0^\infty f(x)dx. \qquad (7.43)$$

Inequalities (7.42) and (7.43) with sharp constants $\left(\frac{p}{p-1}\right)^p$ and e, respectively, follow directly from the general characterization and that $A_{PS} = 1$ in this case. This is not the case when we use e.g. the constant A_{MB} in our characterization. More generally, by using the same technic in the general case it can be proved that if $0 < p \leq q < \infty$, then the inequality

$$\left(\int_0^b \left(\exp\left(\frac{1}{x}\int_0^x \ln f(y)dy\right)\right)^q u(x)dx\right)^{1/q}$$

$$\leq C \left(\int_0^b f^p(x)v(x)dx\right)^{1/p}$$

holds if and only if

$$D := \sup_{0<x<b} x^{-1/p}\left(\int_0^x w(y)dy\right)^{1/q} < \infty,$$

where

$$w(x) := \left[G\left(\frac{1}{v(x)}\right)\right]^{q/p} u(x),$$

where $Gf(x) := \exp\frac{1}{x}\int_0^x \ln f(y)dy$ (see (1.84)). Moreover, $D \leq C \leq e^{1/p}D$, and the constant $e^{1/p}$ cannot be improved in general (see [PeSt1]).

Example 7.28. For example in terms of $A(r)$ defined by (7.37) ($= A_1(r)$ in (7.38) for $1 < r < p$) we have the following estimates of the best constants C in (7.28):

$$\sup_{1<s<p} \left(\frac{\left(\frac{p}{p-s}\right)^p}{\left(\frac{p}{p-s}\right)^p + \frac{1}{s-1}}\right)^{1/p} A(s) \leq C \leq \inf_{1<s<p}\left(\frac{p-1}{p-s}\right)^{1/p'} A(s),$$

$$(7.44)$$

see [KPW1], Theorem 1 and cf. also [We1]. Moreover, for the condition $A_{MB} < \infty$ at the right endpoint we have the standard estimate

$$A_{MB} \le C \le A_{MB} \cdot k(p,q), \qquad (7.45)$$

where $k(p,q) = \left(1 + \frac{q}{p'}\right)^{1/q} \left(1 + \frac{p'}{q}\right)^{1/p'}$ (see pp. 7 and 8). Later on V.M. Manakov [Ma1] improved this constant to

$$k(p,q) = \left(\frac{\Gamma\left(\frac{pq}{q-p}\right)}{\Gamma\left(\frac{q}{q-p}\right)\Gamma\left(\frac{p(q-1)}{q-p}\right)}\right)^{\frac{q-p}{qp}}, \quad p < q.$$

Concerning the constant A_{PS} we have the following estimate (see [PeSt1], Theorem 1):

$$A_{PS} \le C \le p' \, A_{PS}. \qquad (7.46)$$

By using this information and making elementary estimates we can see that by using A_{PS} we get that

$$A_{MB} \cdot (p-1)^{1/q} \le C \le A_{MB} \cdot p'(p-1)^{1/q}, \qquad (7.47)$$

see [We1], p. 129. In particular, this always improves the lower bound in (7.45) for all $p > 2$. In particular, let $p = 3, q = 4$ and choose $s = 1.15$ in the lower bound in (7.44) and we obtain the following estimates:

a) $A_{MB} \le C \le A_{MB} \cdot 1.530348452$,

by using (7.45) with Manakov's constant.

b) $A_{MB} \cdot 1.189207115 \le C$,

by using (7.46).

c) $A_{MB} \cdot 1.396254480 \le C$,

by using the scale of conditions in (7.44).

Summing up, we find that

$$A_{MB} \cdot 1.396254480 \le C \le A_{MB} \cdot 1.530348452,$$

which is a better estimate than those we can find in standard literature.

7.3.3. *A general equivalence theorem and proof of Theorem 7.23*

Theorem 7.29. *For* $-\infty \leq a < b \leq \infty$, α, β *and* s *positive numbers and* f, g *measurable functions which are positive a.e. in* (a, b), *let us denote*

$$F(x) := \int_x^b f(t)dt, \qquad G(x) := \int_a^x g(t)dt \qquad (7.48)$$

and

$$B_1(x; \alpha, \beta) := F^\alpha(x)G^\beta(x),$$

$$B_2(x; \alpha, \beta, s) := \left(\int_x^b f(t)G^{\frac{\beta-s}{\alpha}}(t)dt \right)^\alpha G^s(x),$$

$$B_3(x; \alpha, \beta, s) := \left(\int_a^x g(t)F^{\frac{\alpha-s}{\beta}}(t)dt \right)^\beta F^s(x), \qquad (7.49)$$

$$B_4(x; \alpha, \beta, s) := \left(\int_a^x f(t)G^{\frac{\beta+s}{\alpha}}(t)dt \right)^\alpha G^{-s}(x),$$

$$B_5(x; \alpha, \beta, s) := \left(\int_x^b g(t)F^{\frac{\alpha+s}{\beta}}(t)dt \right)^\beta F^{-s}(x).$$

The numbers $B_1 := \sup_{a<x<b} B_1(x; \alpha, \beta)$ *and*

$$B_i = \sup_{a<x<b} B_i(x; \alpha, \beta, s)(i = 2, 3, 4, 5)$$

are mutually equivalent. The constants in the equivalence relations can depend on α, β *and* s.

Remark 7.30. The proof of Theorem 7.29 is carried out by deriving positive constants c_i and d_i so that

$$c_i \sup_{a<x<b} B_i(x; \alpha, \beta, s) \leq \sup_{a<x<b} B_1(x; \alpha, \beta)$$

$$\leq d_i \sup_{a<x<b} B_i(x; \alpha, \beta, s), i = 2, 3, 4, 5, \quad (7.50)$$

see (7.54) and (7.56)–(7.62) below. This information is useful e.g. for obtaining good estimates of the best constant C in (7.28).

Proof. First we prove that

$$\sup_{a<x<b} B_1(x;\alpha,\beta) \approx \sup_{a<x<b} B_2(x;\alpha,\beta,s), \tag{I}$$

i.e., that (7.50) holds for $i = 2$.

(i) Let $s \le \beta$. Then $\frac{\beta-s}{\alpha} \ge 0$, and since $G(x)$ is increasing, we have that for $t \ge x$

$$G^{\frac{\beta-s}{\alpha}}(t) \ge G^{\frac{\beta-s}{\alpha}}(x). \tag{7.51}$$

Consequently,

$$B_2(x;\alpha,\beta,s) = \left(\int_x^b f(t)G^{\frac{\beta-s}{\alpha}}(t)dt\right)^\alpha G^s(x)$$

$$\ge \left(\int_x^b f(t)dt\right)^\alpha \left(G^{\frac{\beta-s}{\alpha}}(x)\right)^\alpha G^s(x) = F^\alpha(x)G^\beta(x). \tag{7.52}$$

(ii) Let $s > \beta$ and denote

$$W(x) := \int_x^b f(t)G^{\frac{\beta-s}{\alpha}}(t)dt,$$

i.e., $-dW(x) = f(x)G^{\frac{\beta-s}{\alpha}}(x)dx$. Then

$$F^\alpha(x)G^\beta(x) = G^\beta(x)\left(\int_x^b f(t)G^{\frac{\beta-s}{\alpha}}(t)G^{\frac{s-\beta}{\alpha}}(t)W^{\frac{s-\beta}{s}}(t)W^{\frac{\beta-s}{s}}(t)dt\right)^\alpha$$

$$\le \left(\sup_{x<t<b} G^{s-\beta}(t)W^{\frac{(s-\beta)\alpha}{s}}(t)\right)G^\beta(x)$$

$$\times \left(-\int_x^b W^{\frac{\beta-s}{s}}(t)dW(t)\right)^\alpha$$

$$= \left(\sup_{x<t<b} G^s(t)W^\alpha(t)\right)^{\frac{s-\beta}{s}}\left(\frac{s}{\beta}\right)^\alpha G^\beta(x)W^{\frac{\beta}{s}\alpha}(x)$$

$$\leq \left(\frac{s}{\beta}\right)^{\alpha} \left(\sup_{x<t<b} G^s(t)W^{\alpha}(t)\right)^{1-\frac{\beta}{s}} \left(\sup_{x<t<b} G^s(t)W^{\alpha}(t)\right)^{\frac{\beta}{s}}$$

$$= \left(\frac{s}{\beta}\right)^{\alpha} \sup_{x<t<b} B_2(t; \alpha, \beta, s). \qquad (7.53)$$

Hence, for every $s > 0$ it follows from (7.52) and (7.53) that

$$\sup_{a<x<b} B_1(x; \alpha, \beta) \leq \left(\max\left(1, \frac{s}{\beta}\right)\right)^{\alpha} \sup_{a<x<b} B_2(x; \alpha, \beta, s). \qquad (7.54)$$

Also for the proof of the opposite estimate we need to consider two cases.

(iii) Let $s > \beta$. Then we have an inequality opposite to (7.51) and hence

$$B_2(x; \alpha, \beta, s) = G^s(x) \left(\int_x^b f(t) G^{\frac{\beta-s}{\alpha}}(t) dt\right)^{\alpha}$$

$$\leq G^s(x) \left(\int_x^b f(t) dt\right)^{\alpha} G^{\beta-s}(x) = F^{\alpha}(x) G^{\beta}(x). \qquad (7.55)$$

(iv) For $s < \beta$ we have

$$G^s(x) W^{\alpha}(x) = G^s(x) \left(\int_x^b f(t) G^{\frac{\beta-s}{\alpha}}(t) F^{\frac{\beta-s}{\beta}}(t) F^{\frac{s-\beta}{\beta}} dt\right)^{\alpha}$$

$$\leq \left(\sup_{x<t<b} G^{\frac{\beta-s}{\alpha}}(t) F^{\frac{\beta-s}{\beta}}(t)\right)^{\alpha} G^s(x)$$

$$\times \left(\int_x^b F^{\frac{s}{\beta}-1}(t) (-dF(t))\right)^{\alpha}$$

$$= \left(\sup_{x<t<b} G^{\beta}(t) F^{\alpha}(t)\right)^{\frac{\beta-s}{\beta}} \left(\frac{\beta}{s}\right)^{\alpha} G^s(x) F^{\frac{\alpha s}{\beta}}(x)$$

$$\leq \left(\sup_{x<t<b} G^{\beta}(t) F^{\alpha}(t) \right)^{\frac{\beta-s}{\beta}} \left(\frac{\beta}{s} \right)^{\alpha}$$

$$\times \left(\sup_{a<x<b} G^{\beta}(x) F^{\alpha}(x) \right)^{\frac{s}{\beta}}$$

$$\leq \left(\frac{\beta}{s} \right)^{\alpha} \sup_{a<x<b} B_1(x; \alpha, \beta).$$

Therefore, for every $s > 0$ it follows that

$$\sup_{a<x<b} B_2(x; \alpha, \beta, s) \leq \left(\max \left(1, \frac{\beta}{s} \right) \right)^{\alpha} \sup_{a<x<b} B_1(x; \alpha, \beta) \qquad (7.56)$$

and the equivalence (I) on p. 363 follows from (7.54) and (7.56).

Now we consider the equivalence

$$\sup_{a<x<b} B_1(x; \alpha, \beta) \approx \sup_{a<x<b} B_3(x; \alpha, \beta, s). \qquad (\text{II})$$

The proof follows the same idea as the proof of (I); we only have to exchange the roles of F and G. We get

$$\sup_{a<x<b} B_1(x; \alpha, \beta) \leq \left(\max \left(1, \frac{s}{\alpha} \right) \right)^{\beta} \sup_{a<x<b} B_3(x; \alpha, \beta, s), \qquad (7.57)$$

and

$$\sup_{a<x<b} B_3(x; \alpha, \beta, s) \leq \left(\max \left(1, \frac{\alpha}{s} \right) \right)^{\beta} \sup_{a<x<b} B_1(x; \alpha, \beta). \qquad (7.58)$$

Next we prove

$$\sup_{a<x<b} B_1(x; \alpha, \beta) \approx \sup_{a<x<b} B_4(x; \alpha, \beta, s). \qquad (\text{III})$$

If we denote

$$\widetilde{W}(x) = \int_a^x f(t) G^{\frac{\beta+s}{\alpha}}(t) dt$$

so that $B_4(x; \alpha, \beta, s) = G^{-s}(x)\widetilde{W}^\alpha(x)$, and use the fact that $g(t)dt = dG(t)$ and integration by parts, we obtain

$$B_1(x; \alpha, \beta) = G^\beta(x) \left(\int_x^b f(t) G^{\frac{\beta+s}{\alpha}}(t) G^{-\frac{\beta+s}{\alpha}}(t) dt \right)^\alpha$$

$$= G^\beta(x) \left(\int_x^b G^{-\frac{\beta+s}{\alpha}}(t) d\widetilde{W}(t) \right)^\alpha$$

$$\leq G^\beta(x) \left(G^{-\frac{\beta+s}{\alpha}}(b) \widetilde{W}(b) \right.$$

$$\left. + \frac{\beta+s}{\alpha} \int_x^b g(t) G^{-\frac{\beta+s}{\alpha}-1}(t) \widetilde{W}(t) dt \right)^\alpha$$

$$\leq G^\beta(x) \sup_{x<t<b} G^{-s}(t) \widetilde{W}^\alpha(t)$$

$$\times \left(G^{-\frac{\beta}{\alpha}}(b) + \frac{\beta+s}{\alpha} \int_x^b G^{-\frac{\beta}{\alpha}-1}(t) dG(t) \right)^\alpha$$

$$\leq G^\beta(x) \sup_{a<t<b} B_4(t, \alpha, \beta, s)$$

$$\times \left(G^{-\frac{\beta}{\alpha}}(b) + \frac{\beta+s}{\beta} \left(G^{-\frac{\beta}{\alpha}}(x) - G^{-\frac{\beta}{\alpha}}(b) \right) \right)^\alpha$$

$$= \sup_{a<t<b} B_4(t, \alpha, \beta, s) \left[\frac{\beta+s}{\beta} + \left(1 - \frac{\beta+s}{\beta} \right) \left(\frac{G(x)}{G(b)} \right)^{\frac{\beta}{\alpha}} \right]^\alpha$$

$$\leq \left(1 + \frac{s}{\beta} \right)^\alpha \sup_{a<t<b} B_4(t, \alpha, \beta, s).$$

Thus,

$$\sup_{a<x<b} B_1(x; \alpha, \beta) \leq \left(1 + \frac{s}{\beta} \right)^\alpha \sup_{a<x<b} B_4(x, \alpha, \beta, s). \qquad (7.59)$$

To prove the reverse inequality we assume that $\sup_{a<x<b} B_1(x; \alpha, \beta) < \infty$. Then, by using the fact that $f(t)dt =$

$-dF(t)$ and integration by parts, we find

$$B_4(x, \alpha, \beta, s) = G^{-s}(x) \left(\int_a^x G^{\frac{\beta+s}{\alpha}}(t) d\left(-F(t)\right) \right)^\alpha$$

$$= G^{-s}(x) \left(G^{\frac{\beta+s}{\alpha}}(t) F(t) \mid_x^a + \frac{\beta+s}{\alpha} \right.$$

$$\left. \times \int_a^x F(t) G^{\frac{\beta+s}{\alpha}-1} g(t) dt \right)^\alpha$$

$$\leq G^{-s}(x) \left(\sup_{a<t<x} G^\beta(t) F^\alpha(t) \right)$$

$$\times \left(\frac{\beta+s}{\alpha} \int_a^x G^{\frac{s}{\alpha}-1} dG(t) \right)^\alpha$$

$$\leq \left(\frac{\beta+s}{\alpha} \right)^\alpha \sup_{a<t<b} G^\beta(t) F^\alpha(t) G^{-s}(x) \left(\frac{\alpha}{s} G^{\frac{s}{\alpha}}(x) \right)^\alpha$$

$$= \left(\frac{\beta+s}{s} \right)^\alpha \sup_{a<x<b} B_1(x; \alpha, \beta).$$

Hence we have

$$\sup_{a<x<b} B_4(x; \alpha, \beta, s) \leq \left(1 + \frac{\beta}{s} \right)^\alpha \sup_{a<x<b} B_1(x, \alpha, \beta). \qquad (7.60)$$

Now (III) follows by combining (7.59) and (7.60).

Finally, we consider the equivalence

$$\sup_{a<x<b} B_1(x; \alpha, \beta) \approx \sup_{a<x<b} B_5(x; \alpha, \beta, s). \qquad (IV)$$

The proof follows the same ideas as the proof of (III); we only have to exchange the roles of F and G. We have

$$\sup_{a<x<b} B_1(x; \alpha, \beta) \leq \left(1 + \frac{s}{\alpha} \right)^\beta \sup_{a<x<b} B_5(x; \alpha, \beta, s) \qquad (7.61)$$

and

$$\sup_{a<x<b} B_5(x; \alpha, \beta, s) \leq \left(1 + \frac{\alpha}{s} \right)^\beta \sup_{a<x<b} B_1(x; \alpha, \beta). \qquad (7.62)$$

The proof is complete. $\qquad \qquad \square$

Proof of Theorem 7.23. In (7.48) and (7.50) we put $a = 0$, $f(x) = u(x), g(x) = v^{1-p'}(x)$, so that $F(x) = U(x)$, $G(x) = V(x)$, and choose

$$\alpha = \frac{1}{q}, \quad \beta = \frac{1}{p'}.$$

Then the assertion follows from the fact that

$$A_1(s) = \sup_{0 < x < b} B_2\left(x; \frac{1}{q}, \frac{1}{p'}, s\right),$$

$$A_2(s) = \sup_{0 < x < b} B_3\left(x; \frac{1}{q}, \frac{1}{p'}, s\right),$$

$$A_3(s) = \sup_{0 < x < b} B_4\left(x; \frac{1}{q}, \frac{1}{p'}, s\right),$$

$$A_4(s) = \sup_{0 < x < b} B_5\left(x; \frac{1}{q}, \frac{1}{p'}, s\right),$$

are all equivalent with the Muckenhoupt-Bradley constant A_{MB} from (7.30)–(7.31). As mentioned before, the finiteness of A_{MB} is necessary and sufficient for the inequality (7.28) to hold. Moreover, since for the least constant C in (7.28) we have $C \approx A_1$ and it is clear that $C \approx A_i(s)$ the proof is complete.

Remark 7.31. (a) The fact that the conditions (7.39), and also the other mentioned eight scales of conditions can be added to the conditions in Theorem 7.23, are proved in a similar direct way in [GKP2].

(b) The proof of Theorem 7.25 (the case $0 < q < p < \infty$, $p > 1$) is performed by using completely different arguments but also in this case some alternative scales of conditions to characterize (7.28) can be derived by using other known results e.g. those in [SS1] (for details see [PStW1] and cf. also [KPS2]).

Remark 7.32. Some similar scales of (equivalent) conditions characterizing the Hardy inequality with general measures were derived by C. Okpoti *et al.* [OPS1] (the case $1 < p \leq q < \infty$) and [OPS2] (the case $0 < q < p < \infty$, $p > 1$).

7.4. Hardy's Inequalities for "all" Parameters

Together with the Hardy inequality

$$\left(\int_a^b \left(\int_a^x f(t)dt \right)^q u(x)dx \right)^{1/q} \leq C \left(\int_a^b f(x)^p v(x)dx \right)^{1/p} \quad (7.63)$$

we will here consider also the reverse Hardy inequality

$$\left(\int_a^b \left(\int_a^x f(t)dt \right)^q u(x)dx \right)^{1/q} \geq C \left(\int_a^b f(x)^p v(x)dx \right)^{1/p}. \quad (7.64)$$

Up to now we have mostly investigated inequality (7.63) for $1 < p \leq q < \infty$ and for $0 < q < p < \infty$, $q \neq 1$, $1 < p < \infty$. In the case $0 < p < 1$ inequality (7.63) holds only for $f \equiv 0$ (see [KMP], p. 46). Hence, for $p < 1$ we can consider only the reverse inequality (7.64).

In what follows we will consider inequalities (7.63) and (7.64) for general real values of the parameters p, q. But before we show that it is reasonable to consider also other values than $p, q > 1$, let us give an example indicating the connection of our inequalities with "usual" inequalities for sequences, i.e., for arithmetic, geometric and harmonic means.

Example 7.33. Let us take $a = 0$ and $b = \infty$ in (7.63). The expression

$$\frac{1}{x} \int_0^x f(t)dt \quad (7.65)$$

corresponds to the *arithmetic means* $AM_n(\bar{a})$ of a sequence $\bar{a} = \{a_n\}_{n=1}^\infty$:

$$AM_n(\bar{a}) := \frac{1}{n} \sum_{i=1}^n a_i.$$

If we rewrite (7.63) [with $a = 0$ and $b = \infty$] as

$$\left(\int_0^\infty \left(\frac{1}{x} \int_0^x f(t)dt \right)^q x^q u(x)dx \right)^{1/q} \leq C \left(\int_0^\infty f(x)^p v(x)dx \right)^{1/p},$$

then we see that the Hardy inequality is in fact a weighted estimate (with the modified weight $\hat{u}(x) = u(x)x^q$) of the "arithmetic mean" (7.65).

For the *harmonic means* $HM_n(\bar{a})$ of the sequence $\bar{a}$ we have

$$HM_n(\bar{a}) := \frac{n}{\sum\limits_{i=1}^{n} \frac{1}{a_i}} = \left[AM_n \left(\frac{1}{\bar{a}} \right) \right]^{-1}.$$

Consequently, in analogy to (7.65) we can introduce the "harmonic mean"

$$\frac{x}{\int_0^x \frac{1}{f(t)} dt} = \left[\frac{1}{x} \int_0^x \frac{1}{f(t)} dt \right]^{-1} \tag{7.66}$$

and write the Hardy inequality with the "arithmetic mean" (7.65) replaced by the "harmonic mean" (7.66):

$$\left(\int_0^\infty \left(\frac{x}{\int_0^x \frac{1}{f(t)} dt} \right)^q x^{-q} u(x) dx \right)^{1/q} \leq C \left(\int_0^\infty f(x)^p v(x) dx \right)^{1/p},$$

or, if we replace f by $\frac{1}{f}$

$$\left(\int_0^\infty \left(\int_0^x f(t) dt \right)^{-q} u(x) dx \right)^{1/q} \leq C \left(\int_0^\infty f(x)^{-p} v(x) dx \right)^{1/p}. \tag{7.67}$$

Up to now we supposed that $p > 0$, $q > 0$. If we take inequality (7.67) to the power -1 and replace p and q by $-p$ and $-q$, respectively, we obtain that

$$\left(\int_0^\infty \left(\int_0^x f(t) dt \right)^q u(x) dx \right)^{1/q} \geq \tilde{C} \left(\int_0^\infty f(x)^p v(x) dx \right)^{1/p} \tag{7.68}$$

with $\tilde{C} = 1/C$ and this time with $p < 0$ and $q < 0$. Consequently, inequality (7.68) is nothing else but a reverse Hardy inequality with negative values of the parameters p and q (and with $a = 0$).

Remark 7.34. Together with the arithmetic means $AM_n(\bar{a})$ and the harmonic means $HM_n(\bar{a})$ for a sequence $\bar{a}$ we also have the *geometric means* $GM_n(\bar{a})$ defined by

$$GM_n(\bar{a}) = (a_1 a_2 \cdots a_n)^{1/n} = \exp(AM_n(\ln \bar{a})).$$

Correspondingly, we can also introduce, in analogy to the "arithmetic mean" (7.65) and the "harmonic mean" (7.66), the "geometric mean" defined by

$$(Gf)(x) := \exp\left(\frac{1}{x} \int_0^x \ln f(t)dt\right). \tag{7.69}$$

We will deal with inequalities of the type

$$\|Gf\|_{q,u} \le C\|f\|_{p,v}$$

later. A simple example is the previously mentioned Pólya–Knopp's inequality (see (1.80)) and more general results of this type were described in Sec. 1.8.

Now we are prepared to describe criteria, which guarantee the validity of the inequalities (7.63) and (7.64) for values $p, q \in \mathbb{R} \setminus \{0\}$. Let us mention that the results are taken from [KKK1] by A. Kufner *et al.*, where proofs, complementary information and other references can be found.

I. Some notation. For given weight functions u and v on (a, b) we denote

$$U(x) := \int_x^b u(t)dt, \quad V(x) := \int_a^x v(t)^{1-p'}dt, p' = p/(p-1), p \ne 1,$$

$$\tilde{U}(x) := \int_a^x u(t)dt, \quad \tilde{V}(x) := \int_x^b v(t)^{1-p'}dt, \tag{7.70}$$

and, for $\alpha, \beta \in \mathbb{R}$,

$$A_0(x; \alpha, \beta) := U^\alpha(x)V^\beta(x) \quad \text{and} \quad B_0(x; \alpha, \beta) := \tilde{U}^\alpha(x)\tilde{V}^\beta(x). \tag{7.71}$$

We assume that the functions $U, V, \tilde{U}, \tilde{V}$ are finite.

II. The case $1 < p \leq q < \infty$. As it was mentioned in the Introduction, the Hardy inequality (7.63) holds if and only if the function

$$A_{MB}(x) := A_0\left(x; \frac{1}{q}, \frac{1}{p'}\right) \tag{7.72}$$

is finite, i.e., if

$$A_0 := \sup_{a < x < b} A_{MB}(x) < \infty. \tag{7.73}$$

The function $A_{MB}(x)$ is the famous Muckenhoupt-Bradley function, and for the best constant C in (7.63) we have

$$C \approx A_0. \tag{7.74}$$

III. The case $0 < q < p < \infty$, $q \neq 1$, $1 < p < \infty$. As it was mention in the Introduction, the Hardy inequality (7.63) holds if and only if

$$A_1 := \left(\int_a^b U(x)^{r/q} V(x)^{r/q'} v(x)^{1-p'} dx\right)^{1/r} < \infty, \tag{7.75}$$

where $\frac{1}{r} = \frac{1}{q} - \frac{1}{p}$. For the best constant C in (7.63) we again have $C \approx A_1$.

IV. The case $p < 0$, $q > 0$. As it was shown in [KKK2] the reverse Hardy inequality (7.64) holds if and only if the function $A_{MB}(x)$ is finite, and again (7.74) holds.

V. The case $p < 0$, $q < 0$. This case was investigated by P. Beesack and H.P. Heinig already in 1981 (see [BH1]) but a complete exhaustive description was given in 2004 by D.V. Prokhorov [Pr2]. The results are as follows:

(i) For $-\infty < q \leq p < 0$ the reverse Hardy inequality (7.64) holds if and only if

$$B_1 := \sup_{a < x < b} B_0\left(x; -\frac{1}{q}, -\frac{1}{p'}\right) < \infty. \tag{7.76}$$

(ii) For $-\infty < p < q < 0$ the reverse Hardy inequality (7.64) holds if and only if

$$B_2 := \left(\int_a^b \widetilde{U}(x)^{r/p} V(x)^{r/q'} u(x) dx \right)^{1/r} < \infty, \qquad (7.77)$$

where $\frac{1}{r} = \frac{1}{q} - \frac{1}{p}$.

(iii) In both cases we have $C \approx B_i, i = 1, 2$, for the best constant C in (7.64).

VI. The case $0 < p < 1$, $q < 0$. In this case the reverse Hardy inequality (7.64) holds if and only if

$$A_* := \inf_{a<x<b} A_{MB}(x) > 0 \qquad (7.78)$$

with $A_{MB}(x)$ from (7.72), and the best constant C in (7.64) again satisfies $C \approx A_*$. Here, according to (7.71), $A_{MB}(x) = A_0\left(x; \frac{1}{q}, \frac{1}{p'}\right)$, where both parameters $\frac{1}{q}$ and $\frac{1}{p'}$ are negative. Moreover, if we denote

$$\widetilde{A}_{MB}(x) := \frac{1}{A_{MB}(x)} = A_0\left(x; -\frac{1}{q}, -\frac{1}{p'}\right)$$

we can rewrite condition (7.78) as

$$A_0^* := \sup_{a<x<b} \widetilde{A}_{MB}(x) < \infty$$

with positive parameters $\alpha = -\frac{1}{q}$ and $\beta = -\frac{1}{p'}$ in $A_0(x; \alpha, \beta)$.

VII. The case $p \geq 1$, $q < 0$. (i) If $p > 1$, then it was shown in [KKK2] that the Hardy inequality (7.63) holds if and only if

$$B_1 := \sup_{a<x<b} B_0\left(x; -\frac{1}{q}, -\frac{1}{p'}\right) < \infty$$

(see (7.76)). This condition is equivalent to the condition

$$\inf_{a<x<b} B_0\left(x; \frac{1}{q}, \frac{1}{p'}\right) > 0.$$

(ii) If $p = 1$ and if we replace the term $V^{\frac{1}{p'}}(x)$ in $B_0\left(x; \frac{1}{q}, \frac{1}{p'}\right)$ by ess $\sup_{a < t < x} v(t)$, then we obtain a necessary and sufficient condition for the Hardy inequality (7.63) to hold.

VIII. The case $0 < p < 1$, $0 < q < 1$. This case can be handled via duality using the result mentioned in part V.

IX. The case $0 < p < 1$, $q > 1$. This case was investigated in [EGO1]. The corresponding necessary and sufficient condition for the validity of the reverse Hardy inequality (7.64) has the form

$$A := \left(\int_a^b \widetilde{V}(t)^{r/p'} dU(t)^{r/q}\right)^{1/r} + \frac{\widetilde{V}^{1/p'}(a)}{U^{1/q}(a)} < \infty,$$

where $\frac{1}{r} = \frac{1}{q} - \frac{1}{p}$.

Remark 7.35. All results mentioned above can be formulated via duality arguments for the conjugate Hardy and conjugate reverse Hardy inequalities, i.e. for the case when the term $\int_a^x f(t)dt$ in (7.63) and (7.64) is replaced by $\int_x^b f(t)dt$. For example, the counterpart of the Muckenhoupt-Bradley function $A_{MB}(x)$ from (7.72), i.e.,

$$A_{MB}(x) = U^{1/q}(x)V^{1/p'}(x) = \left(\int_x^b u(t)dt\right)^{1/q} \left(\int_a^x v(t)^{1-p'}dt\right)^{1/p'} \tag{7.79}$$

would be

$$\widetilde{A}_{MB}(x) = \widetilde{U}^{1/q}(x)\widetilde{V}^{1/p'}(x) = \left(\int_a^x u(t)dt\right)^{1/q} \left(\int_x^b v(t)^{1-p'}dt\right)^{1/p'}. \tag{7.80}$$

Remark 7.36. In the literature the Hardy inequality (7.63) sometimes appears in the form

$$\left(\int_a^b \left(\int_a^x f(t)v(t)dt\right)^q u(x)dx\right)^{1/q} \leq C \left(\int_a^b f(x)^p dx\right)^{1/p}, \tag{7.81}$$

(see e.g. [Pr2] or [KKK1]). This is, of course, only a formal change since if we introduce instead of the weight function v a new weight function $\hat{v} := v^{-p}$, and write f as $g\hat{v}^{\frac{1}{p}}$, then we have

$$fv = g\hat{v}^{\frac{1}{p}}\hat{v}^{-\frac{1}{p}} = g \quad \text{and} \quad f^p = g^p\hat{v},$$

and we can rewrite (7.81) as

$$\left(\int_a^b \left(\int_a^x g(t)dt\right)^q u(x)dx\right)^{1/q} \leq C \left(\int_a^b g(t)^p\hat{v}(t)dt\right)^{1/p}. \quad (7.82)$$

But this is the classical Hardy inequality (7.63), this time for the function g and with "new" weight functions u and $\hat{v}$.

Furthermore, the Muckenhoupt function $A_{MB}(x)$, which for (7.82) has the form

$$\left(\int_x^b u(t)dt\right)^{1/q} \left(\int_a^x \hat{v}(t)^{1-p'}dt\right)^{1/p'}$$

(see (7.72) with $\hat{v}$ instead of v) can be written in the form

$$\left(\int_x^b u(t)dt\right)^{1/q} \left(\int_a^x v(t)^{p'}dt\right)^{1/p'}.$$

Similarly, we can also rewrite other formulas.

7.5. More on Hardy-type Inequalities for Hardy Operators with Kernel

Inspired by the results in our previous sections 7.3 and 7.4 and some newer results (see e.g. [KPS1]) we shall here present some complements to the results given in Chap. 2 and also raise a number of open questions. Again we consider a general Hardy-type operator K in the form

$$(Kf) := \int_a^x K(x,t)f(t)dt, \ x \in (a,b), \quad (7.83)$$

where $k(x,t) \geq 0$. Without restriction we can put $a = 0$. The most important open question is the following (with no restriction on the kernel $k(x,t)$):

Open question

Find necessary and sufficient conditions on the weights $u(x)$ and $v(x)$ so that

$$\left(\int_0^b \left(\int_0^x k(x,t)f(t)dt \right)^q u(x)dx \right)^{1/q} \leq C \left(\int_0^b f^p(x)v(x)dx \right)^{1/p} \tag{7.84}$$

holds for all measurable functions $f \geq 0$.

We consider now some special cases of kernels.

a) **The case** $k(x,t) \equiv 1$. The newest information was presented in our previous sections 7.3 and 7.4.

b) **The case with product kernels.** By using the results from our Sec. 7.3, a correspondingly fairly complete picture can be given also in the case with product kernels, i.e., when

$$k(x,t) \equiv A(x)B(t), \tag{7.85}$$

with $A(x)$ and $B(x)$ positive a.e.

Then the inequality (7.84) can be rewritten in the form

$$\left(\int_0^b \left(\int_0^x f_0(t)dt \right)^q u_0(x)dx \right)^{1/q} \leq C \left(\int_0^b f_0^p(x)v_0(x)dx \right)^{1/p}, \tag{7.86}$$

i.e., the general Hardy inequality (7.28) with $f(x)$ replaced by $f_0(x) := f(x)B(x)$ and with the new weight functions

$$u_0(x) = [A(x)]^q u(x) \quad \text{and} \quad v_0(x) = [B(x)]^{-p}v(x). \tag{7.87}$$

Hence, by using Theorem 7.23 we have the following general result in this case:

Theorem 7.37. *Let* $1 < p \leq q < \infty$, $0 < s < \infty$, *and define, for the weight functions* $u_0(x)$, $v_0(x)$ *in (7.87), the functions* U_0 *and* V_0 *by*

$$U_0(x) := \int_x^b u_0(t)dt, \quad V_0(x) = \int_0^x v_0^{1-p'}(t)dt. \tag{7.88}$$

Then the general Hardy-type inequality (7.84) *with* $k(x,t) \equiv A(x)B(t)$ *can be characterized by any of the conditions* $B_i(s) < \infty$, $i = 1, 2, 3, 4$ *where* $B_i(s)$ *is equal to* $A_i(s)$ *in Theorem 7.23 with* u, v, U, V, *replaced by* u_0, v_0, U_0, V_0, *defined by* (7.87)–(7.88), *respectively,* $(i = 1, 2, 3, 4)$.

Corollary 7.38. *Let* $1 < p \le q < \infty, s > 0$, *and let* $u_0(x)$, $v_0(x)$, U_0 *and* V_0 *be defined as in* (7.87)–(7.88). *Then the general Hardy inequality* (7.84) *with* $k(x,t) \equiv A(x)B(t)$ *can be characterized by any of the following conditions:*

(a) $B_{MB} := \sup_{0<x<b} (U_0(x))^{1/q} (V_0(x))^{1/p'} < \infty$,

(b) $B_{PS} := \sup_{0<x<b} \left(\int_0^x u_0(t) [V_0(t)]^q \, dt \right)^{1/q} [V_0(x)]^{-1/p} < \infty$,

(c) $B_{PS}^* := \sup_{0<x<b} \left(\int_x^b v_0^{1-p'} [U_0(t)]^{p'} \, dt \right)^{1/p'} [U_0(x)]^{-1/q'} < \infty$,

(d) $B_{TG} := \inf_{h>0} \sup_{0<x<b} \left(\frac{1}{h(x)} \int_0^x u_0(t)(h(t) + V_0(t))^{\frac{q}{p'}+1} dt \right)^{1/q} < \infty$,

(e) $B_{TG}^* := \inf_{h>0} \sup_{0<x<b} \left(\frac{1}{h(x)} \int_x^b v_0^{1-p'}(t)(h(t) \right.$

$\left. + U_0(t))^{\frac{p'}{q}+1} dt \right)^{1/p'} < \infty$,

(f) $B := \sup_{0<x<b} \left(\int_x^b u_0(t) [V_0(t)]^{\frac{q}{p'}-1} dt \, V_0(x) \right)^{1/q} < \infty$.

Remark 7.39. The given first conditions correspond to the previously mentioned conditions by Muckenhoupt-Bradley, Persson-Stepanov in two dual forms and Tomaselli and Gurka in two dual forms. The last condition is obtained by just considering a point in some of the scales in Theorem 7.37 namely $B = A_1(1/q)$, with u_0, v_0, U_0, V_0 instead of u, v, U and V, respectively.

By using the same technique and a special case of Theorem 7.25 we also have the following result for the case $1 < q < p < \infty$:

Theorem 7.40. *Let* $1 < q < p < \infty$, $1/r := 1/q - 1/p$, $0 < s < \infty$ *and define for the weight functions* u_0 *and* v_0 *the functions* U_0 *and* V_0 *by (7.87) and (7.88). Then the general Hardy-type inequality (7.84) with* $k(x,t) \equiv A(x)B(t)$ *and* $b = \infty$ *can be characterized by any of the following conditions:*

(1) $$B_{MR}^{(1)}(s) := \left(\int_0^\infty \left(\int_t^\infty u_0(x) V_0^{q(1/p'-s)}(x) dx \right)^{r/p} \right.$$
$$\left. \times V_0^{q(1/p'-s)+rs}(t) u_0(t) \, dt \right)^{1/r} < \infty,$$

(2) $$B_{PS}^{(1)}(s) := \left(\int_0^\infty \left(\int_0^t u_0(x) V_0^{q(1/p'+s)}(x) dx \right)^{r/p} \right.$$
$$\left. \times u_0(t) V_0^{q(1/p'+s)-rs}(t) \, dt \right)^{1/r} < \infty,$$

(3) $$B_{MR}^{(2)}(s) := \left(\int_0^\infty \left(\int_0^t U_0^{p'(1/q-s)}(x) dV_0(x) \right)^{r/p'} \right.$$
$$\left. \times U_0^{rs-1}(t) u_0(t) \, dt \right)^{1/r} < \infty,$$

(4) $$B_{PS}^{(2)}(s) := \left(\int_0^\infty \left(\int_t^\infty U_0^{q(1/p'+s)}(x) dV_0(x) \right)^{r/p} \right.$$
$$\left. \times U_0^{q(1/p'+s)-rs}(t) \, dV_0(t) \right)^{1/r} < \infty.$$

Remark 7.41. The characterizations $B_{MR}^{(1)}(1/p') < \infty$ and $B_{PS}^{(1)}(1/p) < \infty$ correspond to the original characterizations by

Maz'ja-Rosin and Persson-Stepanov, respectively. Moreover, the first scale in Theorem 7.40 is an extension of the Maz'ja-Rosin condition and the third scale is the dual of this scale. Similarly, the second scale is an extension of the Persson-Stepanov condition and the fourth scale is the dual of this scale.

Remark 7.42. The book [KoMP] contains a lot of complementary results concerning Hardy-type inequalities in the case with product kernels. However, Theorems 7.37 and 7.40 cannot be found there.

Remark 7.43. By using elementary inequalities it is obvious that Theorems 7.37 and 7.40 can also be formulated for the case when the kernel is of the following more general form:

$$k(x,t) = \sum_{n=1}^{N} A_i(x)B_i(t), \qquad (7.89)$$

Hardy-type operators with kernels of the form (7.89) are recently investigated by A. Kufner *et al.* [KKO1], where criteria of boundedness and compactness of the operator are derived.

Open question

Again inspired by the results in Sec. 7.4 it is also of interest to derive alternative conditions, maybe even scales of conditions as before, to characterize (7.84) for kernels satisfying (7.85) or, more generally, (7.89) for other parameters than those considered in our Theorems 7.37 and 7.40.

c) **A sharp sufficient condition.** The following improvement of a result of S. Kaijser *et al.* [KNPW1] was proved in [KPS1]:

Theorem 7.44. *Let* $1 < p \leq q < \infty$, $0 < b \leq \infty$, u *and* v *are weights and* V *is defined by* (7.40). *Let* $k(x,t)$ *be a general non-negative kernel.*
(a) *Then* (7.84) *holds if*

$$A_s := \sup_{0<t<b} \left(\int_t^b k^q(x,t)u(x)V^{(\frac{q(p-s-1)}{p})}(x)dx \right)^{1/q} V^{s/p}(t) < \infty$$

$$(7.90)$$

for any $s < p - 1$.

(b) *The condition* (7.90) *cannot be improved in general for* $s > 0$ *because for product kernels of type* (7.85) *or, more generally,* (7.89) *it is even necessary and sufficient for* (7.84) *to hold.*

(c) *For the best constant* C *in* (7.84) *we have the following estimate*

$$C \le \inf_{s<p-1} \left(\frac{p-1}{p-s-1}\right)^{1/p'} A_s.$$

Proof.   (a) We use Hölder's inequality to obtain that

$$\int_0^x k(x,t)f(t)dt$$

$$= \int_0^x k(x,t)f(t)v^{\frac{1}{p}}(t)\,[V(t)]^{s/p}\left([V(t)]^{-s/p}\,v^{-1/p}(t)\right)dt$$

$$\le \left(\int_0^x k^p(x,t)f^p(t)v(t)[V(t)]^s dt\right)^{1/p}\left(\int_0^x [V(t)]^{-s/(p-1)}dV(t)\right)^{1/p'}$$

$$= \left(\frac{p-1}{p-s-1}\right)^{1/p'}[V(x)]^{\frac{p-s-1}{p}}\int_0^x k^p(x,t)f^p(t)v(t)[V(t)]^s dt.$$

We now use this estimate and Minkowski's integral inequality and find that

$$I := \int_0^b \left(\int_0^x k(x,t)f(t)dt\right)^q u(x)dx$$

$$\le \left(\frac{p-1}{p-s-1}\right)^{q/p'}\int_0^b \left(\int_0^x k^p(x,t)f^p(t)v(t)\,[V(t)]^s dt\right)^{q/p}$$

$$\times u(x)\,[V(x)]^{\frac{p-s-1}{p}q}dx$$

$$\le \left(\frac{p-1}{p-s-1}\right)^{q/p'}\left(\int_0^b f^p(t)v(t)\,[V(t)]^s\right.$$

$$\times \left.\left(\int_t^b k^q(x,t)u(x)\,[V(x)]^{\frac{p-s-1}{p}q}dx\right)^{p/q}dt\right)^{q/p}$$

$$\le \left(\frac{p-1}{p-s-1}\right)^{q/p'}A_s\left(\int_0^b f^p(t)v(t)dt\right)^{q/p}.$$

Hence we have proved that (a) holds with

$$C \le \left(\frac{p-1}{p-s-1} \right)^{1/p'} A_s.$$

Moreover, by taking infimum over all possible $s < p - 1$ we have also proved (c).

(b) According to our discussions in case b), we conclude that it is sufficient to prove the statement for $k(x,t) \equiv 1$ and for this case we see that (7.90) is equivalent to the condition $A_1(s/p) < \infty$ so the proof is complete by just applying Theorem 7.23 for this special case. □

Inspired by the proof above and the discussions in our previous Sec. 7.3 it seems possible to derive other scales of sufficient conditions like those in Theorem 7.44 for general kernels and which are even necessary and sufficient for special kernels e.g. of product type.

Open question

Do there exist other scales of sufficient conditions as those in Theorem 7.44 and which are sharp in the sense that they are also sufficient for special kernels e.g. fitting to our previous Theorems 7.23 and 7.25?

The case with Oinarov kernels

We remind that Oinarov type kernels $k(x,t)$ satisfy the additional conditions (2.25) and (2.26).

Remark 7.45. For the cases $1 < p \le q < \infty$ and $0 < q < p < \infty, p > 1$, we gave some characterizations in Chap. 2, see Theorem 2.10, Theorem 2.15 and Sec. 2.6 but some newer information was presented in [KPS1]. We remind about the fact that for the case $k(x,t) \equiv 1$ both conditions $A_0 < \infty$ and $A_1 < \infty$ in Theorem 2.10 coincide with the previously described Muckenhoupt-Bradley condition $A_{MB} < \infty$. Note that $A_{MB} = A_1(1/p')$ in Theorem 7.23. A similar statement holds concerning the conditions $B_0 < \infty$ and $B_1 < \infty$ in Theorem 2.15 when $k(x,t) \equiv 1$.

More general such reduction of conditions was investigated by A. Kufner [Ku3] (cf. Sec. 4.5).

Inspired by our investigations in our previous sections and the remark above, we also raise the following:

Open question

Derive alternative conditions, maybe even scales of conditions as before, to the conditions $A_0 < \infty$ and $A_1 < \infty$ in Theorem 2.10 and the conditions $B_0 < \infty$ and $B_1 < \infty$ in Theorem 2.15 to characterize the general Hardy inequality (7.84) for the case when $k(x,t)$ is an Oinarov kernel.

This question is open not only for the standard cases $1 < p \leq q < \infty$ and $0 < q < p < \infty$, $p > 1$, but it is also a reasonable question for other parameters p and q discussed in Sec. 7.4.

Remark 7.46. For the case $p = q, u = v, b = \infty$ one result in this direction can be found in [PSU2], Proposition 2.1. In fact, in this case the two conditions $A_0 < \infty$ and $A_1 < \infty$ can be replaced by one condition.

Remark 7.47. In 2007 R. Oinarov [O3] introduced less restrictive classes of kernels O_n^+ and O_n^-, where n is a non-negative integer and in the case $1 < p \leq q < \infty$ he gave a characterization for (7.84) to hold in the class $O_n^+ \bigcap O_n^-$. Also in this more general case this characterization consists of two conditions. Without going into further details we mention that these conditions are of different type than those in Theorem 2.10.

For some further results related to these generalized Oinarov classes, we also refer to the paper [AOP2] and the Ph.D. thesis [Ar1] by L. Arendarenko.

e) **The case with homogeneous kernels.** We consider the integral operator K defined by

$$Kf(x) := \int_0^\infty k(x,t)f(t)dt, \quad x \in \mathbb{R}_+, \tag{7.91}$$

with a kernel $k(t, x)$ which is homogeneous of degree -1, i.e.,

$$k(ax, at) = a^{-1}k(x, t), \quad x, t \in \mathbb{R}_+, a > 0. \tag{7.92}$$

We assume that $k(x, t) \geq 0$ and define the constant

$$\kappa_p := \int_0^\infty k(1, t)t^{-1/p}dt = \int_0^\infty k(x, 1)x^{-1/p'}dx, \quad p \geq 1. \tag{7.93}$$

The following more precise form of some results of Hardy-Littlewood-Pólya [HLP], Theorems 3.19 and 3.36, is known:

Theorem 7.48. *Let the kernel $k(x, t)$ satisfy (7.92) and $p \geq 1$. Then the inequality*

$$\int_0^\infty \left(\int_0^\infty k(x, t)f(t)dt \right)^p dx \leq C \int_0^\infty f^p(x)dx \tag{7.94}$$

holds for some finite constant C if and only if κ_p defined by (7.93) is finite. Moreover, the sharp constant in (7.94) is $C = \kappa_p^p$.

Remark 7.49. A simple proof of Theorem 7.48 can be found in [LPS2], but the result was known much before (cf. our Remark 7.52). In [LPSW1] also the corresponding sharp reversed inequality for the case $0 < p < 1$ was proved and applied (see Theorem 7.73).

The considerations above indicate that it is reasonable to study the following:

Open question

Let $k(x, t)$ be a kernel satisfying (7.92). Find necessary and sufficient conditions on the weights $u(x)$ and $v(x)$ so that the inequality

$$\left(\int_0^\infty \left(\int_0^\infty k(x, t)f(t)dt \right)^q u(x)dx \right)^{1/q} \leq C \left(\int_0^\infty f^p(x)v(x)dx \right)^{1/p}$$

holds for any finite constant C for various parameters p and q, $p \geq 1$, $0 < q \leq \infty$, i.e., in the case $1 < p \leq q < \infty$.

Remark 7.50. In particular, by using such a result we also get a corresponding characterization of the inequality (7.84) under the restriction that the kernel $k(x, t)$ is homogeneous of degree -1 and $b = \infty$. Since we have (at least for the case $p = q$) a characterization of both inequalities (7.84) and (7.94), it seems reasonable that such a result must contain condition(s) with restriction on both the kernel $k(x,t)$ and on the weights $u(x)$ and $v(x)$ like in all previous conditions in this book, cf. also our Theorem 7.44.

We now continue by presenting same unweighted generalizations related to the open question on p. 383. We note that there also exists a corresponding multi-dimensional version of Theorem 7.48 (see Theorem 7.51). We consider the inequality

$$\left(\int_{\mathbb{R}^n} \left| \int_{\mathbb{R}^n} k(x,y)f(y)\,dy \right|^p dx \right)^{\frac{1}{p}}$$

$$\leq C_{k,p} \left(\int_{\mathbb{R}^n} |f(x)|^p dx \right)^{\frac{1}{p}}, \quad 1 \leq p \leq \infty, \qquad (7.95)$$

for multidimensional integral operators

$$\mathbf{K}f(x) := \int_{\mathbb{R}^n} k(x,y)f(y)\,dy \qquad (7.96)$$

with a kernel $k(x,y)$. We assume the following:
1°. the kernel $k(x,y)$ is *homogeneous of degree* $-n$, i.e.,

$$k(tx,ty) = t^{-n}k(x,y), \quad t > 0, \quad x,y \in \mathbb{R}^n, \qquad (7.97)$$

2°. it is *invariant with respect to rotations*, i.e.,

$$k[\omega(x),\omega(y)] = k(x,y), \quad x,y \in \mathbb{R}^n \qquad (7.98)$$

for all rotations $\omega(x)$ in $\mathbb{R}^n$. Let

$$\varkappa_p = \int_{\mathbb{R}^n} |k(\sigma,y)| \frac{dy}{|y|^{\frac{n}{p}}}, \quad \sigma \in \mathbb{S}^{n-1}, \qquad (7.99)$$

where $\mathbb{S}^{n-1}$ is the unit sphere in $\mathbb{R}^n$. In the sequel

$$|\mathbb{S}^{n-1}| = \frac{2\pi^{\frac{n}{2}}}{\Gamma\left(\frac{n}{2}\right)}$$

denotes its surface measure. Because of the invariance condition (7.98), the integral in (7.99) does not depend on the choice of

$\sigma \in \mathbb{S}^{n-1}$ (see details in the book [KS2] by N.K. Karapetyants and S. Samko), so one may choose $\sigma = e_1 = (1, 0, \ldots, 0)$ in (7.99). We also pronounce that the condition (7.97) is crucial in the class of homogeneous kernels. In fact by using a standard dilation argument, we see that (7.95) cannot hold for homogeneity degree of the kernel $k(x, y) \neq -n$.

Theorem 7.51. *Let $1 \leq p \leq \infty$ and the kernel $k(x, y)$ satisfy the assumptions (7.97)–(7.98). If*

$$\varkappa_p < \infty,$$

then the inequality (7.95) holds with $C(k, p) = \varkappa_p$. If $k(x, y) \geq 0$, then the condition $\varkappa_p < \infty$ is also necessary for (7.95) to hold and $\varkappa_p$ is the best constant.

Remark 7.52. The sufficiency part of Theorem 7.51 without the sharp constant was first given in [Mi1] (for $n = 1$ see also [HLP], p. 229). A simpler proof was given in [Sam1]. The necessity of the condition $\varkappa_p < \infty$ for the boundedness and the sharpness of the constant in such a general case was proved in [K1]. A complete proof of Theorem 7.51 in its final form was presented in [KS1], see also its presentation in the book [KS2], p. 70, Theorem 6.4.

For the proof of Theorem 7.51 we need the following lemma.

Lemma 7.53. *Under the assumptions (7.97)–(7.98), the integrals*

$$\varkappa_p = \int_{\mathbb{R}^n} |k(\sigma, y)| \, |y|^{-\frac{n}{p}} dy \, , \quad \sigma \in \mathbb{S}^{n-1}, \tag{7.100}$$

and

$$\kappa_p = \int_{\mathbb{R}^n} |k(x, \theta)| \, |x|^{-\frac{n}{p'}} dx, \quad \theta \in \mathbb{S}^{n-1}, \tag{7.101}$$

where $0 < p \leq \infty$, do not depend on $\sigma \in \mathbb{S}^{n-1}$ and $\theta \in \mathbb{S}^{n-1}$, respectively, and coincide with each other: $\varkappa_p = \kappa_p$.

Proof. By $\omega_x(\eta)$, $\eta \in \mathbb{R}^n$, we denote any rotation in $\mathbb{R}^n$ such that $\omega_x(e_1) = \frac{x}{|x|}$, $e_1 = (1, 0, \ldots, 0)$. Then for $\xi = \omega_x(\eta)$ we have $|\xi| = |\eta|$ and $\xi \cdot \frac{x}{|x|} = \eta \cdot e_1 = \eta_1$.

Making the rotation change of variables $y = \omega_\sigma(z)$ in (7.100), using the rotation invariance $k(\sigma, y) \equiv k(e_1, z)$ of the kernel and the fact that $dy = dz$, we obtain that $\varkappa_p$ does not depend on σ. Similarly, the change of variables $x = \omega_\theta(z)$ in (7.101) shows that κ_p does not depend on θ.

We make use of the fact that $\varkappa_p$ does not depend on σ to rewrite $\varkappa_p$ as follows:

$$\varkappa_p = \frac{1}{|\mathbb{S}^{n-1}|} \int_{\mathbb{S}^{n-1}} d\sigma \int_{\mathbb{S}^{n-1}} d\theta \int_0^\infty |k(\sigma, \rho\theta)| \rho^{n-1-\frac{n}{p}} d\rho .$$

By the homogeneity of the kernel and the change $\rho = \frac{1}{r}$, we get

$$\varkappa_p = \frac{1}{|\mathbb{S}^{n-1}|} \int_{\mathbb{S}^{n-1}} d\sigma \int_{\mathbb{S}^{n-1}} d\theta \int_0^\infty |k(r\sigma, \theta)| r^{n-1-\frac{n}{p'}} dr .$$

Changing the order of integration in σ and θ, we see that the obtained inner integral (in σ and ρ) is equal to the integral defining $\varkappa_p$. Taking into account that $\varkappa_p$ does not depend on θ, we arrive at the equality $\varkappa_p = \kappa_p$. $\qquad\square$

Proof of Theorem 7.51

Sufficiency part. For the operator (7.96), by the Hölder inequality, we obtain

$$|(\mathbf{K}f)(x)| \leq \left\{ \int_{\mathbb{R}^n} |y|^{-\frac{n}{p}} |k(x,y)| dy \right\}^{\frac{1}{p'}}$$

$$\times \left\{ \int_{\mathbb{R}^n} |y|^{\frac{n}{p'}} |k(x,y)| \cdot |f(y)|^p dy \right\}^{\frac{1}{p}} .$$

By the change of variables $y \to |x|y$ in the first integral, the homogeneity of the kernel $k(x,y)$ and Lemma 7.53, we obtain

$$|(\mathbf{K}f)(x)| \leq \frac{\varkappa_p^{\frac{1}{p'}}}{|x|^{\frac{n}{pp'}}} \left\{ \int_{\mathbb{R}^n} |y|^{\frac{n}{p'}} |k(x,y)| \cdot |f(y)|^p dy \right\}^{\frac{1}{p}} .$$

Then

$$\|\mathbf{K}f\|_p \leq \varkappa_p^{\frac{1}{p'}} \left\{ \int_{R^n} |f(y)|^p |y|^{\frac{n}{p'}} dy \int_{\mathbb{R}^n} |k(x,y)| \cdot |x|^{-\frac{n}{p'}} dx \right\}^{\frac{1}{p}}$$

$$= \varkappa_p^{\frac{1}{p'}} \left\{ \int_{R^n} |f(y)|^p dy \int_{R^n} \left| k\left(x, \frac{y}{|y|}\right) \right| \cdot |x|^{-\frac{n}{p'}} dx \right\}^{\frac{1}{p}}$$

$$= \varkappa_p^{\frac{1}{p'}} \kappa_p^{\frac{1}{p}} \|f\|_p$$

by the same Lemma 7.53, we arrive at (7.95) with $C(k,p) = \varkappa_p$.

Necessity part and the sharpness of the constant. Let now the kernel be non-negative. Suppose that (7.95) holds, i.e., the operator $\mathbf{K}$ is bounded. Then

$$\left| \int_{\mathbb{R}^n} (\mathbf{K}f)(x)\psi(x)dx \right| \leq \|\mathbf{K}\| \cdot \|f\|_p \|\psi\|_{p'} \qquad (7.102)$$

for all $f \in L^p(\mathbb{R}^n)$ and $\psi \in L^{p'}(\mathbb{R}^n)$. We choose, for $\varepsilon > 0$,

$$f(x) = 0, \quad \text{if} \quad |x| < 1, \quad \text{and} \quad f(x) = |x|^{-\varepsilon - \frac{n}{p}}, \quad \text{if} \quad |x| \geq 1,$$

and $\psi(x) = [f(x)]^{p-1}$. Substituting this into (7.102), we get

$$\int_{\mathbb{S}^{n-1}} d\sigma \int_{\mathbb{R}^n} k(\sigma, y)|y|^{-\varepsilon - \frac{n}{p}} dy \int_{r > \max(1, |y|^{-1})} r^{-p\varepsilon - 1} dr \leq \|\mathbf{K}\| \cdot \|f\|_p^p.$$
$$(7.103)$$

Direct calculation yields

$$\|f\|_p^p = \frac{|\mathbb{S}^{n-1}|}{p\varepsilon}, \quad \int_{r > \max(1, |y|^{-1})} r^{-p\varepsilon - 1} dr = \frac{1}{p\varepsilon} \left[\max(1, |y|^{-1}) \right]^{-p\varepsilon},$$

so that the inequality (7.103) takes the form

$$\frac{1}{|\mathbb{S}^{n-1}|} \int_{\mathbb{S}^{n-1}} d\sigma \int_{\mathbb{R}^n} k(\sigma, y)|y|^{-\varepsilon - \frac{n}{p}} \left[\max(1, |y|^{-1}) \right]^{-p\varepsilon} dy \leq \|\mathbf{K}\|.$$
$$(7.104)$$

By the rotation invariance of the kernel k, the inner integral in the left-hand side does not depend on σ, so that

$$\int_{\mathbb{R}^n} k(e_1, y)|y|^{-\varepsilon-\frac{n}{p}} \left[\max(1, |y|^{-1})\right]^{-p\varepsilon} dy \leq \|\mathbf{K}\|. \tag{7.105}$$

By the Fatou theorem we may pass to the limit as $\varepsilon \to 0$, which yields the inequality $\varkappa_p \leq \|\mathbf{K}\|$. Together with the inverse inequality proved in the sufficiency part, this gives the equality $\|\mathbf{K}\| = \varkappa_p$ and completes the proof. □

Some applications and consequences of Theorem 7.51

(For proofs and further details see [LPSW1].)

Example 7.54 (Hilbert-type inequality). Let $\lambda > 0, \alpha > 0$ and $1 \leq p < \infty$. Then

$$\int_{\mathbb{R}^n} \left| |x|^{\beta+\lambda\alpha-n} \int_{\mathbb{R}^n} \frac{f(y)dy}{|y|^\beta \left(|x|^\lambda + |y|^\lambda\right)^\alpha} \right|^p dx \leq \varkappa^p \int_{\mathbb{R}^n} |f(x)|^p dx \tag{7.106}$$

holds if and only if $\beta < \frac{n}{p'}$ and $\alpha\lambda > \frac{n}{p'} - \beta$ and

$$\varkappa = \varkappa_{p,\beta} = \int_{\mathbb{R}^n} \frac{|y|^{-\beta-\frac{n}{p}}dy}{(1+|y|^\lambda)^\alpha} = \frac{|\mathbb{S}^{n-1}|}{\lambda} \int_0^\infty \frac{\varrho^{\frac{1}{\lambda}\left(\frac{n}{p'}-\beta\right)-1}}{(1+\varrho)^\alpha} d\varrho$$

$$= \frac{|\mathbb{S}^{n-1}|}{\lambda} B\left(\frac{1}{\lambda}\left(\frac{n}{p'} - \beta\right), \alpha - \frac{1}{\lambda}\left(\frac{n}{p'} - \beta\right)\right)$$

is the sharp constant.

Let now

$$I^\alpha f(x) = \frac{1}{\gamma_n(\alpha)} \int_{\mathbb{R}^n} \frac{f(y)\, dy}{|x - y|^{n-\alpha}}, \quad 0 < \alpha < n, \tag{7.107}$$

be the Riesz potential operator with the normalizing constant

$$\gamma_n(\alpha) = \frac{2^\alpha \pi^{\frac{n}{2}} \Gamma\left(\frac{\alpha}{2}\right)}{\Gamma\left(\frac{n-\alpha}{2}\right)}. \tag{7.108}$$

Example 7.55 (Stein-Weiss inequality). The best constant C for the inequality (see [StW1] and cf. also [StW2])

$$\int_{\mathbb{R}^n} |x|^{\mu} \, |I^{\alpha} f(x)|^p \, dx \le C^p \int_{\mathbb{R}^n} |x|^{\gamma} |f(x)|^p \, dx, \qquad (7.109)$$

valid if and only if $1 \le p < \infty$, $\alpha p - n < \gamma < n(p-1)$, $\mu = \gamma - \alpha p$, can be obtained by means of Theorem 7.51: it is

$$C = 2^{-\alpha} \frac{\Gamma\left(\frac{n(p-1)-\gamma}{2p}\right) \Gamma\left(\frac{n+\gamma-\alpha p}{2p}\right)}{\Gamma\left(\frac{n+\gamma}{2p}\right) \Gamma\left(\frac{n(p-1)+\alpha p-\gamma}{2p}\right)}. \qquad (7.110)$$

(In the case when $\frac{\alpha}{2}$ is an integer, the sharp constant was calculated in [DH1] by other means, for non-integer $\frac{\alpha}{2}$ but $\mu = 0$ we refer to [Y1] and for general μ but $p = 2$ to [E1]).

Example 7.56. The multidimensional Hardy inequalities with power weights

$$\left\| |x|^{\alpha-n} \int_{|y|<|x|} \frac{f(y) \, dy}{|y|^{\alpha}} \right\|_{L^p(\mathbb{R}^n)} \le C_1(p,\alpha) \|f\|_{L^p(\mathbb{R}^n)},$$

$$1 \le p < \infty, \quad \alpha < \frac{n}{p'}, \qquad (7.111)$$

and

$$\left\| |x|^{\beta-n} \int_{|y|>|x|} \frac{f(y) \, dy}{|y|^{\beta}} \right\|_{L^p(\mathbb{R}^n)}$$

$$\le C_2(p,\beta) \|f\|_{L^p(\mathbb{R}^n)}, \quad 1 \le p < \infty, \quad \beta > \frac{n}{p'} \qquad (7.112)$$

hold with the sharp constants in (7.111) and (7.112) given by

$$C_1(p,\alpha) = \frac{|\mathbb{S}^{n-1}|}{\frac{n}{p'} - \alpha} \quad \text{and} \quad C_2(p,\beta) = \frac{|\mathbb{S}^{n-1}|}{\beta - \frac{n}{p'}}, \qquad (7.113)$$

respectively.

Remark 7.57. The best constant for (7.111) was calculated in the non-weighted case $\alpha = 0$ in [CG1], where it was shown that

$$C_1(p, 0) = |B(0, 1)|p', \tag{7.114}$$

where $|B(0, 1)| = \frac{|\mathbb{S}^{n-1}|}{n} = \frac{2\pi^{\frac{n}{2}}}{n\Gamma(\frac{n}{2})}$ is the volume of the unit ball. The weighted case with general weights was studied in [DHK1], but by using this result the sharp constant cannot be obtained.

Example 7.58. The inequality

$$\left\| |x|^{\alpha-n} \int_{a|x|<|y|<b|x|} \frac{f(y)\,dy}{|y|^\alpha} \right\|_{L^p(\mathbb{R}^n)} \le C(p, \alpha) \|f\|_{L^p(\mathbb{R}^n)},$$

$$0 < a < b < \infty, \tag{7.115}$$

holds for all $1 \le p < \infty$ and $\alpha \in \mathbb{R}$, where the sharp constant is equal to

$$C(p, \alpha) = |\mathbb{S}^{n-1}| \left| \frac{b^{\frac{n}{p'}-\alpha} - a^{\frac{n}{p'}-\alpha}}{\frac{n}{p'} - \alpha} \right|, \quad \alpha \ne \frac{n}{p'},$$

with $\dfrac{b^{\frac{n}{p'}-\alpha} - a^{\frac{n}{p'}-\alpha}}{\frac{n}{p'}-\alpha}$ replaced by $\ln\frac{b}{a}$ when $\alpha = \frac{n}{p'}$.

Remark 7.59. Note that (7.115) may be regarded as an extension and unification of (7.111) and (7.112) with the sharp constants given by (7.113). In fact, (7.111) and (7.112) follow by just using (7.115) with $a = 0, b = 1$ (respectively $a = 1, b = \infty$).

Example 7.60. The inequality

$$\left\| |x|^{\alpha+\gamma-n} \int_{|y|<|x|} \frac{f(y)\,dy}{|y|^\alpha(|x|-|y|)^\gamma} \right\|_{L^p(\mathbb{R}^n)} \le C(p, \alpha, \gamma) \|f\|_{L^p(\mathbb{R}^n)},$$

$$1 \le p < \infty, \tag{7.116}$$

holds, where $\alpha < \frac{n}{p'}$ and $\gamma < 1$ and the sharp constant is equal to $C(p, \alpha, \gamma) = |\mathbb{S}^{n-1}| B\left(\frac{n}{p'} - \alpha, 1 - \gamma\right)$.

Finally, we present some limit Pólya-Knopp-type inequalities with sharp constants, which follow from Theorem 7.51 by performing a standard limit procedure and letting $p \to \infty$.

Besides the constant $\varkappa_p$, we also introduce the limit constants

$$\varkappa_\infty = \int_{\mathbb{R}^n} k(e_1, y)\,dy$$

and

$$\varkappa^* = n \frac{\int_{\mathbb{R}^n} k(e_1, y) \ln \frac{1}{|y|}\,dy}{\int_{\mathbb{R}^n} k(e_1, y)\,dy}$$

assuming that $k(x, y) \geq 0$ and strictly positive a.e.

Example 7.61. Let $f(x) \geq 0$, let $\varkappa_\infty < \infty$ and $\varkappa_p < \infty$ for some $p > 1$. If $\varkappa^* < \infty$, then

$$\int_{\mathbb{R}^n} \exp\left(\frac{1}{\varkappa_\infty} \int_{\mathbb{R}^n} k(x, y) \ln f(y)\,dy\right) dx \leq e^{\varkappa^*} \int_{\mathbb{R}^n} f(x)dx \quad (7.117)$$

and the constant $e^{\varkappa^*}$ is sharp.

In the next example we use the notation $\psi(z) = \frac{\Gamma'(z)}{\Gamma(z)}$ for the Euler ψ-function, and define

$$B_n(a, b) := \frac{\gamma_n(a)\gamma_n(b)}{\gamma_n(a+b)},$$

where γ_n is defined by (7.108).

Example 7.62. Let $0 < \alpha < \nu < n$. Then

$$\int_{\mathbb{R}^n} \exp\left(\frac{|x|^{\nu-\alpha}}{B_n(\alpha, n-\nu)} \int_{\mathbb{R}^n} \frac{\ln f(y)\,dy}{|y|^\nu |x-y|^{n-\alpha}}\right) dx \leq e^{\varkappa^*} \int_{\mathbb{R}^n} f(x)dx,$$
$$(7.118)$$

with

$$\varkappa^* = \frac{n}{2}\left[\psi\left(\frac{\nu-\alpha}{2}\right) + \psi\left(\frac{n-\nu+\alpha}{2}\right) - \psi\left(\frac{\nu}{2}\right) - \psi\left(\frac{n-\nu}{2}\right)\right].$$
$$(7.119)$$

The constant $e^{\varkappa^*}$ in (7.118) is sharp.

Remark 7.63. In the case of Newtonian potential, i.e., in the case $\alpha = 2$, the sharp constant in the inequality (7.118) has the following value of $\varkappa^*$:

$$\varkappa^* = \frac{n}{2} \frac{\nu - \frac{n+2}{2}}{(n - \nu)(\nu - 2)}, \quad n \geq 3.$$

This observation follows by using the property $\psi(z + 1) = \psi(z) + \frac{1}{z}$ of the ψ-function.

We also have the following Pólya-Knopp-type inequality generated by the multidimensional Hardy inequality:

Example 7.64. Let $\nu < n$. Then

$$\left\| \exp\left(\frac{n - \nu}{|\mathbb{S}^{n-1}|} |x|^{\nu-n} \int_{|y|<|x|} \frac{\ln f(y)\, dy}{|y|^\nu} \right) \right\|_{L^1(\mathbb{R}^n)} \leq e^{\frac{n}{n-\nu}} \|f\|_{L^1(\mathbb{R}^n)}.$$

$$(7.120)$$

Remark 7.65. (7.120) may be seen as a limit inequality of the Hardy inequality (7.111) with the sharp constant (7.113) in Example 7.56. A similar limit inequality of (7.112) can also be derived from Example 7.56.

Remark 7.66. By letting $\nu = 0$ we obtain another n-dimensional variant of the classical Pólya-Knopp inequality (cf. e.g. (1.80)).

Hardy-Hilbert type inequalities

The main information in this section is taken from [LPS2] and [LPSW1].

As mentioned before Hardy's original motivation when he began his research, which led to the discovery of the original forms of his inequalities, was to find an elementary proof of Hilbert's inequality (see p. 334).

Hilbert's inequality: The inequality

$$\int_0^\infty \int_0^\infty \frac{1}{x+y} f(x)g(y)dxdy$$

$$\leq \frac{\pi}{\sin \frac{\pi}{p}} \left(\int_0^\infty f^p(x)dx \right)^{1/p} \left(\int_0^\infty g^{p'}(y)dy \right)^{1/p'} \quad (7.121)$$

for $p > 1$ is called Hilbert's inequality. It can equivalently be written in the form

$$\int_0^\infty \left(\int_0^\infty \frac{1}{x+y} f(x)dx \right)^p dy \leq \left(\frac{\pi}{\sin \frac{\pi}{p}} \right)^p \left(\int_0^\infty f^p(x)dx \right).$$

$$(7.122)$$

Remark 7.67. Hilbert himself considered only the case $p = 2$ and the corresponding discrete form of (7.121) (see his paper [Hi1] from 1906 and also [Hi2], [Wey1] and the historical description in [KMP]). L^p-spaces with $p \neq 2$ appeared only later (around 1920). Concerning the equivalence of (7.121) and (7.122) see Lemma 7.69 on pp. 294 and 295 for a more general statement.

Hardy's inequality: The first weighted form of Hardy's inequality can equivalently be written in the following way:

$$\int_0^\infty \left(x^{\alpha-1} \int_0^x \frac{f(y)}{y^\alpha} dy \right)^p dx \leq \left(\frac{p}{p-p\alpha-1} \right)^p \left(\int_0^\infty f^p(x)dx \right),$$

$$(7.123)$$

where $p \geq 1, \alpha < \frac{1}{p'}$. The (equivalent) dual form of (7.123) reads:

$$\int_0^\infty \left(x^{\alpha-1} \int_x^\infty \frac{f(y)}{y^\alpha} dy \right)^p dx \leq \left(\frac{p}{1+p\alpha-p} \right)^p \left(\int_0^\infty f^p(x)dx \right),$$

$$(7.124)$$

where $p \geq 1, \alpha > \frac{1}{p'}$.

Hardy-Hilbert-type inequalities for homogeneous kernels: The inequalities (7.121)–(7.124) can all be written on the unified form

$$\int_0^\infty \left(\int_0^\infty k(x,y) f(x)\, dx \right)^p dy \leq C^p \int_0^\infty f^p(x) dx, \quad p \geq 1,$$
(7.125)

with different kernels $k(x,y)$, which are homogeneous of degree -1. A kernel $k(x,y)$ is said to be homogeneous of degree $\lambda, \lambda \in R$, if

$$k(tx,ty) = t^\lambda k(x,y), \quad \text{for all } x,y \in R_+.$$
(7.126)

For the case $\lambda = -1$ see (7.92). It is also well-known that the inequality (7.125) can be equivalently rewritten in the form

$$\int_0^\infty \int_0^\infty k(x,y) f(x) g(y) dx dy$$

$$\leq C \left(\int_0^\infty f^p(x) dx \right)^{\frac{1}{p}} \left(\int_0^\infty g^{p'}(y) dy \right)^{\frac{1}{p'}}, \quad p \geq 1,$$
(7.127)

with the same sharp constant C.

Remark 7.68. There is a huge number of papers devoted to prove (7.125) and (7.127) for concrete kernels $k(x,y)$ other than the classical Hilbert kernel $k(x,y) = 1/(x+y)$. In this connection we refer to the monograph [KPPV1] and the references therein. Moreover, we pronounce that by using a standard dilation argument in (7.125)–(7.127) we see that such kernels must be homogeneous of degree -1. In the papers [LPS2] and [LPSW1] some complements, which unify and extend a great number of these results, were proved.

One key observation to understand this connection is the following result:

Lemma 7.69. a) *Let* $p \geq 1$. *The following statements are equivalent:*

i) *The inequality*

$$\int_0^\infty \int_0^\infty k(x,y) f(x) g(y) dx dy \leq C \|f\|_p \|g\|_{p'}$$
(7.128)

holds for some finite constant C *and all* $f \in L_p$ *and* $g \in L_{p'}$.

ii) *The inequality*

$$\int_0^\infty \left(\int_0^\infty k(x,y) f(x) dx \right)^p dy \le C^p \int_0^\infty f^p(x) dx \qquad (7.129)$$

holds for the same finite constant C *as in* (7.128) *and for all* $f \in L_p$.

b) *Let* $0 < p < 1$. *A similar equivalence as that in a) holds also in this case but with the inequalities in* (7.128) *and* (7.129) *reversed.*

Since $p' < 0$ in case b) we have $\|g\|_{p'} = \left(\int_0^\infty |g(y)|^{p/(p-1)} dy \right)^{\frac{p-1}{p}}$ and we assume that $0 < \|g\|_{p'} < \infty$ here and in the sequel.

Remark 7.70. The statement in a) is well-known and follows from a more general statement in Functional Analysis. However, we give here a simple direct proof, which works also to prove that part b) holds.

Proof of Lemma 7.69. a) Let $p > 1$. Assume that (7.129) holds. Then, by using Hölder's inequality, we find that

$$I_1 = \int_0^\infty \int_0^\infty k(x,y) f(x) g(y) dx dy$$

$$\le \left(\int_0^\infty \left(\int_0^\infty k(x,y) f(x) dx \right)^p dy \right)^{1/p} \left(\int_0^\infty g^{p'}(y) dy \right)^{\frac{1}{p'}}$$

$$\le C \|f\|_p \|g\|_{p'},$$

so (7.128) holds. Now assume that (7.128) holds and choose

$$g(y) = \left(\int_0^\infty k(x,y) f(x) dx \right)^{p-1} \in L_{p'}.$$

With this choice

$$I_1 = \int_0^\infty \left(\int_0^\infty k(x,y) f(x) dx \right)^p dy := I_2.$$

Thus, by (7.128),

$$I_2 \leq C \left\| f \right\|_p \left(\int_0^\infty \left(\int_0^\infty k(x,y)f(x)dx \right)^p dy \right)^{\frac{1}{p'}}$$

$$= C \left\| f \right\|_p I_2^{\frac{1}{p'}}.$$

Hence,

$$I_2 \leq C \left\| f \right\|_p$$

so (7.129) holds.

Let $p = 1$ so $p' = \infty$. By applying (7.128) with $g(y) \equiv 1$ we see that (7.128) implies (7.129). Morover, by using that $g(y) \leq \left\| g \right\|_\infty$, $y \in (0, \infty)$, we find that (7.129) implies (7.128).

b) Hölder's inequality holds in the reversed direction in this case. Hence, the proof of b) only consists of obvious modifications of the proof of a).   □

Theorem 7.71. *Let $p \geq 1$, the kernel $k(x,y)$ satisfy (7.92) and κ_p be the constant defined by (7.93). Then the following three statements are equivalent:*

(i) *The constant κ_p is finite.*
(ii) *The inequality*

$$\int_0^\infty \int_0^\infty k(x,y)f(x)g(y)dxdy \leq C \left\| f \right\|_p \left\| g \right\|_{p'} \qquad (7.130)$$

holds for some finite constant C for all $f \in L_p$ and $g \in L_{p'}$.
(iii) *The inequality*

$$\int_0^\infty \left(\int_0^\infty k(x,y)f(x)dx \right)^p dy \leq C^p \int_0^\infty f^p(x)dx \qquad (7.131)$$

holds for the same finite constant C as in (7.130) and for all $f \in L_p$.

Moreover, the constant $C = \kappa_p$ is sharp in both (7.130) and (7.131).

Remark 7.72. The proof of (7.131) under the condition $\kappa_p < \infty$ was already given in the book [HLP], Theorem 3.19. Apart from the original proof in [HLP], this sufficiency part may be derived, via a change of variables, from the Young theorem for convolutions in R, for details see [KS1] and [KS2]. In this way the sharpness of the constant is derived from the fact that the Young inequality $\|h * f\|_p \leq \|h\|_1 \|f\|_p$ holds with the sharp constants $\|h\|_1$ when h is non-negative.

For the case $0 < p < 1$ it is expected that the inequalities (7.130) and (7.131) hold in the reversed direction but now with the natural restrictions

$$I_1 = \int_0^\infty \int_0^\infty k(x,y)f(x)g(y)dxdy < \infty \qquad (7.132)$$

and

$$I_2 = \int_0^\infty \left(\int_0^\infty k(x,y)f(x)dx \right)^p dy < \infty \qquad (7.133)$$

so the reversed inequalities (7.130) and (7.131) make sense. We also need the following minor techniqual condition:

$$\lim_{\varepsilon \to 0+} \int_0^{\varepsilon_0} k(1,y)y^{-1/p}y^{(p-1)\varepsilon}dy = \int_0^{\varepsilon_0} k(1,y)y^{-1/p}dy \qquad (7.134)$$

for some $\varepsilon_0 > 0$.

Theorem 7.73. *Let $0 < p < 1$ and the kernel $k(x,y)$ satisfy (7.92). Moreover, assume that (7.132)–(7.134) hold. Then all the statements in Theorem 7.71 hold with inequalities (7.130) and (7.131) holding in reversed direction.*

Here we use the same convention concerning $\|g\|_{p'}$ as before, see sentence after Lemma 7.69.

Remark 7.74. For the proof of the fact that $\kappa_p < \infty$ implies the equivalent reversed conditions (7.130) and (7.131) we do not need the restriction (7.134).

Remark 7.75. By using a standard dilation argument it is seen that the inequalities considered in Theorem 7.71 can hold if and only if $\lambda = -1$. However, by changing the norms in the left-hand sides in

(7.130) and (7.131) to power-weighted norms we can from our result obtain a similar result for homogeneous kernels of any degree λ. In order to be able to compare with a result in [Zh1] (see Remark 7.77) we formulate this result as follows:

Theorem 7.76. *Let $p \geq 1$ and $\alpha, \beta \in \mathbb{R}$. Let the kernel $k_{\lambda_0}(x, y)$ satisfy* (7.126) *for $\lambda_0 = -1 + \alpha + \beta$, and define*

$$\kappa_{p,\beta} = \int_0^\infty k_{\lambda_0}(1, y) y^{-\beta - (1/p)} dy. \tag{7.135}$$

Then the following four conditions are equivalent:

i*) *The constant $\kappa_{p,\beta}$ is finite.*
ii*) *The inequality*

$$\int_0^\infty \int_0^\infty k_{\lambda_0}(x, y) f(x) g(y) dx dy \leq C \|f\|_{p,x^\alpha} \|g\|_{p',x^\beta} \tag{7.136}$$

holds for some finite constant C for all $f \in L_{p,x^\alpha}$ and $g \in L_{p',x^\beta}$.
iii*) *The inequality*

$$\int_0^\infty \left(y^{-\beta} \int_0^\infty k(x, y) f(x) dx \right)^p dy \leq C^p \int_0^\infty f^p(x) x^{\alpha p} dx \tag{7.137}$$

holds for the same finite constant C as in (7.136) *and all $f \in L_{p,x^\alpha}$.*
iv*) *The constant $C = \kappa_{p,\beta}$ (defined by* (7.135)*) is sharp in both* (7.136) *and* (7.137).

Remark 7.77. By choosing $\lambda = -\lambda_0$, $\alpha = 1 - \frac{\lambda}{r} - \frac{1}{p}$, $\beta = 1 - \frac{\lambda}{s} - \frac{1}{p'}$ $(= -\frac{\lambda}{s} + \frac{1}{p})$ with $s > 1$, $\frac{1}{r} + \frac{1}{s} = 1$ we can compare with Theorem 2.1 in [Zh1]. For the case $p > 1$, $\lambda_0 > 0$ the equivalence in ii*) and iii*) were already established in this theorem and also the sharpness in iv*) for these cases. However, the necessity pointed out in i*) was not explicitly pointed out in [Zh1].

Remark 7.78. By using our Theorem 7.73 and making similar calculations as in the proof of Theorem 7.76 we can obtain a similar complement and strengthening of Theorem 2.2 in [Zh1] yielding for $0 < p < 1$ and kernels of any homogeneity $\lambda_0 \in \mathbb{R}$.

In order to cover even more direct applications we finally also state another consequence (but also formal generalization) of Theorem 7.71. We consider here (skew-symmetric) kernels with the following generalized homogeneity of order -1

$$k(t^a x, t^b y) = t^{-1} k(x, y), \quad a, b \neq 0. \tag{7.138}$$

Theorem 7.79. *Let $p \geq 1$ and let the kernel $k(x, y)$ satisfy (7.138) with (generalized duality) condition $\frac{a}{p'} + \frac{b}{p} = 1$ and define*

$$\kappa_{p,\beta}(a, b) := \left(\frac{a}{b}\right)^{\frac{1}{p'}} \int_0^\infty k(1, t) \, t^{\frac{1}{b}\left[\left(\frac{b-2}{p}+1\right)\right]-1} dt.$$

Then the following conditions are equivalent:

(i) *The constant $\kappa_{p,\beta}(a, b)$ is finite.*
(ii) *The inequality*

$$\int_0^\infty \int_0^\infty k(x, y) f(x) g(y) dx dy \leq C \|f\|_p \|g\|_{p'} \tag{7.139}$$

holds for some finite constant C for all $f \in L_p$ and $g \in L_{p'}$.
(iii) *The inequality*

$$\int_0^\infty \left(\int_0^\infty k(x, y) f(x) dy\right)^p dx \leq C^p \int_0^\infty f^p(x) dx \tag{7.140}$$

holds for the same finite constant C as in (7.139) and all $f \in L_p$.
(iv) *The sharp constant in both (7.139) and (7.140) is $C = \kappa_{p,\beta}(a, b)$.*

Remark 7.80. By using a similar proof as that of Theorem 7.79 we can obtain a similar consequence (and formal extension) also of our Theorem 7.73.

Examples of inequalities covered by the results above

Example 7.81 (cf. (7.121)). Let $f(x, y) = \frac{1}{x+y}$ and $p > 1$. Then, Theorem 7.71 guarantees that the following equivalent inequalities hold

$$\int_0^\infty \left(\int_0^\infty \frac{f(y)}{x+y} dy \right)^p dx \le \kappa_p^p \int_0^\infty f^p(x)\, dx$$

and

$$\int_0^\infty \int_0^\infty \frac{f(x)f(y)}{x+y} dx dy \le \kappa_p \left(\int_0^\infty f^p(x)\, dx \right)^{\frac{1}{p}} \left(\int_0^\infty g^{p'}(y)\, dy \right)^{\frac{1}{p'}}$$

with the sharp constant

$$\kappa_p = \int_0^\infty \frac{y^{-\frac{1}{p}}}{1+y} dy = \frac{\pi}{\sin \frac{\pi}{p}}.$$

In a similar way we can get a great number of the so-called Hardy-Hilbert type inequalities by using other related kernels of homogeneous type -1. For example, if $\lambda p' > 1$ we have the following equivalent inequalities

$$\int_0^\infty \left(x^{\lambda-1} \int_0^\infty \frac{f(y)}{x^\lambda + y^\lambda} dy \right)^p dx \le \kappa_p^p \int_0^\infty f^p(x)\, dx \qquad (7.141)$$

and

$$\int_0^\infty \int_0^\infty \frac{x^{\lambda-1} f(y) g(x)}{x^\lambda + y^\lambda} dx dy$$
$$\le \kappa_p \left(\int_0^\infty f^p(y)\, dy \right)^{\frac{1}{p}} \left(\int_0^\infty g^{p'}(x)\, dx \right)^{\frac{1}{p'}} \qquad (7.142)$$

with sharp constant

$$\kappa_p = \int_0^\infty \frac{y^{-\frac{1}{p}}}{1+y^\lambda} dy = \frac{1}{\lambda} \int_0^\infty \frac{y^{\frac{1}{\lambda p'}-1}}{1+y} dy = \frac{\pi}{\lambda \sin \frac{\pi}{\lambda p'}}.$$

Example 7.82 (cf. (7.123) and (7.124)). Let $k(x,y) = x^{\alpha-1} y^{-\alpha}$, $0 < y \le x$, $k(x,y) = 0$, $y > x$. Then, Theorem 7.71 implies the following equivalent inequalities

$$\int_0^\infty \left(x^{\alpha-1} \int_0^x \frac{f(y)}{y^\alpha} dy \right)^p dx \leq \kappa_p^p \int_0^\infty f^p(x)\, dx \qquad (7.143)$$

and

$$\int_0^\infty \int_0^x \frac{x^{\alpha-1} f(y) g(x)}{y^\alpha} dy dx$$

$$\leq \kappa_p \left(\int_0^\infty f^p(y)\, dy \right)^{\frac{1}{p}} \left(\int_0^\infty g^{p'}(x)\, dx \right)^{\frac{1}{p'}}$$

with the sharp constant

$$\kappa_p = \int_0^1 y^{-\alpha} y^{-\frac{1}{p}} dy = \frac{p}{p-1-\alpha p}, \quad \alpha < \frac{1}{p'}, \quad 1 \leq p \leq \infty.$$

Instead by using the kernel $k(x,y) = x^{\alpha-1} y^{-\alpha}$, $y \geq x$, $k(x,y) = 0$, $0 < y < x$ Theorem 7.71 implies the equivalent inequalities

$$\int_0^\infty \left(x^{\alpha-1} \int_x^\infty \frac{f(y)}{y^\alpha} dy \right)^p dx \leq \kappa_p^p \int_0^\infty f^p(x)\, dx \qquad (7.144)$$

and

$$\int_0^\infty \int_x^\infty \frac{x^{\alpha-1} f(y) g(x)}{y^\alpha} dy dx$$

$$\leq \kappa_p \left(\int_0^\infty f^p(y)\, dy \right)^{\frac{1}{p}} \left(\int_0^\infty g^{p'}(x)\, dx \right)^{\frac{1}{p'}}$$

with the sharp constant

$$\kappa_p = \frac{p}{\alpha p - p + 1}, \quad \alpha > \frac{1}{p'}, \quad 1 \leq p \leq \infty.$$

Remark 7.83. The inequality (7.143) is an equivalent form of the first weighted version of Hardy's original inequality proved by Hardy himself in 1928 (see [Ha5]). (7.144) is sometimes called the dual form of (7.143), in fact these inequalities are in a sense equivalent.

In our next example we unify and generalize Examples 7.81 and 7.82 by presenting a scale of inequalities between those in these examples (a genuine Hardy-Hilbert-type inequality).

Example 7.84. Apply Theorem 7.71 with the kernel

$$k(x,y) = \frac{x^{\alpha+\beta-1}}{y^{\alpha}(x+y)^{\beta}}, \quad 0 < y \le x, \quad \text{and} \quad k(x,y) = 0, \quad y > x.$$

We find that the (Hardy-Hilbert-type) inequality

$$\int_0^{\infty} \left(x^{\alpha+\beta-1} \int_0^{ax} \frac{f(y)}{y^{\alpha}(x+y)^{\beta}} dy \right)^p dx \le \kappa_p^p \int_0^{\infty} f^p(x) \, dx, \quad (7.145)$$

where $0 < a \le \infty$, holds with the sharp constant

$$\kappa_p = \int_0^a \frac{y^{-\frac{1}{p}}}{y^{\alpha}(x+y)^{\beta}} dy = \int_0^{\frac{a}{1+a}} t^{-\alpha-\frac{1}{p}}(1-t)^{\alpha+\frac{1}{p}+\beta-2} dx$$

$$= B_{\frac{a}{1+a}}\left(\frac{1}{p'} - \alpha, \alpha + \beta - \frac{1}{p'} \right),$$

where $0 < a \le \infty$, $1 \le p \le \infty$,

$$\begin{cases} \alpha < \dfrac{1}{p'}, \quad \beta \in R, \quad \text{if} \quad a < \infty, \\[2mm] \alpha < \dfrac{1}{p'}, \quad \alpha + \beta > \dfrac{1}{p'}, \quad \text{if} \quad a = \infty, \end{cases}$$

and $B_z(u,v)$ denotes the incomplete Beta-function

$$B_z(u,v) = \int_0^z t^{1-u}(1-t)^{v-1} dt \quad 0 < z \le 1.$$

Remark 7.85. Concerning (7.145) note especially that

(*) if $a = 1, \beta = 0$, we obtain the Hardy inequality (7.143) in Example 7.82,

(**) if $a = \infty$, $\beta = 1, \alpha = 0$ we get the Hilbert inequality in Example 7.81,

(***) in all (Hardy like) cases $\beta = 0$ we have the sharp constant

$$\frac{a^{\frac{1}{p'}-\alpha}}{\frac{1}{p'}-\alpha}, \quad \alpha < \frac{1}{p'}.$$

The following example is a dual counterpart to Example 7.84.

Example 7.86. Apply Theorem 7.71 with the kernel

$$k(x,y) = \frac{x^{\alpha+\beta}}{y^\alpha(x+y)^\beta}, \ y \geq x \quad \text{and} \quad k(x,y) = 0, \quad 0 < y \leq x,$$

and we find that

$$\int_0^\infty \left| x^{\alpha+\beta} \int_{ax}^\infty \frac{f(y)\,dy}{y^{1+\alpha}(x+y)^\beta} \right|^p dx \leq \varkappa_p^p \int_0^\infty |f(x)|^p\,dx,$$

where $0 \leq a < \infty$, with the sharp constant

$$\varkappa_p = \int_a^\infty \frac{dy}{y^{1+\alpha+\frac{1}{p}}(1+y)^\beta} = \int_0^{\frac{1}{1+a}} t^{\alpha+\beta+\frac{1}{p}-1}(1-t)^{-\alpha-\frac{1}{p}-1}\,dt)$$

$$= B_{\frac{1}{1+a}}\left(\alpha+\beta+\frac{1}{p},-\alpha-\frac{1}{p}\right),$$

where $0 \leq a < \infty$, $1 \leq p \leq \infty$, and

$$\begin{cases} -\beta < \alpha + \dfrac{1}{p}, & \beta \in \mathbb{R}, \quad \text{if } a > 0, \\[2mm] -\beta < \alpha + \dfrac{1}{p} < 0, & \beta > 0, \quad \text{if } a = 0. \end{cases}$$

Example 7.87 (Hardy-Littlewood inequality [HL1]). It yields that

$$\int_0^\infty \frac{1}{x^\alpha}\left| \int_0^x \frac{f(y)\,dy}{(x-y)^{1-\alpha}} \right|^p dx \leq \varkappa_p^p \int_0^\infty |f(x)|^p\,dx$$

with the sharp constant

$$\varkappa_p = \int_0^1 \frac{dy}{y^{\frac{1}{p}}(1-y)^{1-\alpha}} = B\left(\alpha,\frac{1}{p'}\right), \quad \alpha > 0, \ 1 < p < \infty.$$

The following example is also a particular case of Theorem 7.71:

Example 7.88 (Unifying Examples 7.82 and 7.87). It yields that

$$\int_0^\infty x^{\alpha+\beta-1} \left| \int_0^x \frac{f(y)\,dy}{y^\alpha (x-y)^\beta} \right|^p dx \le \varkappa_p^p \int_0^\infty |f(x)|^p \, dx$$

with the sharp constant

$$\varkappa_p = \int_0^1 \frac{dy}{y^{\alpha+\frac{1}{p}} (1-y)^\beta}$$

$$= B\left(1-\beta, \frac{1}{p'} - \alpha\right), \quad \alpha < \frac{1}{p'}, \quad \beta < 1, \quad 1 < p < \infty.$$

Example 7.89. Let $\alpha > 0$, $p > 1$, and λ, μ satisfy that

$$\frac{1}{\lambda p'} + \frac{1}{\mu p} = \alpha, \quad 2 - p < \frac{1}{\alpha\mu} < 2.$$

Then the following inequalities hold and are equivalent:

(i) $\int_0^\infty \int_0^\infty \left(\frac{1}{x^\lambda + y^\mu}\right)^\alpha f(x) g(y) \, dx\,dy \le C \|f\|_p \|g\|_{p'}$ for all $f \in L_p$ and $g \in L_{p'}$.

(ii) $\int_0^\infty \left(\int_0^\infty \left(\frac{1}{x^\lambda + y^\mu}\right)^\alpha f(x) \, dy\right)^p dx \le C^p \int_0^\infty f^p(x) \, dx$ for all $f \in L_p$.

The sharp constant C in both (i) and (ii) is

$$C = \frac{1}{|\lambda|^{\frac{1}{p'}} |\mu|^{\frac{1}{p}}} B(a_0, a_1),$$

with $a_0 = \frac{1}{p}\left(2\alpha - \frac{1}{\mu}\right)$ and $a_1 = \alpha - a_0$.

In fact, the proof follows by just using Theorem 7.79 with $a = \frac{1}{\lambda\alpha}$, $b = \frac{1}{\mu\alpha}$ and making some straightforward calculations.

Remark 7.90. In the classical Hilbert case $\alpha = \lambda = \mu = 1$ we obtain that

$$C = B\left(\frac{1}{p}, \frac{1}{p'}\right) = \frac{\pi}{\sin\frac{\pi}{p}}$$

so that (i) coincides with the classical form (7.121) of Hilbert's inequality.

In addition to the constant κ_p defined in (7.93), we also introduce the constants

$$\kappa_\infty := \int_0^\infty k(1, y)\, dy$$

and

$$\kappa^* := \frac{\int_0^\infty k(1, y) \ln\frac{1}{y} dy}{\int_0^\infty k(1, y)\, dy}$$

assuming that $k(x, y) \geq 0$ and may be zero only on a set of measure zero.

The following limit geometric mean inequality follows from our Theorem 7.51:

Theorem 7.91. *Let $f(x) \geq 0$ and let $\kappa_p < \infty$ for some $p > 1$. If $\kappa^* < \infty$, then*

$$\int_0^\infty \exp\left(\frac{1}{\kappa_\infty} \int_0^\infty k(x, y) \ln f(y) dy\right) dx \leq e^{\kappa^*} \int_0^\infty f(x) dx$$

$$(7.146)$$

and the constant e^{κ^*} is sharp.

Example 7.92 (Generated by a weighted Hardy inequality). Take $k(x, y) = \frac{x^{a-1}}{y^\alpha}$ when $y \leq x$ and $k(x, y) = 0$ otherwise, where $\alpha < 1$. Then $\varkappa_\infty = \frac{1}{1-\alpha}$ and

$$\varkappa^* = (1 - \alpha) \int_0^1 y^{-\alpha} \ln\frac{1}{y} dy = (1 - \alpha) \int_0^\infty t e^{-(1-\alpha)t} dt = \frac{1}{1 - \alpha}$$

and (7.146) turns into

$$\int_0^\infty \exp\left((1-\alpha)\, x^{\alpha-1} \int_0^x \frac{\ln f(y)dy}{y^\alpha}\right) dx \le e^{\frac{1}{1-\alpha}} \int_0^\infty f(x)dx$$

with $e^{\frac{1}{1-\alpha}}$ as the sharp constant. For $\alpha = 0$ this is the classical Pólya-Knopp inequality (see (7.43) or (1.80)).

Example 7.93 (Generated by the weighted Hilbert inequality). Take $k(x,y) = \left(\frac{x}{y}\right)^\alpha \frac{1}{x+y}$ where $0 < \alpha < 1$. Then

$$\varkappa_\infty = \int_0^\infty \frac{dy}{y^\alpha(1+y)} = \frac{\pi}{\sin \alpha\pi}.$$

To calculate $\varkappa^*$ we differentiate the last equality in α and get

$$\int_0^\infty \frac{\ln \frac{1}{y}dy}{y^\alpha(1+y)} = -\frac{\pi^2 \cos \alpha\pi}{\sin^2 \alpha\pi}$$

so that $\varkappa^* = -\pi \cot \alpha\pi$ and (7.146) turns into the sharp inequality

$$\int_0^\infty \exp\left(\frac{\pi x^\alpha}{\sin \alpha\pi} \int_0^\infty \frac{\ln f(y)}{y^\alpha(x+y)}dy\right) dx \le e^{-\pi \cot \alpha\pi} \int_0^\infty f(x)dx.$$

Example 7.94 (Generated by the Hardy-Littlewood inequality). Take $k(x,y) = \frac{1}{x^\alpha(x-y)^{1-\alpha}}$ when $y < x$ and $k(x,y) = 0$ otherwise, where $\alpha > 0$. Then $\varkappa_\infty = \frac{1}{\alpha}$. Via integration by parts and some additional tricks it may be shown that

$$\int_0^\infty k(1,y) \ln \frac{1}{y}dy = \int_0^\infty \frac{\ln \frac{1}{y}}{(1-y)^{1-\alpha}}dy = \frac{\psi(1+\alpha) - \psi(1)}{\alpha},$$

where $\psi(z) = \frac{\Gamma'(z)}{\Gamma(z)}$ is the Euler psi function and we find that (7.146) turns into Pólya-Knopp-type inequality

$$\int_0^\infty \exp\left(\frac{\alpha}{x^\alpha} \int_0^\infty \frac{\ln f(y)}{(x-y)^{1-\alpha}}dy\right) dx \le e^{\frac{\psi(1+\alpha)-\psi(1)}{\alpha}} \int_0^\infty f(x)dx.$$

$$(7.147)$$

Note that in the case $\alpha = 1$ the inequality (7.147) turns into the classical Pólya-Knopp inequality (see (7.43)) with the sharp constant e in view of the property $\psi(2) = \psi(1) + 1$ of the psi-function.

7.6. Hardy-type Inequalities in Other Function Spaces

7.6.1. *Hardy-type inequalities for Orlicz, Lorentz, rearrangement invariant and general Banach function spaces*

Some more information concerning the Hardy inequality in other Banach function spaces can be found in [KMP], which in a sense may be regarded as a complement of this book. Here we just mention some typical results in the cases A–D below.

A. The Hardy inequality in Orlicz spaces

An Orlicz function Φ is a continuous increasing unbounded function $\Phi : [0, \infty) \to [0, \infty)$ such that $\Phi(0) = 0$. Convex Orlicz functions are called Young functions. A Young function Φ is called an N-function if moreover

$$\lim_{t \to 0^+} \frac{\Phi(t)}{t} = 0 \ \text{ and } \ \lim_{t \to \infty} \frac{\Phi(t)}{t} = \infty.$$

For an Orlicz function Φ, a σ-finite measure space (Ω, Σ, μ) and weight w on Ω the weighted Orlicz space $L_w^\Phi = L_w^\Phi(\mu)$ is defined as the space of all classes of μ measurable functions $f : \Omega \to \mathbb{R}$ such that the modular

$$\varrho_{\Phi,w}\left(\frac{f}{\lambda}\right) := \int_\Omega \Phi\left(\frac{|f(x)|}{\lambda} w(x) d\mu(x)\right)$$

is finite for some $\lambda > 0$. If Φ is a Young function, then the weighted Orlicz space L_w^Φ is a Banach function space with the Luxemburg-Nakano norm

$$\|f\|_{\Phi,w} := \inf\left\{\lambda > 0 : \varrho_{\Phi,w}\left(\frac{f}{\lambda}\right) \le 1\right\}.$$

Moreover, the complementary function Φ^* to Φ is given by

$$\Phi^*(u) := \sup_{v>0}[uv - \Phi(v)].$$

One main result here is the following (see [KMP], Theorem 14):

Theorem 7.95. *Let Φ_1, Φ_2 be N-functions and suppose that u, v_0, v_1 are locally integrable weights and μ a positive regular Borel measure on $(0, \infty)$. The following conditions are equivalent:*

(i) *The weak type modular inequality*

$$\Phi_2^{-1}\left[\int_{\{x>0:|\int_0^x f(t)dt|>\lambda\}} \Phi_2(\lambda u(x))d\mu(x)\right]$$

$$\leq \Phi_1^{-1}\left[\int_0^\infty \Phi_1(A|f(x)|v_1(x))v_0(x)dx\right]$$

holds for some $A > 0$.

(ii) *There is a constant $B > 0$ such that*

$$\int_0^r \Phi_1^*\left(\frac{\alpha(\lambda, r)}{B\lambda v_0(x)v_1(x)}\right) v_0(x)dx \leq \alpha(\lambda, r) < \infty$$

for every $\lambda > 0$ and every $r > 0$, where $\alpha(\lambda, r) = \Phi_1 \circ \Phi_2^{-1}\left[\int_r^\infty \Phi_2(\lambda u(x))d\mu(x)\right]$. Moreover, $\frac{A}{2} \leq B \leq 4A$.

If, in addition, the composition $\Phi_2^{-1} \circ \Phi_1$ is a countable super-additive function on $(0, \infty)$, then (i) and (ii) are also equivalent to the following conditions:

(iii) *The strong type modular inequality*

$$\Phi_2^{-1}\left[\int_0^\infty \Phi_2\left(u(x)\left|\int_0^x f(t)dt\right|\right)d\mu(x)\right]$$

$$\leq \Phi_1^{-1}\left[\int_0^\infty \Phi_1(C|f(x)|v_1(x))v_0(x)dx\right]$$

holds for some $C > 0$.

(iv) *There is a constant $D > 0$ such that*

$$\Phi_2^{-1}\left[\int_r^\infty \Phi_2\left(\frac{u(x)}{D}\left\|\frac{\chi_{(0,r)}}{v_0 v_1 \varepsilon}\right\|_{\Phi_1^*,\varepsilon v_0}\right)d\mu(x)\right] \leq \Phi_1^{-1}\left(\frac{1}{\varepsilon}\right)$$

for every $\varepsilon > 0$ and $r > 0$.
Moreover, $D \leq C \leq 8D$ and $A \leq C \leq 8A$.

B. The Hardy inequality in Lorentz spaces

By using the information in Chap. 6 we get much information concerning the mapping properties of Hardy-type operations in standard weighted Lorentz spaces $\Lambda^p(w)$ as defined by (6.1). Here we just give some complementary information from the book [KMP].

For $0 < p < \infty, 0 < q \leq \infty$ and a weight w on Ω, the Lorentz space $L^{p,q}(w)$ is defined by

$$L^{p,q}(w) := \left\{ f \in L^0(\Omega) : \right.$$

$$\left. \|f\|_{L^{p,q}(w)} := \left(\frac{q}{p}\int_0^\infty t^{q/p-1}f_w^*(t)^q dt\right)^{1/q} < \infty\right\}$$

if $0 < q < \infty$, and by

$$L^{p,\infty}(w) := \left\{ f \in L^0(\Omega) : \|f\|_{L^{p,\infty}(w)}(w) := \sup_{t>0} t^{1/p}f_w^*(t) < \infty\right\}$$

if $q = \infty$, where

$$f_w^*(t) := \inf\{y > 0 : \lambda_f^w(y) \leq t\}, \quad \lambda_f^w(y) := \int_{\{x\in\Omega:|f(x)|>y\}} w(x)dx.$$

One typical result here is the following theorem by Sawyer [Sa1]:

Theorem 7.96. *Let $0 < p < \infty, 0 < q < \infty, 1 < r < \infty, 1 \leq s \leq \infty$, and let u, v be weights. Assume that the estimate*

$$\left\|\int_0^x f(t)dt\right\|_{L^{p,q}(u)} \leq C\|f\|_{L^{r,s}(v)} \tag{7.148}$$

holds for all $f \in L^{r,s}(v)$. Then

$$\sup_{x>0} \|\chi_{(x,\infty)}\|_{L^{p,q}(u)} \left\| \frac{\chi_{(0,x)}}{v} \right\|_{L^{r',s'}(v)} < \infty \qquad (7.149)$$

($v > 0$ a.e. on $(0,x)$) if $\int_x^\infty u(t)dt > 0$. Conversely, (7.149) implies (7.148) if and only if $q \geq \max\{r, s\}$.

C. The Hardy inequality in rearrangement invariant (r.i.) spaces

The r.i. space X on $I = (a, b), a < b \leq \infty$, has the lower and upper Boyd indices p_X, q_X defined by

$$p_X := \lim_{s \to \infty} \frac{\ln s}{\ln h(s, X)} = \sup_{s>1} \frac{\ln s}{\ln h(s, X)}$$

and

$$q_X := \lim_{s \to 0^+} \frac{\ln s}{\ln h(s, X)} = \inf_{0<s<1} \frac{\ln s}{\ln h(s, X)},$$

respectively, where $h(s, X) = \|D_s\|_{X \to X}$ and $D_s f(x) = f(x/s)\chi_{(0,b)}(x/s)$.

Moreover, we define the modified Hardy operator H_θ as follows:

$$H_\theta f(x) := \frac{1}{x^\theta} \int_0^x t^{\theta-1} f(t)dt, \ \theta = 1/p, 1 \leq p < \infty.$$

One important result here is the following (see [KMP], Theorem 16).

Theorem 7.97. *Let X be an arbitrary r.i. space on $I = (0, b), 0 < b \leq \infty$. Then the Hardy operator H_θ is bounded in X if and only if the lower Boyd index p_X satisfies $p_X > 1/\theta$.*

We also define the non-linear Hardy-type operators $H^{(p,r)}$ and $H_{(q,r)}$ as follows:

$$H^{(p,r)} f(t) := t^{-1/p} \left(\int_0^t f^*(s)^r ds^{r/p} \right)^{1/r}$$

and

$$H_{(q,r)}f(t) := t^{-1/q} \left(\int_t^\infty f^*(s)^r ds^{r/q} \right)^{1/r},$$

respectively, with suitable modifications for infinite indices.

The main result reads (see [KMP], Theorem 17):

Theorem 7.98. *Let X be a r.i. quasi-Banach space on $I = (0,1)$ or $I = (0,\infty)$.*

(i) *For $0 < p, r < \infty$ the operator $H^{(p,r)}$ is bounded in X if and only if $p_X > p$.*

(ii) *For $0 < q \leq \infty$ and $0 < r < \infty$ the operator $H_{(q,r)}$ is bounded in X if and only if $q_X < q$.*

D. The Hardy inequality in general Banach function spaces

In this case very little is known. We only refer to the short information in [KMP], see e.g. Theorem 18 and the paper [Ber1].

In the following two subsections we will present and discuss some new results for Morrey-type and Hölder-type spaces, respectively.

7.6.2. Hardy-type Inequalities in Morrey-type spaces

The one-dimensional case

The classical Morrey space $L^{p,\lambda}(\mathbb{R}_+), p \geq 1, 0 \leq \lambda < 1$ is defined by the norm $\|f\|_{p,\lambda}$ defined by

$$\|f\|_{p,\lambda} = \sup_{x>0, r>0} \left(\frac{1}{r^\lambda} \int_{(x-r)_+}^{x+r} |f(y)|^p dy \right)^{\frac{1}{p}}.$$

The dilation operator is defined by $\Pi_\varrho f(x) = f(\varrho x)$, $x > 0$. The Morrey space $L^{p,\lambda}(\mathbb{R}_+)$ has the following property

$$\|\Pi_\varrho f\|_{p,\lambda} := \frac{1}{\varrho^{\frac{1-\lambda}{p}}} \|f\|_{p,\lambda}, \ \varrho > 0. \tag{7.150}$$

For the Hardy-type operators

$$H_\gamma f(x) = x^{\gamma-1} \int_0^x \frac{f(t)dt}{t^\gamma}, \quad \mathcal{H}_\gamma f(x) = x^\gamma \int_x^\infty \frac{f(t)dt}{t^{1+\gamma}}, \quad x > 0,$$

$$(7.151)$$

with power weights the following theorems hold:

Theorem 7.99. *Let $1 \le p < \infty, 0 \le \lambda < 1$ and $\lambda + p > 1$. Then the inequality*

$$\|H_\gamma\|_{p,\lambda} \le C\|f\|_{p,\lambda} \qquad (7.152)$$

holds for some $C : 1 \le C < \infty$ if and only if $\gamma < \frac{\lambda}{p} + \frac{1}{p'}$. Moreover, under this condition $C = \frac{p}{\lambda+p-1-\gamma p}$ is the best constant.

Proof.   Let $\gamma < \frac{\lambda}{p} + \frac{1}{p'}$. We have $H_\gamma f(x) = \int_0^1 \frac{f(xt)}{t^\gamma}dt$. Hence $\|H\|_{p,\lambda} \le \int_0^1 \frac{\|f(\cdot t)\|_{p,\lambda}}{t^\gamma}dt$ and then by (7.150) we get

$$\|H_\gamma\|_{p,\lambda} \le \int_0^1 \frac{dt}{t^{\gamma+\frac{1-\lambda}{p}}}\|f(t)\|_{p,\lambda} = \frac{p}{\lambda+p-1-\gamma p}\|f(t)\|_{p,\lambda},$$

which proves (7.152). The constant $C = \frac{p}{\lambda+p-1-\gamma p}$ is the best possible. The case $\lambda = 0$ is well known (see e.g. Theorem 7.10). In the case $\lambda > 0$ it suffices to observe that for the function $f_0 = x^{\frac{\lambda-1}{p}} \in L^{p,\lambda}(\mathbb{R}_+)$ the inequality (7.152) turns into equality, since $H_\gamma f_0(x) = \frac{p}{\lambda+p-1-\gamma p}f_0(x)$.

The necessity of the condition $\gamma < \frac{\lambda}{p} + \frac{1}{p'}$ follows from the fact that $H_\gamma f_0$ just does not exist for the function f_0 if $\gamma \ge \frac{\lambda}{p} + \frac{1}{p'}$. $\square$

Theorem 7.100. *Let $1 \le p < \infty$ and $0 \le \lambda < 1$ and $\lambda + p > 1$. Then the inequality*

$$\|\mathcal{H}_\gamma\|_{p,\lambda} \le C\|f\|_{p,\lambda} \qquad (7.153)$$

holds if and only if $\gamma > \frac{\lambda-1}{p}$, and under this condition $C = \frac{p}{\gamma p+1-\lambda}$ is the best constant.

We omit the proof of this theorem since it is similar to that of Theorem 7.99.

Open question

Theorems 7.99 and 7.100 correspond to the case of power weighted Hardy-type inequalities. The weighted boundedness in the general setting, i.e., with a complete characterization of admissible weights, like the Muckenhoupt class A_p for Lebesgue spaces, is an open question for Morrey spaces.

Remark 7.101. We remark that the class of weights for Morrey-type spaces must differ essentially from the Muckenhoupt class A_p of such weights for the Lebesgue spaces $L^p(\Omega)$. It should already depend not only on p, but also on λ. As an illustrating example, note that, as shown in [S2] in the one-dimensional case, the singular operator (Hilbert transform) is bounded in the classical (local or global) Morrey spaces with power weights $|x - x_0|^\alpha$, $x_0 \in \mathbb{R}$, if and only if

$$\lambda - 1 < \alpha < \lambda + p - 1. \tag{7.154}$$

An open question is to find the characterization of the whole class of admissible weights, denote it say, as $A_{p,\varphi}$ or $A_{p,\lambda}$ in the case of classical Morrey spaces, similar to the Muckenhoupt class A_p. A certain candidate for the class $A_{p,\lambda}$ was introduced in [S5], where its necessity was shown for the boundedness of the Hilbert transform. For definitions and basic information about weighted local Morrey spaces $\mathcal{L}^{p,\varphi}_{\{x_0\}}(\Omega, \mu)$ we refer to the paper [S4]. Here, for two weighted local Morrey spaces $\mathcal{L}^{p,\varphi}_{\{x_0\}}(\Omega, u)$ and $\mathcal{L}^{p,\varphi}_{\{x_0\}}(\Omega, v)$ general type sufficient conditions and necessary conditions imposed on the functions φ and ψ and the weights u and v for the boundedness of the maximal operator from $\mathcal{L}^{p,\varphi}_{\{x_0\}}(\Omega, u)$ to $\mathcal{L}^{p,\varphi}_{\{x_0\}}(\Omega, v)$, with some "logarithmic gap" between the sufficient and necessary conditions were obtained. Both the conditions formally coincide if we omit a certain logarithmic factor in these conditions. We also refer to the paper [EKM1], where there was considered a version of weighted Morrey spaces.

The multi-dimensional case

In this section we present some boundedness results concerning multi-dimensional Hardy-type operators between generalized Morrey spaces $\mathcal{L}^{p,\varphi}(\mathbb{R}^n, w)$ with different p defined by an almost increasing function $\varphi(r)$ and radial type weight $w(|x|)$. We obtain sufficient conditions in terms of some integral inequalities imposed on φ and w, for such boundedness. In some cases the obtained conditions are also necessary. We also present the results on the boundedness in terms of numerical characteristics of φ and w, the so-called Matuszewska-Orlicz-type lower and upper indices.

We also present some results on boundedness of multi-dimensional weighted Hardy-type operators in local vanishing Morrey spaces.

We study the weighted generalized Hardy-type operators

$$H_w^\alpha f(x) = |x|^{\alpha-n} w(|x|) \int_{|y|<|x|} \frac{f(y)dy}{w(|y|)},$$

$$\mathcal{H}_w^\alpha f(x) = |x|^\alpha w(|x|) \int_{|y|>|x|} \frac{f(y)dy}{|y|^n w(|y|)}, \qquad (7.155)$$

where $x \in \mathbb{R}^n$ and $\alpha \geq 0$. The one-dimensional case includes the versions

$$H_w^\alpha f(x) = x^{\alpha-1} w(x) \int_0^x \frac{f(t)dt}{w(t)},$$

$$\mathcal{H}_w^\alpha f(x) = x^\alpha w(x) \int_x^\infty \frac{f(t)dt}{tw(t)}, \quad x > 0 \qquad (7.156)$$

adjusted for the half-axis $\mathbb{R}_+^1$, so that in the sequel $\mathbb{R}^n$ with $n = 1$ may be read either as $\mathbb{R}$ or $\mathbb{R}_+$. The notation $H^\alpha = H_w^\alpha|_{w\equiv 1}$ will also be used in the sequel.

Let Ω be an open set in $\mathbb{R}^n$, $\Omega \subseteq \mathbb{R}^n$ and $\ell = \text{diam } \Omega$, $0 < \ell \leq \infty$, $B(x,r) = \{y \in \mathbb{R}^n : |x - y| < r\}$ and $\widetilde{B}(x,r) = B(x,r) \cap \Omega$. Let also $\varphi(r)$ be a non-negative function on $[0, \ell]$ such that

$$\inf_{\delta < r < \ell} \varphi(r) > 0 \qquad (7.157)$$

for every $\delta > 0$. Let also $1 \leq p < \infty$. We will use the notation

$$\mathfrak{M}_{p,\varphi}(f; x, r) := \frac{1}{\varphi(r)} \int_{\widetilde{B}(x,r)} |f(y)|^p \, dy. \qquad (7.158)$$

We introduce the vanishing generalized Morrey spaces, global and local, by the following definition.

Definition 7.102. The generalized vanishing Morrey spaces $V\mathcal{L}^{p,\varphi}(\Omega)$ and $V\mathcal{L}^{p,\varphi}_{\mathrm{loc};x_0}(\Omega)$ are defined as the spaces of functions $f \in L^p_{\mathrm{loc}}(\Omega)$ such that

$$\lim_{r\to 0}\sup_{x\in\Omega}\mathfrak{M}_{p,\varphi}(f;x,r) = 0 \tag{7.159}$$

and

$$\lim_{r\to 0}\mathfrak{M}_{p,\varphi}(f;x_0,r) = 0, \tag{7.160}$$

respectively, where $x_0 \in \Omega$. Equipped with the norms

$$\|f\|_{p,\varphi} := \sup_{x\in\Omega,r>0}\left(\frac{1}{\varphi(r)}\int_{\widetilde{B}(x,r)}|f(y)|^p\,dy\right)^{\frac{1}{p}} < \infty, \tag{7.161}$$

$$\|f\|_{p,\varphi;\mathrm{loc}} := \sup_{r>0}\left(\frac{1}{\varphi(r)}\int_{\widetilde{B}(x_0,r)}|f(y)|^p\,dy\right)^{\frac{1}{p}} < \infty, \tag{7.162}$$

they are Banach spaces (closed subspaces of the corresponding Morrey spaces, when $\varphi(0) = 0$).

Besides the modular $\mathfrak{M}_{p,\varphi}(f;x,r)$ we will also use its least non-decreasing dominant

$$\widetilde{\mathfrak{M}}_{p,\varphi}(f;x,r) := \sup_{0<t<r}\mathfrak{M}_{p,\varphi}(f;x,t),$$

which may be equivalently used in the definition of the vanishing spaces, since

$$\lim_{r\to 0}\sup_{x\in\Omega}\mathfrak{M}_{p,\varphi}(f;x,r) = 0 \iff \lim_{r\to 0}\sup_{x\in\Omega}\widetilde{\mathfrak{M}}_{p,\varphi}(f;x,r) = 0.$$

Obviously,

$$V\mathcal{L}^{p,\varphi}(\Omega) \subset V\mathcal{L}^{p,\varphi}_{\mathrm{loc};x_0}(\Omega).$$

The spaces $V\mathcal{L}^{p,\varphi}(\Omega)$, $V\mathcal{L}^{p,\varphi}_{\mathrm{loc};x_0}(\Omega)$ will be called vanishing *global* and *local Morrey spaces*, respectively.

Let w denote a weight function on Ω. Everywhere in the sequel we assume that

$$\lim_{r \to 0} \frac{r^n}{\varphi(r)} = 0, \qquad (7.163)$$

and additionally

$$\sup_{0 < r < \infty} \frac{r^n}{\varphi(r)} < \infty \quad \text{in the case } \Omega \text{ is unbounded}, \qquad (7.164)$$

which makes the spaces $V\mathcal{L}^{p,\varphi}(\Omega)$, $V\mathcal{L}^{p,\varphi}_{\mathrm{loc};x_0}(\Omega)$ non-trivial, because bounded functions with compact support belong then to these spaces. Note that the condition

$$\sup_{0 < r < \infty} \frac{r^n}{\varphi(r)} < \infty, \qquad (7.165)$$

means the similar non-triviality of the crresponding non-vanishing spaces.

Definition 7.103. Let $0 < \ell < \infty$.
1) By $W = W([0, \ell])$ we denote the class of continuous and positive functions φ on $(0, \ell]$ such that there exists finite or infinite limit $\lim_{x \to 0} \varphi(x)$;
   2) by $W_0 = W_0([0, \ell])$ we denote the class of almost increasing functions $\varphi \in W$ on $(0, \ell)$;
   3) by $\overline{W} = \overline{W}([0, \ell])$ we denote the class of functions $\varphi \in W$ such that $x^a \varphi(x) \in W_0$ for some $a = a(\varphi) \in \mathbb{R}$;
   4) by $\underline{W} = \underline{W}([0, \ell])$ we denote the class of functions $\varphi \in W$ such that $\frac{\varphi(t)}{t^b}$ is almost decreasing for some $b \in \mathbb{R}$.

Definition 7.104. Let $0 < \ell < \infty$.
1) By $W_\infty = W_\infty([\ell, \infty])$ we denote the class of functions φ which are continuous and positive and almost increasing on $[\ell, \infty)$ and which have the finite or infinite limit $\lim_{x \to \infty} \varphi(x)$,
2) by $\overline{W}_\infty = \overline{W}_\infty([\ell, \infty))$ we denote the class of functions $\varphi \in W_\infty$ such that $x^a \varphi(x) \in W_\infty$ for some $a = a(\varphi) \in \mathbb{R}$.

By $\overline{W}(\mathbb{R}^1_+)$ we denote the set of functions on $\mathbb{R}_+$ whose restrictions onto $(0,1)$ are in $\overline{W}([0,1])$ and restrictions onto $[1,\infty)$ are in $\overline{W}_\infty([1,\infty))$. Similarly, the set $\underline{W}(\mathbb{R}_+)$ is defined.

It is known that the property of a function to be almost increasing or almost decreasing after the multiplication (division) by a power function is closely related to the notion of the so-called Matuszewska-Orlicz indices. We refer for instance to [KSa1], [S1], [M1] for the properties of the indices of such a type:

$$m(\varphi) = \sup_{0<x<1} \frac{\ln\left[\limsup_{h\to 0} \frac{\varphi(hx)}{\varphi(h)}\right]}{\ln x}, \quad M(\varphi) = \sup_{x>1} \frac{\ln\left[\limsup_{h\to 0} \frac{\varphi(hx)}{\varphi(h)}\right]}{\ln x}$$

$$(7.166)$$

and

$$m_\infty(\varphi) = \sup_{x>1} \frac{\ln\left[\liminf_{h\to\infty} \frac{\varphi(xh)}{\varphi(h)}\right]}{\ln x}, \quad M_\infty(\varphi) = \inf_{x>1} \frac{\ln\left[\limsup_{h\to\infty} \frac{\varphi(xh)}{\varphi(h)}\right]}{\ln x}.$$

$$(7.167)$$

Theorem 7.105. *Let* $1 \le p < \infty, 1 \le q < \infty$ *and* φ *satisfy the conditions* (7.163)–(7.164) *and let*

$$w \in \overline{W}(\mathbb{R}_+), \quad w(2t) \le cw(t), \quad \frac{\varphi^{\frac{1}{p}}}{w} \in \underline{W}(\mathbb{R}_+). \tag{7.168}$$

Then the Hardy-type operator H^α_w *is bounded from* $\mathcal{L}^{p,\varphi}_{loc;0}(\mathbb{R}^n)$ *to* $\mathcal{L}^{q,\varphi}_{loc;0}(\mathbb{R}^n)$, *and acts from* $V\mathcal{L}^{p,\varphi}_{loc;0}(\mathbb{R}^n)$ *to* $V\mathcal{L}^{q,\varphi}_{loc;0}(\mathbb{R}^n)$ *if* $\sup_{r>0} \mathbb{W}(r) < \infty$, *where*

$$\mathbb{W}(r) := \frac{1}{\varphi(r)} \int_0^r s^{n-1+q\left(\alpha-\frac{n}{p}\right)} \varphi^{\frac{q}{p}}(s) \left(1 + \frac{1}{V(s)} \int_0^s \frac{V(t)}{t} dt\right)^q ds$$

$$(7.169)$$

and

$$V(t) = \frac{t^{\frac{n}{p'}} \varphi^{\frac{1}{p}}(t)}{w(t)}. \tag{7.170}$$

Under this condition,

$$\|H_w^\alpha f\|_{\mathcal{L}_{loc;0}^{q,\varphi}} \le C \sup_{r>0} \mathbb{W}^{\frac{1}{q}}(r)\|f\|_{\mathcal{L}_{loc;0}^{p,\varphi}} \tag{7.171}$$

and

$$\mathfrak{M}_{q,\varphi}\left(H_w^\alpha f;0,r\right) \le C\widetilde{\mathfrak{M}}^{\frac{q}{p}}(f;0,r) \sup_{r>0} \mathbb{W}(r). \tag{7.172}$$

The following corollary gives sufficient conditions for the boundedness of the operator H_w^α in terms of the Matuszewska-Orlicz indices of the function φ and the weight w.

Corollary 7.106. *Let $1 \le p \le q < \infty$ (with $q = p$ admitted in the case $\alpha = 0$) and the conditions (7.163) -(7.164) and (7.168) be satisfied. Then the operator H_w^α is bounded from $V\mathcal{L}_{loc;0}^{p,\varphi}(\mathbb{R}^n)$ to $V\mathcal{L}_{loc;0}^{q,\varphi}(\mathbb{R}^n)$, if*

$$\min\{m(V), m_\infty(V)\} > 0 \quad and \quad \min\{m(\varphi), m_\infty(\varphi)\} > 0 \tag{7.173}$$

and

$$\varphi(r) \le Cr^{n-\frac{\alpha pq}{q-p}} \tag{7.174}$$

(the latter condition required in the case $\alpha > 0$). The condition $\min\{m(V), m_\infty(V)\} > 0$ is guaranteed by the inequalities

$$M(w) < \frac{n}{p'} + \frac{m(\varphi)}{p}, \quad M_\infty(w) < \frac{n}{p'} + \frac{m_\infty(\varphi)}{p}.$$

In the case of the classical (localized) Morrey space, i.e., $\varphi(r) = r^\lambda, 0 < \lambda < n$, and the power weight $w(r) = r^\mu$, the conditions (7.173)–(7.174) reduce to

$$1 \le p < \frac{n-\lambda}{\alpha}, \quad \frac{1}{q} = \frac{1}{p} - \frac{\alpha}{n-\lambda}, \quad \mu < \frac{n}{p'} + \frac{\lambda}{p}; \tag{7.175}$$

conditions (7.175) are also necessary for the operator H_w^α to be bounded from $\mathcal{L}_{loc;0}^{p,\lambda}(\mathbb{R}^n)$ to $\mathcal{L}_{loc;0}^{q,\lambda}(\mathbb{R}^n)$ and acting $V\mathcal{L}_{loc;0}^{p,\lambda}(\mathbb{R}^n)$ to $V\mathcal{L}_{loc;0}^{q,\lambda}(\mathbb{R}^n)$.

Let now

$$\mathcal{W}(r) := \frac{1}{\varphi(r)} \int_0^r w^q(s) s^{q\alpha+n-1} \left(\int_s^\infty \frac{\mathcal{V}(t)}{t} \, dt \right)^q ds, \qquad (7.176)$$

where

$$\mathcal{V}(t) = \frac{1}{w(t)} \left[\frac{\varphi(t)}{t^n} \right]^{\frac{1}{p}}. \qquad (7.177)$$

Theorem 7.107. *Let* $1 \leq p < \infty, 1 \leq q < \infty$ *and* φ *satisfy the conditions* (7.163)–(7.164) *and*

$$\frac{\varphi^{\frac{1}{p}}}{w} \in \overline{W}(\mathbb{R}_+^1), \quad or \quad w \in \overline{W}(\mathbb{R}_+^1) \quad and \quad w(2t) \leq Cw(t). \quad (7.178)$$

Then the Hardy-type operator $\mathcal{H}_w^\alpha$ *is bounded from* $\mathcal{L}_{loc;0}^{p,\varphi}(\mathbb{R}^n)$ *to* $\mathcal{L}_{loc;0}^{q,\varphi}(\mathbb{R}^n)$, *and acts from* $V\mathcal{L}_{loc;0}^{p,\varphi}(\mathbb{R}^n)$ *to* $V\mathcal{L}_{loc;0}^{q,\varphi}(\mathbb{R}^n)$, *if*

$$\int_0^\infty \frac{\mathcal{V}(t)}{t} dt < \infty \qquad (7.179)$$

and

$$\sup_{r>0} \mathcal{W}(r) < \infty. \qquad (7.180)$$

Under the condition (7.180) *it yields that*

$$\|\mathcal{H}_w^\alpha f\|_{\mathcal{L}_{loc;0}^{q,\varphi}} \leq C \sup_{r>0} \mathcal{W}^{\frac{1}{q}}(r) \|f\|_{\mathcal{L}_{loc;0}^{p,\varphi}}. \qquad (7.181)$$

As before, we provide also sufficient conditions for the boundedness of the operator $\mathcal{H}_w^\beta$ in terms of the Matuszewska-Orlicz indices.

Corollary 7.108. *Let* $1 \leq p \leq q < \infty$ *(with* $q = p$ *admitted in the case* $\alpha = 0$ *and* φ *satisfy the conditions* (7.163)–(7.164) *and* (7.178)). *Then the operator* $\mathcal{H}_w^\alpha$ *is bounded from* $V\mathcal{L}_{loc;0}^{p,\varphi}(\mathbb{R}^n)$ *to* $V\mathcal{L}_{loc;0}^{q,\varphi}(\mathbb{R}^n)$, *if*

$$\max\{M(\mathcal{V}), M_\infty(\mathcal{V})\} < 0 \quad and \quad \min\{m(\varphi), m_\infty(\varphi)\} > 0, \qquad (7.182)$$

and the condition (7.174) holds; the assumption $\max\{M(\mathcal{V}),$ $M_\infty(\mathcal{V})\} < 0$ *is guaranteed by the conditions*

$$m(w) > \frac{M(\varphi) - n}{p}, \quad m_\infty(w) > \frac{M_\infty(\varphi) - n}{p}.$$

In the case $\varphi(r) = r^\lambda, 0 < \lambda < n,$ *and* $w(r) = r^\mu,$ *the conditions (7.182) and (7.174) reduce to*

$$1 \le p < \frac{n - \lambda}{\alpha}, \quad \frac{1}{q} = \frac{1}{p} - \frac{\alpha}{n - \lambda}, \quad \mu > \frac{n - \lambda}{p}; \qquad (7.183)$$

conditions (7.183) are also necessary for the operator $\mathcal{H}_w^\alpha$ *to be bounded from* $\mathcal{L}_{loc;0}^{p,\lambda}(\mathbb{R}^n)$ *to* $\mathcal{L}_{loc;0}^{q,\lambda}(\mathbb{R}^n)$ *and acting from* $V\mathcal{L}_{loc;0}^{p,\lambda}(\mathbb{R}^n)$ *to* $V\mathcal{L}_{loc;0}^{q,\lambda}(\mathbb{R}^n).$

The proofs of Theorems 7.105 and 7.107 and corollaries to them can be found in [PeS2] and [S3]. In these papers some complementary results and applications are also presented.

Concerning the boundedness of Hardy-type operators in local vanishing Morrey spaces on metric measure spaces we refer to [LPSa1].

Necessary and sufficient conditions for boundedness of a Hardy-type operator from a weighted Lebesgue space to a Morrey-type space was proved by V.I. Burenkov and R. Oinarov [BO1]. We also refer to [BG1] and the review article [Bu1] and the references given there.

7.6.3. *Hardy-type inequalities in Hölder-type spaces*

Hölder spaces on unbounded sets can be defined with compactification of $\mathbb{R}^n$ at infinity (see Definition 7.112) or without.

By $C^\lambda(\Omega),$ $0 < \lambda \le 1,$ where Ω is an open set in $\mathbb{R}^n, \Omega \subseteq \mathbb{R}^n,$ $n \ge 1,$ we denote the class of bounded Hölder continuous functions, defined by the seminorm

$$[f]_\lambda := \sup_{\substack{x, x+h \in \Omega \\ |h| < 1}} \frac{|f(x + h) - f(x)|}{|h|^\lambda} < \infty.$$

Equipped with the norm

$$\|f\|_{C^\lambda} = \sup_{x \in \Omega} |f(x)| + [f]_\lambda$$

$C^\lambda(\Omega)$ is a Banach space. We deal with the case $\Omega = B_R$, where $B_R = B(0, R) := \{x \in \mathbb{R}^n : |x| < R\}$, $0 < R \leq \infty$.

We consider the Hardy-type operators

$$H^\alpha f(x) = |x|^{\alpha - n} \int_{|y| < |x|} f(y) dy \quad \text{and}$$

$$\mathcal{H}^\alpha f(x) = |x|^\alpha \int_{|y| > |x|} \frac{f(y)}{|y|^n} dy, \ \alpha > 0,$$

where $x \in B_R$ with $0 < R \leq \infty$ for the operator H^α and $R = \infty$ for the operator $\mathcal{H}^\alpha$. We write $H = H^\alpha$ and $\mathcal{H} = \mathcal{H}^\alpha$ in the case $\alpha = 0$.

The operator $H^\alpha, \alpha = 0$, may be considered in both settings, with and without compacification of $\mathbb{R}^n$ at infinity but a consideration of $\mathcal{H}^\alpha$ requires the compactification of $\mathbb{R}^n$ at infinity, due to the needed convergence of integrals at infinity. We provide details for the operator $H^\alpha, \alpha \geq 0$, without compactification and for both the operators H and $\mathcal{H}$ with compactification. We also show that in the setting of the spaces with compactification we may consider only the case $\alpha = 0$. For further details see E. Burtseva *et al.* [BLPSa1].

Denote also

$$C_0^\lambda(B_R) = \{f \in C^\lambda(B_R) : f(0) = 0\}.$$

We need the following lemma:

Lemma 7.109. *Let*

$$g(r) = \frac{1}{r^n} \int_{|y| < r} f(y) dy, \ 0 < r < R,$$

where $f \in C^\lambda(B_R)$, $0 < \lambda \leq 1, 0 < R \leq \infty$. Then

$$|g'(r)| \leq C_{n,\lambda} \frac{[f]_\lambda}{r^{1-\lambda}}, \ 0 < r < R, \tag{7.184}$$

where $C_{n,\lambda}$ depends only on n and λ.

Proof. Passing to polar coordinates (γ, σ), we have

$$g(r) = \frac{1}{r^n} \int_0^r t^{n-1} \Phi(t) dt, \quad \Phi(t) = \int_{\mathbb{S}^{n-1}} f(t\sigma) d\sigma.$$

Hence,

$$g'(r) = -\frac{n}{r^{n+1}} \int_0^r t^{n-1}\Phi(t)dt + \frac{\Phi(r)}{r} = \frac{n}{r^{n+1}} \int_0^r t^{n-1}[\Phi(r) - \Phi(t)]dt.$$

Therefore,

$$|g'(r)| \leq \frac{n}{r^{n+1}} \int_0^r t^{n-1}|\Phi(r) - \Phi(t)|dt.$$

It is easily seen that

$$|\Phi(r) - \Phi(t)| \leq [f]_\lambda |\mathbb{S}^{n-1}|(r-t)^\lambda.$$

Consequently,

$$|g'(r)| \leq \frac{n|\mathbb{S}^{n-1}|[f]_\lambda}{r^{n+1}} \int_0^r t^{n-1}(r-t)^\lambda dt$$

$$= n\frac{|\mathbb{S}^{n-1}|[f]_\lambda}{r^{1-\lambda}} \int_0^1 s^{n-1}(1-s)^\lambda ds,$$

and we arrive at (7.184). $\qquad\square$

In the following theorem we also deal with the space $\tilde{C}_0^\lambda$ consisting of functions f for which $[f]_\lambda < \infty$ and $f(0) = 0$. This space contains functions unbounded at infinity. Note that $[f]_\lambda$ is a norm in this space.

Now we are in a position to prove the following theorem.

Theorem 7.110. *Let $\alpha \geq 0$, $\lambda > 0$ and $\lambda + \alpha \leq 1$. In the case $\alpha = 0$ the Hardy operator H^α is bounded in $C^\lambda(B_R)$ and $[Hf]_\lambda \leq C[f]_\lambda$. In the case $\alpha > 0$ the operator H^α is bounded from $\tilde{C}_0^\lambda(B_R)$ into $\tilde{C}_0^{\lambda+\alpha}(B_R)$, $0 < R \leq \infty$.*

Proof. Let first $\alpha = 0$. For

$$Hf = H^\alpha f|_{\alpha=0}$$

we have

$$Hf(x) = |x|^{-n} \int_{|y| \leq |x|} f(y)dy = \int_{B(0,1)} f(|x|y)dy$$

so that

$$|Hf(x+h) - Hf(x)| \le \int_{B(0,1)} |f(|x+h|y) - f(|x|y)| dy$$

$$\le [f]_\lambda \int_{B(0,1)} ||x+h| - |x||^\lambda |y|^\lambda dy =: A.$$

Since by triangle inequality $||x+h| - |x||^\lambda \le |h|^\lambda$, $\lambda \in [0,1]$, for all $x, x+h \in \mathbb{R}^n$, we obtain that

$$A \le [f]_\lambda \int_{B(0,1)} |h|^\lambda |y|^\lambda dy \le [f]_\lambda |h|^\lambda \int_{B(0,1)} |y|^\lambda dy = C|h|^\lambda [f]_\lambda.$$

Thus, $|Hf(x+h) - Hf(x)| \le C|h|^\lambda [f]_\lambda$ and therefore $[Hf]_\lambda \le C[f]_\lambda$, with C not depending on x and h.

The boundedness of H in supremum norm is evident.

Let now $\alpha > 0$ and $f \in \tilde{C}_0^\lambda(B_R)$. We have

$$H^\alpha f(x) = |x|^\alpha g(|x|), \; g(r) = \frac{1}{r^n} \int_{B(0,r)} f(y) dy = \int_{B(0,1)} f(ry) dy.$$

Hence, by the triangle inequality,

$$|H^\alpha f(x+h) - H^\alpha f(x)| \le ||x+h|^\alpha - |x|^\alpha| |g(|x+h|)| + |g(|x+h|)$$

$$-g(|x|)| |x|^\alpha$$

$$\le C[f]_\lambda ||x+h|^\alpha - |x|^\alpha| |x+h|^\lambda$$

$$+|g(|x+h|) - g(|x|)| |x|^\alpha =: \Delta_1 + \Delta_2,$$

$$(7.185)$$

where we used the fact that $f(0) = 0$ and consequently

$$|g(|x+h|)| = |Hf(|x+h|)| \le C|x+h|^\lambda [f]_\lambda$$

according to the case $\alpha = 0$ in the last passage.

We consider separately the cases $|x+h| \le 2|h|$ and $|x+h| \ge 2|h|$.

The case $|x+h| \le 2|h|$.

In this case we also have $|x| \le 3|h|$.

By (7.186) we have

$$\Delta_1 \le C[f]_\lambda |h|^\alpha |x+h|^\lambda \le C_1[f]_\lambda |h|^{\lambda+\alpha}$$

and

$$\Delta_2 \le C[g]_\lambda |h|^\lambda |x|^\alpha \le C_1[f]_\lambda |h|^{\lambda+\alpha}.$$

The case $|x+h| \ge 2|h|$.
We have

$$\Delta_1 \le C[f]_\lambda |x+h|^{\lambda+\alpha} \left| 1 - \left(\frac{|x|}{|x+h|} \right)^\alpha \right|.$$

Since $|1 - t^\alpha| \le |1-t|$ for all $0 < t \le 1$, we obtain

$$\Delta_1 \le C[f]_\lambda \frac{||x+h| - |x||}{|x+h|^{1-\lambda-\alpha}} \le C[f]_\lambda |h|^{\lambda+\alpha}.$$

For Δ_2 we use the mean value theorem and obtain

$$\Delta_2 \le C \left| g'(\xi) \right| \, ||x+h| - |x|| \, |x|^\alpha \le C|g'(\xi)||h||x|^\alpha$$

with ξ between $|x|$ and $|x+h|$.

If $|x| \le |x+h|$, then, by Lemma 7.109, we get

$$\Delta_2 \le C \frac{[f]_\lambda}{|\xi|^{1-\lambda}} |x|^\alpha |h| \le C \frac{[f]_\lambda}{|x|^{1-\lambda-\alpha}} |h| \le C[f]_\lambda |h|^{\lambda+\alpha}$$

because $|x| \ge |x+h| - |h| \ge |h|$. Finally, when $|x| \ge |x+h|$, we have

$$\Delta_2 \le C \frac{[f]_\lambda}{|\xi|^{1-\lambda}} |x|^\alpha |h| \le C \frac{[f]_\lambda}{|x+h|^{1-\lambda}} |x|^\alpha |h|$$

$$= C \frac{[f]_\lambda}{|x+h|^{1-\lambda-\alpha}} \left(\frac{|x|}{|x+h|} \right)^\alpha |h|,$$

where $\frac{|x|}{|x+h|} \le \frac{|h|}{|x+h|} + \frac{|x+h|}{|x+h|} \le \frac{3}{2}$. Therefore,

$$\Delta_2 \le C[f]_\lambda |h|^{\lambda+\alpha}.$$

It remains to gather the estimates for Δ_1 and Δ_2 in (7.185) and use the definition of Hölder space. Since obviously $H^\alpha f(0) = 0$ the proof is complete. $\qquad\square$

We define the generalized Hölder space $C^{\omega(\cdot)}(\Omega)$ as the set of functions continuous in Ω having the finite norm

$$\|f\|_{C^{\omega(\cdot)}} = \sup_{x \in \Omega} |f(x)| + [f]_{\omega(\cdot)}$$

with the seminorm

$$[f]_{\omega(\cdot)} = \sup_{\substack{x, x+h \in \Omega \\ |h| < 1}} \frac{|f(x+h) - f(x)|}{\omega(|h|)},$$

where $\omega : [0,1] \to \mathbb{R}_+$ is a non-negative increasing function in $C([0,1])$ such that $\omega(0) = 0$ and $\omega(t) > 0$ for $0 < t \leq 1$. Such spaces are known in the literature, see for instance [BLPSa1] and the references given therein.

Let also $C_0^{\omega(\cdot)}(B_R) = \{f \in C^{\omega(\cdot)}(B_R) : f(0) = 0\}$.

Following the same lines as in proof of Theorem 7.110 one can prove the following its generalization.

Theorem 7.111. *Let $\omega \in C([0,1])$ be positive on $[0,1]$, increasing and such that $\omega(0) = 0$ and $\frac{\omega(t)}{t^{1-\alpha}}$ is almost decreasing. In the case $\alpha = 0$ the operator $H = H^0$ is bounded in $C^{\omega(\cdot)}(B_R)$. When $\alpha > 0$, it is bounded from $\tilde{C}_0^{\omega(\cdot)}(B_R)$ into $\tilde{C}_0^{\omega_\alpha(\cdot)}(B_R)$, where $\omega_\alpha(t) = t^\alpha \omega(t)$.*

Now we pass to Hölder type spaces with compactification of $\mathbb{R}^n$ at infinity.

Definition 7.112. Let $\dot{\mathbb{R}}^n$ denote the compactification of $\mathbb{R}^n$ by a single infinite point. Let $0 \leq \lambda < 1$. We say that f belongs to $C^\lambda(\dot{\mathbb{R}}^n)$ if

$$|f(x) - f(y)| \leq C \frac{|x-y|^\lambda}{(1+|x|)^\lambda (1+|y|)^\lambda}.$$

The set $C^\lambda(\dot{\mathbb{R}}^n)$ is a Banach space with respect to the norm

$$\|f\|_{C^\lambda(\dot{\mathbb{R}}^n)} = \|f\|_{C(\dot{\mathbb{R}}^n)} + \sup_{x,y} |f(x) - f(y)| \left(\frac{(1+|x|)(1+|y|)}{|x-y|} \right)^\lambda.$$

It may be shown that $C^\lambda(\dot{\mathbb{R}}^n)$ is the subspace of $C^\lambda(\mathbb{R}^n)$ invariant with respect to the inversion change of variables $x_* = \frac{x}{|x|^2}$, i.e.,

$$C^\lambda(\dot{\mathbb{R}}^n) = \{f : f \in C^\lambda(\mathbb{R}^n) \text{ and } f_* \in C^\lambda(\mathbb{R}^n)\},$$

where $f_* = f(x_*)$.

In the setting of the spaces $C^\lambda(\dot{\mathbb{R}}_n)$ we consider only the case $\alpha = 0$, see Remark 7.115.

Theorem 7.113. *Let $0 \le \lambda < 1$. Then the operator H is bounded in $C^\lambda(\dot{\mathbb{R}}^n)$.*

To formulate the corresponding result for the operator $\mathcal{H}$ we need the following subspaces:

$$C_0^\lambda(\dot{\mathbb{R}}^n) = \{f \in C^\lambda(\dot{\mathbb{R}}^n) : f(0) = 0\},$$

$$C_\infty^\lambda(\dot{\mathbb{R}}^n) = \{f \in C^\lambda(\dot{\mathbb{R}}^n) : f(\infty) = 0\},$$

$$C_{0,\infty}^\lambda = C_0^\lambda \cap C_\infty^\lambda.$$

Theorem 7.114. *Let $0 < \lambda < 1$. Then the operator $\mathcal{H}$ is bounded from $C_{0,\infty}^\lambda(\dot{\mathbb{R}}^n)$ to $C_\infty^\lambda(\dot{\mathbb{R}}^n)$.*

For the proofs of these theorems we refer to [BLPSa1].

Remark 7.115. When $\alpha > 0$, Theorems 7.113 and 7.114 may not be extended to the setting $C^\lambda(\dot{\mathbb{R}}^n) \longrightarrow C^{\lambda+\alpha}(\dot{\mathbb{R}}^n)$, in which we require the Hölder behavior of functions also at the infinite point, in contrast to Theorem 7.110. The function $f_0 = \dfrac{1}{(1+x^\lambda)} \in C_\infty^\lambda(\dot{\mathbb{R}}_+)$ provides the corresponding counterexample for both the operators H^α and $\mathcal{H}^\alpha$. For example, for the operator H^α we have

$$H^\alpha f_0(x) = \frac{x^{\alpha-1}}{1-\lambda}[(1+x)^{1-\lambda} - 1].$$

Hence, when $x \to \infty$ we obtain that $H^\alpha f_0(x) \sim cx^{\alpha-\lambda}$, while the inclusion $H^\alpha f_0(x) \in C_\infty^\lambda(\dot{\mathbb{R}}_+)$ requires the behavior $|H^\alpha f_0(x)| \le c(1+x)^{-\alpha-\lambda}$.

Corresponding generalizations of Theorems 7.113 and 7.114 may also be formulated in terms of the generalized Hölder spaces $C^\omega(\dot{\mathbb{R}}^n)$, $C^\omega_\infty(\dot{\mathbb{R}}^n)$, $C^\omega_0(\dot{\mathbb{R}}^n)$ and $C^\omega_{\infty,0}(\dot{\mathbb{R}}^n)$ defined below. For details, see [BLPSa1].

Definition 7.116. The generalized space $C^\omega(\dot{\mathbb{R}}^n)$ is defined by the condition

$$|f(x) - f(y)| \leq C\omega\left(\frac{|x-y|}{(1+|x|)(1+|y|)}\right),$$

where $\omega(h)$ is an increasing function. The subspaces $C^\omega_\infty(\dot{\mathbb{R}}^n)$, $C^\omega_0(\dot{\mathbb{R}}^n)$ and $C^\omega_{\infty,0}(\dot{\mathbb{R}}^n)$ of the space $C^\omega(\dot{\mathbb{R}}^n)$ are defined by the conditions $f(\infty) = 0$, $f(0) = 0$ and $f(0) = f(\infty) = 0$, respectively.

7.6.4. Hardy-type inequalities in $L^{p(\cdot)}$ spaces and further results

We will finalize this section by mention some other known results. First we consider the case with variable $L^{p(\cdot)}$ spaces on $(0, \infty)$. These spaces are defined via the modular

$$I_p(f) := \int_0^\infty |f(x)|^{p(x)} dx,$$

where $p(x)$ is a measurable function with values in $[1, \infty)$ and the norm of f in $L^{p(\cdot)}$ defined by

$$\|f\|_{L^{p(\cdot)}} := \inf\left\{\lambda : I_p\left(\frac{f}{\lambda}\right) \leq 1\right\}.$$

By $\mathcal{M}_{0,\infty} = \mathcal{M}_{0,\infty}(\mathbb{R}_+)$ we denote the set of all measurable bounded functions $p(x) : \mathbb{R}_+ \to \mathbb{R}_+$, which satisfy the following conditions:

(i) $0 < p_- \leq p(x) \leq p_+ < \infty, x \in \mathbb{R}_+$,
(ii) there exists $p(0) = \lim_{x\to 0_+} p(x)$ and $|p(x) - p(0)| \leq \frac{A}{\ln\frac{1}{x}}$, $0 < x \leq 1/2$,
(iii) there exists $p(\infty) = \lim_{x\to\infty} p(x)$ and $|p(x) - p(\infty)| \leq \frac{A}{\ln x}$, $x \geq 2$.

By $\mathcal{P}_{0,\infty} = \mathcal{P}_{0,\infty}(\mathbb{R}_+)$ we denote the subset of functions $p(x) \in \mathcal{M}_{0,\infty}$ with $\inf_{x \in \mathbb{R}_+} p(x) \geq 1$.

The following result from 2007 is due to L. Diening and S. Samko (see [DS1]).

Theorem 7.117. *Let* $p \in \mathcal{P}_{0,\infty}$ *and* $\mu \in \mathcal{M}_{0,\infty}$ *and*

$$0 \leq \mu(0) < \frac{1}{p(0)} \quad \text{and} \quad 0 \leq \mu(\infty) < \frac{1}{p(\infty)}.$$

Let also $q(x)$ *be any function in* $\mathcal{P}_{0,\infty}$ *such that*

$$\frac{1}{q(0)} = \frac{1}{p(0)} - \mu(0) \quad \text{and} \quad \frac{1}{q(\infty)} = \frac{1}{p(\infty)} - \mu(\infty).$$

Then the Hardy-type inequalities

$$\left\| x^{\alpha + \mu(x) - 1} \int_0^x \frac{f(y)}{y^\alpha} dy \right\|_{L^{q(\cdot)}} \leq C \|f\|_{L^{p(\cdot)}}$$

and

$$\left\| x^{\beta + \mu(x)} \int_x^\infty \frac{f(y)}{y^{\beta + 1}} dy \right\|_{L^{q(\cdot)}} \leq C \|f\|_{L^{p(\cdot)}}$$

are valid if and only if α *and* β *satisfy*

$$\alpha < \min\left\{ \frac{1}{p'(0)}, \frac{1}{p'(\infty)} \right\} \quad \text{and} \quad \beta > \max\left\{ -\frac{1}{p(0)}, -\frac{1}{p(\infty)} \right\},$$

respectively.

Remark 7.118. Theorem 7.117 is a generalization of Hardy's inequalities in power weighted case in two dual forms with parameters $1 < p \leq q < \infty$. For the case $q(\cdot) = p(\cdot)$ we also refer to a result by T. Kopaliani [Kop1], where even a power weighted multidimensional version (in differential form) was proved.

Open question

To characterize the weights $u(x)$ and $v(x)$ so that Theorem 7.117 holds with these weights instead of the special case of power weights.

This question is open also for other possible cases than those corresponding to the case $1 < p \leq q < \infty$.

Some more information concerning integral operators (e.g. Hardy-type operators) in non-standard Banach spaces can be found in the books [KMRS] and [LE1]. In addition to the case of variable $L^{p(\cdot)}$ spaces some corresponding, but less complete, results of this type for variable exponent Morrey spaces, variable exponent grand Lebesgue spaces etc. are also given. Moreover, we refer to the paper [LPS1], where some weighted Hardy-type inequalities in generalized complementary Morrey spaces were proved.

7.7. More on Multidimensional Hardy-type Inequalities

The main information in this section can be found in [Pe2] by L.E. Persson. For proofs and complementary information see the Ph.D. theses [We1] by A. Wedestig and [U1] by E. Ushakovo and also the review article [PeS1].

Some two-dimensional results

We first remind the following two-dimensional inequality, which was proved by E.T. Sawyer in 1985 (see [Sa2]):

Theorem 7.119. *Let $1 < p \leq q < \infty$ and u and v be weights on $\mathbb{R}^2_+$. Then the inequality*

$$\left(\int_0^\infty \int_0^\infty \left(\int_0^{x_1} \int_0^{x_2} f(t_1, t_2) dt_1 dt_2 \right)^q u(x_1, x_2)\, dx_1 dx_2 \right)^{\frac{1}{q}}$$

$$\leq C \left(\int_0^\infty \int_0^\infty f^p(x_1, x_2) v(x_1, x_2)\, dx_1 dx_2 \right)^{\frac{1}{p}} \qquad (7.186)$$

holds for all non-negative and measurable functions on $\mathbb{R}^2_+$, if and only if the following three conditions are satisfied:

$$\sup_{(y_1, y_2) \in \mathbb{R}^2_+} \left(\int_{y_1}^\infty \int_{y_2}^\infty u(x_1, x_2) dx_1 dx_2 \right)^{\frac{1}{q}}$$

$$\times \left(\int_0^{y_1} \int_0^{y_2} v(x_1, x_2)^{1-p'} dx_1 dx_2 \right)^{\frac{1}{p'}} < \infty, \qquad (7.187)$$

$$\sup_{(y_1,y_2)\in\mathbb{R}_+^2} \frac{\left(\int_0^{y_1}\int_0^{y_2}\left(\int_0^{x_1}\int_0^{x_2}v(t_1,t_2)^{1-p'}dt_1dt_2\right)^q u(x_1,x_2)dx_1dx_2\right)^{\frac{1}{q}}}{\left(\int_0^{y_1}\int_0^{y_2}v(x_1,x_2)^{1-p'}dx_1dx_2\right)^{\frac{1}{p}}} < \infty$$

(7.188)

$$\sup_{(y_1,y_2)\in\mathbb{R}_+^2} \frac{\left(\int_{y_1}^{\infty}\int_{y_2}^{\infty}\left(\int_{x_1}^{\infty}\int_{x_2}^{\infty}u(t_1,t_2)dt_1dt_2\right)^{p'}v(x_1,x_2)^{1-p'}dx_1dx_2\right)^{\frac{1}{q}}}{\left(\int_{y_1}^{\infty}\int_{y_2}^{\infty}u(x_1,x_2)dx_1dx_2\right)^{\frac{1}{q'}}} < \infty.$$

(7.189)

All three conditions (7.187)–(7.189) are independent and no one may be removed.

Remark 7.120. Note that (7.187) corresponds to the Muckenhoupt-Bradley condition (7.30), (7.188) corresponds to the condition $A_{PS} < \infty$ in (7.34) and (7.189) corresponds to the dual condition $A_{PS}^* < \infty$ of (7.34). According to Theorem 7.23 and Remark 7.24 all these conditions are equivalent in the one-dimensional case but as seen above it is not so in the two-dimensional case.

One of the recent progresses related to Theorem 7.119 was obtained in A. Wedestig's Ph.D's thesis [We1] from 2004. It was shown there that in the case where the weight $v(x_1, x_2)$ on the right-hand side of (7.186) has the form of the product $v_1(x_1)v_2(x_2)$, then only one condition appears (but this condition is not unique and can in fact be given in infinite many equivalent forms). Namely, the following statement holds:

Theorem 7.121. *Let $1 < p \le q < \infty$ and let u be a weight on $\mathbb{R}_+^2$ and v_1 and v_2 be weights on $\mathbb{R}_+$. Then the inequality*

$$\left(\int_0^{\infty}\int_0^{\infty}\left(\int_0^{x_1}\int_0^{x_2}f(t_1,t_2)dt_1dt_2\right)^q u(x_1,x_2)\,dx_1dx_2\right)^{\frac{1}{q}}$$

$$\le C\left(\int_0^{\infty}\int_0^{\infty}f^p(x_1,x_2)v_1(x_1)v_2(x_2)\,dx_1dx_2\right)^{\frac{1}{p}}$$

(7.190)

holds for all non-negative and measurable functions f on $\mathbb{R}^2_+$ *if and only if*

$$A_W(s_1, s_2) := \sup_{(t_1, t_2) \in \mathbb{R}^2_+} (V_1(t_1))^{\frac{s_1 - 1}{p}} (V_2(t_2))^{\frac{s_2 - 1}{p}}$$

$$\times \left(\int_{t_1}^{\infty} \int_{t_2}^{\infty} u(x_1, x_2)(V_1(x_1))^{q\frac{p - s_1}{p}} \right.$$

$$\left. \times (V_2(x_2))^{q\frac{p - s_2}{p}} dx_1 dx_2 \right)^{\frac{1}{q}} < \infty$$

holds for some $s_1, s_2 \in (1, p)$ *(and, hence, for all* $s_1, s_2 \in (1, p)$*), where* $V_i(t_i) := \int_0^{t_i} v_i(\xi)^{1 - p'} d\xi$, $i = 1, 2$. *Moreover, for the best constant* C *in* (7.190) *it yields that* $C \approx A_W(s_1, s_2)$.

A limit result of Theorem 7.121 is the following two-dimensional Pólya-Knopp-type inequality, which was also proved in the same Ph.D. thesis [We1]:

Theorem 7.122. *Let* $0 < p \le q < \infty$ *and* u *and* v *be weights on* $\mathbb{R}^2_+$. *Then the inequality*

$$\left(\int_0^{\infty} \int_0^{\infty} \left[\exp \left(\frac{1}{x_1 x_2} \int_0^{x_1} \int_0^{x_2} \log f(t_1, t_2) dt_1 dt_2 \right) \right]^q \right.$$

$$\left. \times u(x_1, x_2) dx_1 dx_2 \right)^{\frac{1}{q}}$$

$$\le C \left(\int_0^{\infty} \int_0^{\infty} f^p(x_1, x_2) v(x_1, x_2) dx_1 dx_2 \right)^{\frac{1}{p}} \qquad (7.191)$$

holds for all non-negative and measurable functions f on $\mathbb{R}^2_+$ *if and only if*

$$\sup_{y_1 > 0, y_2 > 0} y_1^{\frac{s_1 - 1}{p}} y_2^{\frac{s_2 - 1}{p}} \left(\int_{y_1}^{\infty} \int_{y_2}^{\infty} x_1^{-\frac{s_1 q}{p}} x_2^{-\frac{s_2 q}{p}} w(x_1, x_2) dx_1 dx_2 \right)^{\frac{1}{q}} < \infty,$$

holds for some $s_1 > 1, s_2 > 1$ *(and thus for all* $s_1 > 1, s_2 > 1$*) and where*

$$w(x_1, x_2) := u(x_1, x_2) \left[\exp \left(\frac{1}{x_1 x_2} \int_0^{x_1} \int_0^{x_2} \log \frac{1}{v(t_1, t_2)} dt_1 dt_2 \right) \right]^{\frac{q}{p}}.$$

Remark 7.123. Observe that this limit inequality indeed holds for all weights (and not only for product weights on the right-hand side) and also for $0 < p \leq 1$. The reason for this comes from the useful technical details when we perform the limit procedure, e.g. that we first do a substitution so we only need to use the case when the weight in the right-hand side in (7.190) is equal to 1. Also here we have a good estimate of the best constant C in (7.191).

Remark 7.124. The corresponding statements as those in Theorems 7.121 and 7.122 hold also for any dimension n. However, in our next subsection we will present some results mainly from the Ph.D. thesis [U1] of E. Ushakova from 2006, where also the case with product weights on the left-hand side was considered. The proofs there are completely different from those in [We1] before and the obtained characterizations are different.

Some more multidimensional results

In the sequel we assume that f is a non-negative and measurable function.

Let $x = (x_1, \ldots, x_n), t = (t_1, \ldots, t_n) \in \mathbb{R}_+^n, n \in \mathbb{Z}_+$ and $1 < p \leq q < \infty$. We consider the n-dimensional Hardy-type operator

$$(H_n f)(x) = \int_0^{x_1} \cdots \int_0^{x_n} f(t) dt_1 \cdots dt_n$$

and study the inequality

$$\left(\int_{\mathbb{R}_+^n} (H_n f)^q (x) u(x) dx \right)^{\frac{1}{q}} \leq C \left(\int_{\mathbb{R}_+^n} f^p(x) v(x) dx \right)^{\frac{1}{p}}. \quad (7.192)$$

Sometimes we assume that one of the involved weight functions v and u is of product type, i.e., that

$$u(x) = u_1(x_1) u_2(x_2) \cdots u_n(x_n), \quad (LP)$$

or

$$v(x) = v_1(x_1) v_2(x_2) \cdots v_n(x_n). \quad (RP)$$

Moreover,

$$U(t) = U(t_1, \cdots, t_n) := \int_{t_1}^{\infty} \cdots \int_{t_n}^{\infty} u(x) dx_1 \cdots dx_n$$

and

$$V(t) = V(t_1, \cdots, t_n) := \int_0^{t_1} \cdots \int_0^{t_n} (v(x))^{1-p'} dx_1 \cdots dx_n.$$

The next statement gives a necessary condition for (7.192) to hold with help of some n-dimensional versions of the constants A_{MB} and A_{PS} (see (7.30) and (7.34)).

Theorem 7.125. *Let $1 < p \le q < \infty$ and assume that (7.192) holds for all non-negative and measurable functions f on $\mathbb{R}_+^n$ with a finite constant C, which is independent on f. Then*

$$A_{MB}^{(n)} := \sup_{t_i > 0} (U(t_1, \ldots, t_n))^{1/q} (V(t_1, \cdots, t_n))^{1/p'} < \infty$$

and

$$A_{PS}^{(n)} := \sup_{t_i > 0} (V(t_1, \cdots, t_n))^{-1/p}$$
$$\times \left(\int_{t_1}^{\infty} \cdots \int_{t_n}^{\infty} u(x) V^q(x) dx_1 \cdots dx_n \right)^{1/q} < \infty.$$

Our next result is that in the case of product weights on the right-hand side we get a complete characterization of (7.192).

Theorem 7.126. *Let $1 < p \le q < \infty$ and the weight v be of product type (RP). Then (7.192) holds for all non-negative and measurable functions f on $\mathbb{R}_+^n$ with some finite constant C, which is independent on f, if and only if $A_{MB}^{(n)} < \infty$ or $A_{PS}^{(n)} < \infty$. Moreover, $C \approx A_{MB}^{(n)} \approx A_{PS}^{(n)}$ with constants of equivalence only depending on the parameters p and q and the dimension n.*

Note that here it yields that $V(t_1, \ldots, t_n) = V_1(x_1) V_2(x_2) \ldots V_n(x_n)$, where $V_i(t_i) := \int_0^{t_i} (v_i(x_i))^{1-p'} dx_i, i = 1, \ldots, n$. For a proof we refer to the mentioned Ph.D. thesis [U1].

We can also consider the case when u is of product type (LP) and where we need the dual of the constants $A_{MB}^{(n)}$ and $A_{PS}^{(n)}$:

$$A_{MB}^{*(n)} := \sup_{t_i>0} (U_1(t_1)\ldots U_n(t_n))^{1/q} (V(t_1,\ldots,t_n))^{1/p'} < \infty,$$

and

$$A_{PS}^{*(n)} := \sup_{t_i>0} (U_1(t_1)\cdots U_n(t_n))^{-1/q'}$$

$$\times \left(\int_{t_1}^\infty \cdots \int_{t_n}^\infty v^{1-p'}(x)\,(U_1(x_1)\cdots U_n(x_n))^{p'}\,dx_1\cdots dx_n\right)^{1/p'}.$$

Theorem 7.127. *Let $1 < p \le q < \infty$ and the weight u be of product type (LP). Then (7.192) holds for all non-negative and measurable functions f on $\mathbb{R}_+^n$ with some finite constant C, which is independent of f, if and only if $A_{MB}^{*(n)} < \infty$ or $A_{PS}^{*(n)} < \infty$. Moreover, $C \approx A_{MB}^{*(n)} \approx A_{PS}^{*(n)}$ with constants of equivalence only depending on the parameters p and q and the dimension n.*

Also the case $1 < q < p < \infty$ can be considered and the following multidimensional versions of the usual Mazya-Rosin and Persson-Stepanov constants in one dimension can be defined:

$$B_{MR}^{(n)} := \left(\int_{\mathbb{R}_+^n} (U(t))^{r/q}\,(V_1(t_1))^{r/q'}\cdots(V_n(t_n))^{r/q'}\,dV_1(t_1)\cdots dV_n(t_n)\right)^{1/r},$$

$$B_{PS}^{(n)} := \left(\int_{\mathbb{R}_+^n} \left(\int_0^{t_1}\cdots\int_0^{t_n} u(x)(V_1(x_1)\cdots V_n(x_n))^q dx\right)^{r/q}\right.$$

$$\left.\times (V_1(t_1)\cdots V_n(t_n))^{-r/q}\,dV_1(t_1)\cdots dV_n(t_n)\right)^{1/r}.$$

Here, as usual, $1/r = 1/q - 1/p$. For technical reasons we also need the following additional condition:

$$V_1(\infty) = \cdots = V_n(\infty) = \infty.$$

Theorem 7.128. *Let* $1 < q < p < \infty$ *and* $1/r = 1/q - 1/p$. *Assume that the weight* v *is of product type* (RP). *Then* (7.192) *holds for all non-negative and measurable functions* f *on* $\mathbb{R}_+^n$ *with some finite constant* C, *which is independent on* f, *if and only if* $B_{MR}^{(n)} < \infty$, *or* $B_{PS}^{(n)} < \infty$. *Moreover*, $C \approx B_{MR}^{(n)} \approx B_{PS}^{(n)}$ *with constants of equivalence depending only on* p *and* q *and the dimension* n.

Remark 7.129. Also for $1 < p < q < \infty$ the case when the left-hand side is of product type can be considered and a theorem similar to Theorem 7.128 can be proved by using some dual forms of the constants $B_{MR}^{(n)}$ and $B_{PS}^{(n)}$ (see [U1]).

We finalize this section by shortly discussing some limit multi-dimensional (Pólya-Knopp-type) inequalities. Consider the inequality

$$\left(\int_{\mathbb{R}_+^n} (G_n f)^q (x) u(x) dx \right)^{1/q} \le C \left(\int_{\mathbb{R}_+^n} f^p(x) v(x) dx \right)^{1/p} , \quad (7.193)$$

where the n-dimensional geometric mean operator G_n is defined by

$$(G_n f)(x) = \exp \left(\frac{1}{x_1 \cdots x_n} \int_0^{x_1} \cdots \int_0^{x_n} \ln f(t_1, \ldots, t_n) \, dt_1 \ldots dt_n \right).$$

We denote

$$A_G^{(n)} := \sup_{t_i > 0} (t_1 \cdots t_n)^{-1/p} \left(\int_0^{t_1} \cdots \int_0^{t_n} w(x) dx \right)^{1/q}$$

with

$$w(x) := ((G_n v)(x))^{-q/p} u(x).$$

Theorem 7.130. *Let* $0 < p \le q < \infty$. *Then* (7.193) *holds for all non-negative and measurable functions on* $\mathbb{R}_+^n$ *if and only if* $A_G^{(n)} < \infty$. *Moreover*, $C \approx A_G^{(n)}$ *with constants of equivalence depending only on the parameters* p *and* q *and the dimension* n.

Remark 7.131. The proof in [U1] shows that Theorem 7.130 may be regarded as a natural limit case of Theorem 7.126 characterized by the condition $A_{PS}^{(n)} < \infty$. For $n = 2$ we get another characterization than that in Theorem 7.122. Note also that in this case the limit result holds in a wider range of parameters and for general weights.

Remark 7.132. A similar limit (geometric mean) result as that in Theorem 7.130 can also be derived for the case $0 < q < p < \infty$ now as a limiting case of Theorem 7.128.

References

The following six books can serve as standard reference sources for integral inequalities (with and without weights). They are mentioned in the text in the abbreviated forms [HLP], [KoMP], [KMRS], [KMP], [MPF] and [OK]:

[HLP]   HARDY, G.H., LITTLEWOOD, J.E. and PÓLYA, G., *Inequalities*, 2nd ed., Cambridge Univ. Press 1952 (first ed. 1934). MR0944909

[KoMP] KOKILASHVILI, V., MESKHI, A. and PERSSON, L.E., *Weighted Norm Inequalities for Integral Transforms with Product Kernels*, Nova Scientific Publishers, Inc., New York, 2009. MR2807671

[KMRS] KOKILASHVILI, V., MESKHI, A., RAFEIRO, H. and SAMKO, S., *Integral Operators in Non-standard Function Spaces. Vol. I. Variable Exponent Lebesgue and Amalgam Spaces*, Birkhäuser, 2015. MR3559401

[KMP]   KUFNER, A., MALIGRANDA, L. and PERSSON, L.E., *The Hardy Inequality-About its History and Some Related Results*, Vydavatelsky Servis Publishing House, Pilsen, 2007. MR2351524

[MPF]   MITRINOVIĆ, D.S., PEČARIČ, J.E. and FINK, A. M., *Inequalities involving functions and their integrals and derivatives*, Kluwer Academic Publishers Group, 1991. MR1190927

[OK]   OPIC, B. and KUFNER, A., *Hardy-type inequalities*, Pitman Research Notes in Mathematics Series, Longman Scientific & Technical, Harlow, 1990. MR1069756

FURTHER REFERENCES:

ADAMS, R. A.:

[A1] *Sobolev Spaces*, Second Edition, *Pure and Applied Mathematics*, Vol. 65, Elsevier/Academic Press, MR2424078

ADAMS, R., ARONSZAJN, N. and SMITH, K.T.:

[AAS1] Theory of Bessel potentials, Part II, *Ann. Inst. Fourier* (Grenoble) **17** (1967), 1–135. MR0228702

ALZER, H.:

[Al1] A refinement of Carleman's inequality, *J. Approx. Theory* **95** (1998), 497–499. MR1657699

ANDERSEN, K. and MUCKENHOUPT, B.:

[AM1] Weighted weak type Hardy inequalities with applications to Hilbert transforms and maximal functions, *Studia Math.* **72** (1982) (1), 9–26. MR0665888

APPELL, J. and KUFNER, A.:

[AK1] On the two-dimensional Hardy operator in Lebesgue spaces with mixed norms, *Analysis* **15** (1995), 91–98. MR1322131

ARENDARENKO, L.:

[Ar1] Estimates for Hardy-type Integral Operators in Weighted Lebesgue Spaces, Doctoral thesis, Department of Mathematics, Luleå University of Technology, Sweden, 2012.

ARENDARENKO, L., OINAROV, R. and PERSSON, L.E.:

[AOP1] Some new Hardy-type integral inequalities on cones of monotone functions, In: *Advances in Harmonic Analysis and Operator Theory* (Eds: A. Almeida, L. Castro and F.O. Speck), *Operator Theory: Advances and Applications*, **229**, Birkhäuser, Basel, 2013, 77–89. MR3060408

[AOP2] On the boundedness of some classes of integral operators in weighted Lebesgue space, *Eurasian Math. J.* **3** (2012), 5–17. MR3060408

ARIÑO, M. and MUCKENHOUPT, B.:

[ArM1] Maximal functions on classical Lorentz spaces and Hardy's inequality with weights for non-increasing functions, *Trans. Amer. Math. Soc.* **320** (1990) (2), 727–735. MR0989570

ARONSZAJN, N. and SMITH, K.T.:

[AS1] Theory of Bessel potentials, Technical report 26, Department of Mathematics, University of Kansas, USA, 1961.

ASEKRITOVA, I., KRUGLJAK, N., MALIGRANDA, L. and PERSSON, L.E.:

[AKMP1] Distribution and rearrangement estimates of the maximal function and interpolation, *Studia Math.* **124** (1997) (2), 107–131. MR1447618

BARZA, S.:

[B1] Weighted Multidimensional Integral Inequalities and Applications, Doctoral thesis, Department of Mathematics, Luleå University of Technology, Sweden, 1999.

BARZA, S., HEINIG, H. P. and PERSSON, L.E.:
[BHP1] Duality theorem over the cone of monotone functions and sequences in higher dimensions, *J. Inequal. Appl.* **7** (2002) (1), 79–108. MR1923569

BARZA, S., PERSSON, L.E., and SAMKO, N.:
[BPS1] Some new sharp limit Hardy-type inequalities via convexity, *J. Inequal. Appl.* 2014, 2014:6. MR3336919

BARZA, S., PERSSON, L.E. and SORIA, J.:
[BPSo1] Sharp weighted multidimensional integral inequalities for monotone functions, *Math. Nachr.* **210** (2000), 43–58. MR1738936

BARZA, S., PERSSON, L.E. and STEPANOV, V.D.:
[BPSt1] On weighted multidimensional embeddings for monotone functions, *Math. Scand.* **88** (2001), 303–319. MR1839578

BATUEV, E.N. and STEPANOV, V.D.:
[BS1] Weighted inequalities of Hardy type (Russian), *Sibirsk. Mat. Zh.* **30** (1989), (1) 13–22; translation in *Siberian Math. J.* **30** (1989) (1), 8–16. MR0995015

BECKENBACH, E. F. and BELLMAN, R.:
[BB1] *Inequalities*, Ergebnisse der Mathematik und ihrer Grenzgebiete, Vol **30**, Springer, Berlin, 1961. MR0158038

BEESACK, P. and HEINIG, H.P.:
[BH1] Hardy's inequalities with indices less than 1, *Proc. Amer. Math. Soc.* **83** (1981) (3), 532–536. MR0627685

BENNETT, C.:
[Ben1] Intermediate spaces and the class LLogL, *Ark. Mat.* **11** (1973), 215–228. MR0352966

BENNETT, G.:
[Be1] Some elementary inequalities III, *Quart. J. Math. Oxford Ser.* **42** (1991) (166), 149–174. MR1107279

[Be2] *Factorizing the classical inequalities*, Mem. Amer. Math. Soc. **120**, No. 576, 1996. MR1317938

BENNETT, C., DE VORE, R. and SHARPLEY, R.:
[BDS1] Weak-L^∞ and BMO, *Ann. of Math.* **113** (1981), 601–611. MR0621018

BENNETT, C. and SHARPLEY, R.:
[BeS1] *Interpolation of Operators*, Academic Press, 1988. MR0928802

BEREZHNOI, E.I.:
[Ber1] Inequalities with weights for Hardy operators in function spaces, in: Function Spaces (Poznan, 1989), *Teubner-Texte Math.*, Stuttgart **120** (1991), 75–79. MR1155160

BERGH, J. and LÖFSTRÖM, J.:
[BL1] *Interpolation Spaces — An Introduction,* Springer Verlag, Berlin–Heidelberg–New York, 1976. MR0482275

BERGH, J., BURENKOV, V.I. and PERSSON, L.E.:
[BBP1] Best constants in reversed Hardy's inequalities for quasimonotone functions, *Acta. Sci. Math. (Szeged)* **59** (1994) (1–2), 221–239. MR1285442

[BBP2] On some sharp reversed Hölder and Hardy-type inequalities, *Math. Nachr.* **169** (1994), 19–29. MR1292795

BLISS, G.A.:
[Bl1] An integral inequality, *J. London Math. Soc.* **5** (1930), 40–46.

BLOOM, S. and KERMAN, R.:
[BK1] Weighted norm inequalities for operators of Hardy-type, *Proc. Amer. Math. Soc.* **113** (1991) (1), 135–141. MR1059623

BRADLEY, J.S.:
[Br1] Hardy inequalities with mixed norms, *Canad. Math. Bull.* **21** (1978)(4), 405–408. MR0523580

BREZIS, H. and WAINGER, S.:
[BW1] A note on limiting cases of Sobolev embeddings and convolution inequalities, *Commun. Partial Differ. Equ.* **5** (1980), 773–789. MR0579997

BRUDNYĬ, Yu. A. and KRUGLJAK, N.:
[BrK1] *Interpolation functors and interpolation spaces*, North-Holland Publishing Co., Amsterdam, 1991. MR1107298

BURENKOV, V.I.:
[Bu1] Recent progress in studying the boundedness of classical operators of real analysis in general Morrey-type spaces II, *Eurasian Math J.* **4** (2013) (1), 21–45. MR3118889

BURENKOV, V.I. and EVANS, W.D.:
[BE1] Weighted Hardy-type inequalities for differences and the extension problem for spaces with generalized smoothness, *J. London Math. Soc.* **57** (1998) (1), 209–230. MR1624754

BURENKOV, V.I., EVANS, W.D. and GOLDMAN, M.L.:
[BEG1] On weighted Hardy and Poincaré-type inequalities for differences, *J. Inequal. Appl.* **1** (1997), 1–10. MR1731738

BURENKOV, V.I. and GOLDMAN, M.L.:
[BG1] Necessary and sufficient conditions for boundedness of the maximal operator from Lebesgue spaces to Morrey-type spaces, *Math. Inequal. Appl.* **17** (2014) (2), 401–418. MR3235019

BURENKOV, V.I., JAIN P. and TARARYKOVA, T.:
[BJT1] On Hardy-Steklov and geometric Steklov operators, *Math. Nachr.* **280** (2007) (11), 1244–1256. MR2337342

BURENKOV, V.I. and OINAROV, R.:
[BO1] Necessary and sufficient conditions for boundedness of the Hardy-type operator from a weighted Lebesgue space to a Morrey-type space, *Math. Inequal. Appl.* **16** (2013) (1), 1–19. MR3060376

BURTSEVA, E., LUNDBERG, S., PERSSON, L.E and SAMKO, N.:
[BLPSa1] Hardy-type operators in some Hölder type spaces. *J. Math. Inequal.* 2017, to appear.

CAFFARELLI, L., KOHN, R. and NIRENBERG, L.:
[CKN1] First order interpolation inequalities with weights, *Compositio Math.* **53** (1984) (3), 259–275. MR0768824

CARLEMAN, T.:
[C1] Sur les fonctions quasi-analytiques, in: *Conférences faites au cinquième congrès des mathématiciens scandinaves* (Helsingfors 1923), 181–196.

CARLESON, L.:
[Ca1] A proof of an inequality of Carleman, *Proc. Amer. Math. Soc.* **5** (1954), 932–933. MR0065601

CARRO, M.J. and HEINIG, H.P.:
[CH1] Modular inequalities for the Calderon operator, *Tohoku Math. J.* **52** (2000) (1), 31–46. MR1740541

CARRO, M.J. and SORIA, J.:
[CS1] Boundedness of some integral operators, *Canad. J. Math.* **45** (1993) (6), 1155–1166. MR1247539
[CS2] Weighted Lorentz spaces and the Hardy operator, *J. Funct. Anal.* **112** (1993) (2), 480–494. MR1213748
[CS3] The Hardy-Littlewood maximal function and weighted Lorentz spaces, *J. London Math. Soc.* **55** (1997), 146–158. MR1455455

CARTON-LEBRUN, C., HEINIG, H.P. and HOFMANN, S.C.:
[CHH1] Integral operators on weighted amalgams, *Studia Math.* **109** (1994) (2), 133–157. MR1269772

CERDÁ, J. and MARTÍN, J.:
[CM1] Weighted Hardy inequalities and Hardytransforms with weights, *Studia Math.* **139** (2000), 189–196. MR1772116

CHEN, T. and SINNAMON, G.:
[ChS1] Generalized Hardy operators and normalizing measures, *J. Inequal. Appl.,* **7** (2002) (6), 829–866. MR1949628

CHRIST, M. and GRAFAKOS, L.:
[CG1] Best constants for two nonconvolution inequalities, *Proc. Amer. Math. Soc.* **123** (1995) (6), 1687–1693. MR1239796

ČIŽMEŠIJA, A. and PEČARIĆ, J.E.:
[CP1] Mixed means and Hardy's inequality, *Math. Inequal. Appl.* **1** (1998), 491–506. MR1646729
[CP2] Classical Hardy's and Carleman's inequalities and mixed means, in: *Survey on Classical Inequalities* (Ed: T.M. Rassias), Kluwer Academic Publishers, Dordrecht-Boston-London, 2000, 27–65. MR1894716
[CP3] Some new generalisations of inequalities of Hardy and Levin-Cochran-Lee type, *Bull. Austral Math. Soc.* **63** (2001), 105–113. MR1812314

ČIŽMEŠIJA, A., PEČARIĆ, J.E. and PERSSON, L.E.:
[CPP1] Strengthened Hardy and Pólya-Knopp's inequalities, *J. Approx. Theory* **125** (2003), 74–84. MR2016841

COCHRAN, J.A. and LEE, C-S.:
[CL1] Inequalities related to Hardy's and Heinig's, *Math. Proc. Cambridge Philos. Soc.* **96** (1984), 1–7. MR0743695

DAVIES, E.B. and HINZ, A.M.:
   [DH1] Explicit constants for Rellich inequalities in $L_p(\Omega)$, *Math. Z.* **227** (1998) (3), 511–523. MR1612685
DIENING, L. and SAMKO, S.:
   [DS1] Hardy inequality in variable exponent Lebesgue spaces, *Frac. Calc. Appl. Anal.* **10** (2007) (1), 1–18. MR2348863
DOLBEAULT, J., ESTEBAN, M. and LOSS, M.:
   [DEL1] Rigidity versus symmetry breaking via nonlinear flows in cylinders and Euclidean spaces, *Invent. Math.* **206** (2016) (2), 397–440. MR3570296
DOLBEAULT, J., ESTEBAN, M., TARANTELLO, G. and TERTIKAS, A.:
   [DETT1] Radial symmetry and symmetry breaking for some interpolation inequalities, *Calc. Var.* **42** (2011), 461–485. MR2846263
DRÁBEK, P., HEINIG, H.P. and KUFNER, A.:
   [DHK1] Higher-dimensional Hardy inequality, in: *General Inequalities* 7 (Oberwolfach, 1995), *Internal Ser. Numer. Math.* **123** (1997), Birkhäuser, Basel, 1997, 3–16. MR1457264
DRÁBEK, P. and KUFNER, A.:
   [DK1] The Hardy inequality and Birkhoff interpolation, *Bayreuth Math. Schr.* **47** (1994), 99–104. MR1285205
EILERTSEN, S.:
   [E1] On weighted fractional integral inequalities, *J. Funct Anal.* **185**(1) (2001), 342–366. MR1853761
ERIDANI, A., KOKILASHVILI, V. and MESKHI, A.:
   [EKM1] Morrey spaces and fractional integral operators. *Expos. Mathem.* **27** (2009) (3), 227–239. MR2555369
EVANS, D., GOGATISHVILI, A. and OPIC, B.:
   [EGO1] The reverse Hardy inequality with measures, *Math. Inequal. Appl.* **11** (2007) (1), 43–74. MR2376257
GARCÍA-HUIDOBRO, M., KUFNER, A., MANÁSEVICH, R. and YARUR, C.S.:
   [GKMY1] Radial solutions for a quasilinear equation via Hardy inequalities, *Adv. Differential Equations* **6** (2001), 1517–1540. MR1858431
GENEBASHVILI, I.Z., GOGATISHVILI, A.S. and KOKILASHVILI, V.M.:
   [GGK1] Solution of two-weight problems for integral transforms with positive kernels, *Georgian Math. J.* **3** (1996) (4), 319–342. MR1397815
GODUNOVA, E.K.:
   [G1] Inequalities based on convex functions, *Amer. Math. Soc. Trans.* II Ser **88** (1970), 55–66. Translation from *Izv. Vyssh. Uchebn. Zaved. Mat.* **4**(47) (1965), 45–53. MR0193199
GOGATISHVILI, A., JOHANSSON, M., OKPOTI, C.A. and PERSSON, L.E.:
   [GJOP1] Characterizations of embeddings in Lorentz spaces, *Bull. Austral. Math. Soc.* **76** (2007), 69–92. MR2343440
GOGATISHVILI, A., KUFNER, A. and PERSSON L.E.:
   [GKP1] Some new scales of weight characterications of the class Bp, *Acta Math. Hungar* **123** (2009) (4), 365–377. MR2506756

[GKP2]   Some new scales of characterizations of Hardy's inequality, *Proc. Est. Acad. Sci.* **59** (2010) (1), 7–18. MR2647852

GOGATISHVILI, A., KUFNER, A., PERSSON, L.E. and WEDESTIG, A.:

[GKPW1] An equivalence theorem for integral conditions related to Hardy's inequality with applications *Real Anal. Exchange* **29** (2004) (2), 867–880. MR2083821

GOGATISHVILI, A.S. and LANG. J.:

[GL1]   The generalized Hardy operator with kernel and variable integral limits in Banach function spaces, *J. Inequal. Appl.* **4** (1999), 1–16. MR1733113

GOLDMAN, M.L.:

[G1]   On integral inequalities on the set of functions with some properties of monotonicity, Function Spaces, Differential Operators and Nonlinear Analysis (Friedrichroda, 1992), 274–279, Teubner-Texte Math, 133, Teubner, Stuttgart, 1993. MR1242589

GOLDMAN, M.L., HEINIG, H.P. and STEPANOV, V.D.:

[GHS1]   On the principle of duality in Lorentz spaces, *Canad. J. Math.* **48** (1996) (5), 959–979. MR1414066

GRISVARD, P.:

[Gr1]   Espaces intermédiaires entre espaces de Sobolev avec poids, *Ann. Scuola Norm. Sup. Pisa* **17** (1963), 255–296. MR0164236

GUPTA, B., JAIN, P., PERSSON, L.E. and WEDESTIG, A.:

[GJPW1]   Weighted geometric mean inequalities over cones in $\mathbf{R}^N$, *JIPAM, J. Inequal. Pure Appl. Math.* **4** (2003) (4), Article 68, 12 pp. (electronic). MR2051569

GURKA, P.:

[Gu1]   Generalized Hardy's inequality, *Časopis Pěst. J. Mat.* **109** (1984) (2), 194–203. MR0744875

GURKA, P. and OPIC, B.:

[GO1]   Hardy inequality of fractional order, *Banach J. Math. Anal.* **2** (2008) (2), 9–15. MR2391243

[GO2]   Sharp Hardy inequalities of fractional order involving slowly varying functions, *J. Math. Anal. Appl.* **386** (2012), 728–737. MR2855769

HALPERIN, I.:

[H1]   Function spaces, *Canad. J. Math.* **5** (1953), 273–288. MR0056195

HARDY, G.H.:

[Ha1]   Notes on some points in the integral calculus, LI, *Messenger of Math.* **48** (1919), 107–112.

[Ha2]   Note on a Theorem of Hilbert, *Math. Z.* **6** (1920), 314–317.

[Ha3]   Notes on some points in the integral calculus, LX, *Messenger of Math.* **54** (1925), 150–156.

[Ha4]   Prolegomena to a chapter on inequalities, *J. London Math. Soc.* **4** (1929), 61–78.

[Ha5]   Notes on some points in the integral calculus, *Messenger Math.*, LXIV. Further inequalities between integrals. *Messenger Math.* **57** (1928), 12–16.

HARDY, G.H., and LITTLEWOOD, J.E.:
[HL1] Some properties of fractional integrals, *Math. Z.* **27** (1928) (4), 565–606. MR1544927

HEINIG, H.P.:
[He1] Some extensions of Hardy's inequality, *SIAM J. Math. Anal.* **6** (1975), 698–713. MR0376988

HEINIG, H.P., KERMAN, R. and KRBEC, M.:
[HKK1] Weighted exponential inequalities, *Georgian Math. J.* **8** (2001) (1), 69–86. MR1828685

HEINIG, H.P. and KUFNER, A.:
[HK1] Weighted Friedrichs inequalities in amalgams, *Czechoslovak Math. J.* **43** (1993) (2), 285–308. MR1211731
[HK2] Hardy operators on monotone functions and sequences in Orlicz spaces, *J. London. Math. Soc.* **53** (1996) (2), 256–270. MR1373059

HEINIG, H.P., KUFNER, A. and PERSSON, L.E.:
[HKP1] On some fractional order Hardy inequalities, *J. Inequal. Appl.* **1** (1997), 25–46. MR1731740
[HKP2] Some generalizations and refinements of the Hardy inequality, in: *Recent Progress in Inequalities — A volume dedicated to Professor D.S. Mitrinović* (Ed: G.V. Milovanović), Math. Appl. 430, Kluwer Acad. Publ. 1998, 271–288. MR1609959

HEINIG, H.P. and MALIGRANDA, L.:
[HM1] Interpolation with weights in Orlicz spaces, *Boll. Un. Mat. Ital.* B(7) **8** (1994), 37–55. MR1274318
[HM2] Weighted inequalities for monotone and concave functions, *Studia Math.* **116** (1995), 133–270. MR1354136

HEINIG, H.P. and SINNAMON, G.:
[HS1] Mapping properties of integral averaging operators, Studia Math. **129** (1998), 157–177. MR1608162

HERZ, C.:
[Her1] The Hardy-Littlewood maximal theorem, in: *Symposium on Harmonic Analysis* (University of Warwick), 1968, 1–27.

HILBERT, D.:
[Hi1] Grundzüge einer allgemeinen Theorie der linearen Intergraleichungen, *Göttingen Nachr.* (1906), 157–227.
[Hi2] Grundzüge einer allgemeinen therie der linearen Intergraleichungen, Teubner, Leipzig, 1912.

JAIN, P. and GUPTA, B.:
[JG1] Compactness of Hardy-Steklov operators, *J. Math. Anal. Appl.* **288** (2003) (2), 680–691. MR2020189

JAIN, P., JAIN, P.W. and GUPTA, B.:
[JJG1] On a conjecture by Kufner and Persson, *Rocky. Mountain J. Math.* **37** (2007) (6), 1941–1951. MR2382635

JAIN, P., PERSSON, L.E. and WEDESTIG, A.:
[JPW1] From Hardy to Carleman and general mean-type inequalities, in: *Function Spaces and Applications* (Eds: D.E. Edmunds *et al.*), CRC

Press (New York)/ Narosa Publishing House (New Delhi), 2000, 117–130. MR1974797

[JPW2] Carleman-Knopp type inequalities via Hardy's inequality, *Math. Inequal. Appl.* **4** (2001), 343–355. MR1841067

JAKOVLEV, G.N.:

[J1] Boundary properties of a certain class of functions (Russian), *Trudy Mat. Inst. Steklov* **60** (1961), 325–349. MR0144205

[J2] Boundary properties of functions of the class $W_p^{(l)}$ in regions with corners (Russian), *Dokl. Akad. Nauk. SSSR* **140** (1961), 73–76. MR0136988

JOHANSSON, M.:

[Jo1] Carleman type Inequalities and Hardy-type Inequalities for Monotone Functions, Doctoral thesis, Department of Mathematics, Luleå University of Technology, Sweden, 2007.

JOHANSSON, M., PERSSON, L.E. and WEDESTIG, A.:

[JPW1] Carleman's inequality-history, proofs and some new generalizations. *JIPAM, J. Inequal. Pure Appl Math.* **4** (2003), no. 3, Article 53, 19 pp. MR2044402.

JOHANSSON, M., STEPANOV, V. and USHAKOVA, E.:

[JSU1] Hardy inequality with three measures on montone functions, *Math. Inequal. Appl.* **11** (2008) (3), 393–413. MR2431205

KAIBLINGER, N., MALIGRANDA, L. and PERSSON, L.E.:

[KaMP1] Norms in weighted L_2-spaces and Hardy operators, in: *Lecture Notes in Pure and Appl. Math.* **213** (Eds.: H. Hudzik and L. Skrzypczak), Marcel Dekker, New York, 2000, 205–216. MR1772126

KAIJSER, S., NIKOLOVA, L., PERSSON, L.E. and WEDESTIG, A.:

[KNPW1] Hardy-type inequalities via convexity, *Math. Inequal. Appl.* **8** (2005) (3), 403–417. MR2148234

KAIJSER, S., PERSSON, L.E. and ÖBERG, A.:

[KPO1] On Carleman's and Knopp's inequalities, *J. Approx. Theory,* **117** (2002), 140–151. MR1920723

KALYBAY, A.:

[Kal1] A new development of Nikolskii-Lizorkin and Hardy-type inequalities with applications, Doctoral thesis, Department of Mathematics, Luleå University of Technology, 2006.

KALYBAY, A. and PERSSON, L.E.:

[KP1] Three weights higher order Hardy inequalities, *J. Funct. Spaces Appl.* **4** (2006) (2), 163–191. MR2227044

KALYBAY, A., PERSSON, L.E. and TEMIRKHANOVA; A.:

[KPT1] A new discrete Hardy-type inequality with kernels and monotone functions, *J. Inequal. Appl.* 2015, 2015:321. MR3405707

KARAPETYANTS, N.K.:

[K1] Necessary conditions for boundedness of an operator with a nonnegative quasihomogeneous kernel (Russian), *Mat. Zametki* **30** (1981) (5), 787–794. MR0640078

KARAPETYANTS, N.K. and SAMKO, S.:

[KS1] Multidimensional integraloperators with homogeneous kernels, *Frac. Calc. Appl. Anal.* **2** (1999) (1), 67–96. MR1679630

[KS2] *Equations with Involutive Operators*, Birkhäuser, Boston, 2001. MR1836251

KARAPETYANTS, N.K. and SAMKO, N.:

[KSa1] Weighted theorems on fractional integrals in the generalized Hölder spaces via indices m_ω and M_ω, *Fract. Calc. Appl. Anal.* **7** (2004) (4), 437–458. MR2251526

KATO, T.:

[Ka1] *Perturbation Theory for Linear Operators,* Springer Verlag, New York, 1966. MR0203473

KILGORE, T.:

[Ki1] A weighted inequality for derivatives on the half-line, *J. Inequal. Appl.* **4** (1997) (4), 357–367. MR1732632

KNOPP, K.:

[Kn1] Über Reihen mit positiven Gliedern, *J. London Math. Soc.* **3** (1928), 205–211. MR1574991

KOKILASHVILI, V.M.:

[Ko1] On Hardy's inequality in weighted spaces (Russian), *Soobshch. Akad. Nauk Gruzin. SSR* **96** (1979) (1), 37–40. MR0564755

[Ko2[New aspects in the weight theory and applications, in: *Function Spaces, Differential Operators and Nonlinear Analysis* (Ed: J.Rákosník), Prometheus, Prague, 1996, 51–70. MR1480929

KOKILASHVILI, V.M. and MESKHI, A.:

[KM1] Boundedness and compactness criteria for some classical integral transforms, in: *Lecture Notes in Pure and Appl. Math.* **213** (Eds.: H. Hudzik and L. Skrypczak), Marcel Dekker, New York, 2000, 279–296. MR1772132

KOPALIANI, T.:

[Kop1] On the Hardy inequality on $L^{p(\cdot)}(R^n)$ spaces, *Frac. Calc. Appl. Anal.* **12** (2009) (4), 423–432. MR2598189

KRNIĆ, M., PEČARIĆ, J., PERIĆ, I. and VUCOVIĆ, P.:

[KPPV1] *Recent Advances in Hilbert-Type Inequalities, Monographs in Inequalities* **3**, Element, Zagreb, 2012. MR3025815

KRUGLJAK, N., MALIGRANDA, L. and PERSSON, L.E.:

[KrMP1] The failure of the Hardy inequality and interpolation of intersections, *Ark. Mat.* **37** (1999) (2), 323–344. MR1714765

[KrMP2] On an elementary approach to the fractional Hardy inequality, *Proc. Amer. Math. Soc.* **128** (2000), 727–734. MR1676324

KRULIC, K.:

[Kru1] Generalizations and Refinements of Hardy's Inequalities, Doctoral thesis, Department of Mathematics, University of Zagreb, Croatia, 2010.

KUFNER, A.:

[Ku1] *Weighted Sobolev Spaces,* John Wiley & Sons, Inc., New York, 1985. MR0802321

[Ku2] Higher order Hardy inequalities, *Bayreuth. Math. Schr.* **44** (1993), 105–146. MR1224775

[Ku3] Some remarks concerning the Hardy inequality, in: *Function Spaces, Differential Operators and Nonlinear Analysis* (Eds: H. J. Schmeisser and H. Triebel), *Teubner Texte Math.* **133**, 1993, 290–294. MR1242592

[Ku4] Hardy's inequalities and related topics, in: *Function Spaces, Differential Operators and Nonlinear Analysis* (Ed: J. Rákosník), Prague, 1996, 89–99. MR1480931

[Ku5] Fractional order inequalities of Hardy type, in: *Analytic and Geometric Inequalities and Applications* (Eds: T.M. Rassias and H.M. Srivastava), *Math. Appl. 478 Kluwer Acad. Publ.*, Dordrecht, 1999, 183–189. MR1785869

[Ku6] Some comments to the Hardy inequality, in: *Function Spaces and Applications* (Eds: D.E. Edmunds *et al.*), CRC Press (New York)/ Narosa Publishing House (New Delhi), 2000, 143–152. MR1974799

[Ku7] The critical exponent for weighted Sobolev imbeddings, *Acta Appl. Math.* **65** (2001) (1–3), 273–281. MR1843796

KUFNER, A. and HEINIG, H.P.:

[KH1] The Hardy inequality for higher-order derivatives, *Trudy Mat. Inst. Steklov* **192** (1990), 105–113 (Russian); translation in *Proc. Steklov Inst. Math.* **192** (1992), 113–121. MR1097892

KUFNER, A., KULIEV, K. and KULIEVA, G.:

[KKK1] Conditions characterizing the Hardy and reverse Hardy inequalities, *Math. Inequal. Appl.* **12** (2009) (4), 693–700. MR2598260

[KKK2] The Hardy inequality with one negative parameter, *Banach J. Math. Anal.* **2** (2008) (2), 76–84. MR2404105

KUFNER, A., KULIEV, K. and OINAROV, R.:

[KKO1] Some criteria for boundedness and compactness of the Hardy operator with some special kernels, *J. Inequal. Appl.* **2013**, 2013:310, 15 pp. MR3088528

KUFNER, A., KULIEV, K. and PERSSON, L.E.:

[KKP1] Some higher order Hardy inequalities, *J. Inequal. Appl.* **2012**, 2012:69, 14 pp. MR2931013

KUFNER, A. and LEINFELDER, H.:

[KL1] On overdetermined Hardy inequalities, *Math. Bohem.* **123** (1998) (3), 279–293. MR1645446

KUFNER, A., MALIGRANDA, L: and PERSSON, L.E.:

[KMPe1] The prehistory of the Hardy inequality, *Amer. Math. Monthly* **113** (2006) (8), 715–732. MR2256532

KUFNER, A. and PERSSON, L.E.:

[KuP1] Hardy inequalities of fractional order via interpolation, in: *Inequalities and Applications, World Sci. Ser. Appl. Anal.*, 3 (1994), 417–430. MR1299571

[KuP2] A note on fractional order Hardy inequalities, in: *Boundary Value Problems, Special Functions and Fractional Calculus* (Minsk 1996), Beloruss. *Gos. Univ., Minsk*, 1996, 190–194. MR1428941

KUFNER, A:, PERSSON, L.E. and SAMKO, N.:

[KPS1] Hardy-type inequalities with kernels: the current status and some new results, *Math. Nachr.* **290** (2017) (1), 57–65. MR3604622

[KPS2] Some new scales of weight characterizations of Hardy-type inequalities, *Oper. Theory Adv. Appl.* **228**, Birkhäuser/Springer, Basel AG, 2013, 261–274. MR3025499

KUFNER, A:, PERSSON, L.E. and WEDESTIG, A.:

[KPW1] A study of some constants characterizing the weighted Hardy inequality, Orlicz centenary volume, *Banach Center Publ.* **64** (2004), 135–146. MR2099465

KUFNER, A. and SIMADER, C.G.:

[KSi1] Hardy inequalities for overdetermined classes of functions, *Z. Anal. Anwendungen* **16** (1997) (2), 387–403. MR1459965

KUFNER, A. and SINNAMON, G.:

[KSin1] Overdetermined Hardy inequalities, *J. Math. Anal. Appl.* **213** (1997), 468–486. MR1470864

KUFNER, A. and TRIEBEL, H.:

[KT1] Generalizations of Hardy's inequality, *Confer. Sem. Mat. Univ. Bari* **156** (1978), 1–21. MR0541051

KUFNER, A. and WANNEBO, A.:

[KW1] Some remarks to the Hardy inequality for higher order derivatives. in: *General Inequalities* **6** (Oberwolfach 1990), *Internat. Ser. Numer. Math.* **103**, Birkhäuser, Basel, 1992, 33–48. MR1212994

LAI, Q.:

[L1] Two weight mixed Φ-inequalities for the Hardy operator and the Hardy-Littlewood maximal operator, *J. London Math. Soc.* **46** (1992) (2), 301–318. MR1182486

[L2] Two weight Φ-inequalities for the Hardy operator, Hardy-Littlewood maximal operator, and fractional integrals, *Proc. Amer. Math. Soc.* **118** (1993), 129–142. MR1123665

[L3] Weighted modular inequalities for Hardy-type operators, *Proc. London Math. Soc.* **79** (1999), 649–672. MR1710168

LANG, J. and EDMUNDS, D.:

[LE1] Eigenvalues, embeddings and generalized trigonometric functions, *Lecture Notes in Mathematics*, **2016**, Springer, Heidelberg, 2011. MR2796520

LEINDLER, L.:

[Le1] Generalization of inequalities of Hardy and Littlewood, *Acta Sci. Math. (Szeged)* **31** (1970), 279–285. MR0320618

LEVINSON, N.:
    [Lev1] Generalizations of an inequality of Hardy, *Duke Math. J.* **31** (1964), 389–394. MR0171885
LIONS, J.L. and MAGENES, E.:
    [LM1] *Non-homogeneous Boundary Value Problems and Applications.* Vol. I., Grundlehren der Mathematischen Wissenschaften, Band 181, Springer Verlag, New York-Heidelberg, 1972. MR0350178
LOMAKINA, E. and STEPANOV, V.D.:
    [LS1] On the Hardy-type integral operators in Banach function spaces, *Publ. Math.* **42** (1998) (1), 165–194. MR1628166
LORENTZ, G.G.:
    [Lo1] Some new functional spaces, *Ann. of Math.* **51** (1950), 37–55. MR0033449
LORENTZ, R.A.:
    [Lor1] *Multivariate Birkhoff interpolation,* Lecture Notes in Mathematics 1516, Springer-Verlag 1992. MR1222648
LOVE, E.R.:
    [Lov1] Inequalities related to those of Cochran and Lee, *Math. Proc. Cambridge Philos. Soc.* **99** (1986) (3), 395–408. MR0830353
    [Lov2] Some inequalities for geometric means, in: *General Inequalities* **5** (Oberwolfach 1986), *Internat. Schriftenreihe Numer. Math.* 80, Birkhäuser, Basel-Boston, 1987. MR1018137
    [Lov3] Inequalities related to Knopp's inequality, *J. Math. Anal. Appl* **137** (1989) (1), 173–180. MR0981930
LUKKASSEN, D., PERSSON, L.E. and SAMKO, N.:
    [LPSa1] Hardy-type operators in locally vanishing Morrey spaces on fractal sets, *Frac. Calc Appl. Anal.* **18** (2015) (5), 1252–1276. MR3417092
LUKKASSEN, D., PERSSON, L.E. and SAMKO, S.:
    [LPS1] Weighted Hardy operators in complementary Morrey spaces, *J. Funct. Spaces Appl.* (2012), Art. ID 283285. MR2997727
    [LPS2] Some sharp inequalities for integral operators with homogeneous kernels, *J. Inequal. Appl.* **2016**, 2016:114, 18 pp. MR3484026
LUKKASSEN, D., PERSSON, L.E., SAMKO, S. and WALL, P.:
    [LPSW1] Some sharp inequalities for multidimensional integral operators with homogeneous kernel: an overview and new results, *Math. Inequal. Appl.* **19** (2016), 551–564. MR3458764
MALIGRANDA, L.:
    [M1] Indices and interpolation, *Dissertationes Math. (Rozprawy Mat.),* **234**, 1985. MR0820076
    [M2] Weighted inequalities for quasi-monotone functions, *J. London Math. Soc.* **57** (1998), 363–370. MR1644209
MANAKOV, V.M.:
    [Ma1] On the best constant in weighted inequalities for the Riemann-Liouville integrals, *Bull. London. Math. Soc.* **24** (1992) (5), 442–448. MR1173940

MARTÍN-REYES, F.J.:

[Mar1] Weights, one-sided operators, singular integrals and ergodic theorems, in: *Nonlinear Analysis, Function Spaces and Applications,* Vol. 5 (Eds: M. Krbec, A. Kufner, B. Opic and J. Rákosník), Prometheus, Prague, 1994, 103–137. MR1322311

MARTÍN-REYES, F.J. and SAWYER, E.:

[MS1] Weighted inequalities for the Riemann-Liouville fractional integrals of order one and greater, *Proc. Amer. Math. Soc.* **106** (1989) (3), 727–733. MR0965246

MAZ'JA, V.G.:

[Maz1] *Sobolev Spaces,* Springer-Verlag, Springer Series in Soviet Mathematics, Springer-Verlag, Berlin 1985. MR0817985

MAZ'JA, V.G. and VERBITSKY, I.E.:

[MV1] Boundedness and compactness criteria for the one-dimensional Schrödinger operator, in: *Function Spaces, Interpolation Theory and Related Topics* (Eds.: M. Cwikel *et al.*), Walter de Gruyter, Berlin-New York, 2002, 369–382. MR1943294

MIKHAILOV. L.G.:

[Mi1] Some multi-dimensional integral equations with homogeneous kernels (Russian), *Dokl. Akad. Nauk. SSSR* **176** (1967), 263–265. MR0220107

MILMAN, M. and SAGHER, Y.:

[MiS1] An interpolation theorem,   *Ark. Mat.* **22** (1984) (1), 33–38. MR0735876

MUCKENHOUPT, B.:

[Mu1] Hardy's inequality with weights, *Studia Math.* **44** (1972), 31–38. MR0311856

[Mu2] Weighted norm inequalities for the Hardy maximal function, *Trans. Amer. Math. Soc.* **165** (1972), 207–226. MR0312139

[Mu3] Weighted norm inequalities for classical operators, in: *Harmonic analysis in Euclidean spaces, Proc. Sympos. Pure Math. 35, Amer. Math. Soc.*, Providence, R.I., 1979, 69–83. MR0545240

MYASNIKOV, E.A., PERSSON, L.E. and STEPANOV, V.D.:

[MPS1] On the best constants in certain integral inequalities for monotone functions,   *Acta. Sci. Math.(Szeged)* **59** (1994) (3-4), 613–624. MR1317179

NASYROVA, M.:

[N1] Overdetermined weighted Hardy inequalities on semiaxis, in: *Function Spaces and Applications* (Eds: D.E. Edmunds *et al.*), CRC Press (New York)/Narosa Publishing House (New Delhi), 2000, 201–231. MR1974802

[N2] Weighted Inequalities Involving Hardy-type and Limiting Geometric Mean Operators, Doctoral thesis, Department of Mathematics, Luleå University of Technology, Sweden, 2002.

NASYROVA, M., PERSSON, L.E. and STEPANOV, V.D.:

[NPS1] On weighted inequalities with geometric mean operator by the Hardy-type integral transform, *JIPAM J. Inequal. Pure Appl. Math.* **3** (2002) (4), Article 48, 16 pp. MR1923347

NASYROVA, M. and STEPANOV, V.D.:
[NS1] On weighted Hardy inequalities on semiaxis for functions vanishing at the endpoints, *J. Inequal. Appl.* **1** (1997) (3), 223–238. MR1731339
[NS2] On maximal overdetermined Hardy's inequality of second order on a finite interval, *Math. Bohem.* **124** (1999) (2–3), 293–302. MR1780698

NASYROVA, M. and USHAKOVA, E.:
[NU1] Hardy-Steklov operators and embedding inequalities of Sobolev type, arXiv 1503.06518v2, 2015.

NURSULTANOV, E. and TIKHONOV, S.:
[NT1] Convolution inequalities in Lorentz spaces, *J. Fourier Anal. Appl.* **17** (2011) (3), 486–505. MR2803945

OINAROV, R.:
[O1] Weighted inequalities for a class of integral operators (Russian), *Dokl. Akad. Nauk. SSSR* **319** (1991) (5), 1076–1078; translation in *Soviet Math. Dokl.* **44** (1992) (1), 291–293. MR1152890
[O2] Two-sided estimates for the norm of some classes of integral operators (Russian), *Trudy Mat. Inst. Steklov* **204** (1993), 240–250; translation in *Proc. Steklov Inst. Math.* **204** (1994) (3), 205–214. MR1320028
[O3] Boundedness and compactness of Volterra-type integral operators, *Sibirsk Mat. Zh.* **48** (2007) (5), 1100–1105, transl in *Siberian Math. J.* **48** (2007) (5), 884–896. MR2364630

OKPOTI, C. and PERSSON, L.E. and SINNAMON, G.:
[OPS1] An equivalence theorem for some integral conditions with general measures related to Hardy's inequality, *J. Math. Anal. Appl.* **326** (2007) (1), 398–413. MR2277791
[OPS2] An equivalence theorem for some integral conditions with general measures II, *J. Math. Anal. Appl.* **337** (2008) (1), 219–230. MR2358669

OPIC, B. and GURKA, P.:
[OG1] Weighted inequalities for geometric means, *Proc. Amer Math. Soc.* **120** (1994) (3), 771–779. MR1169043

PALMIERI, G.:
[P1] Un approcio alia teoria degli spazi di traccia relativi agli spazi di Orlicz-Sobolev (Italian), *Boll. Un. Mat Ital.* B(5) **16** (1979) (1), 100–119. MR0536530

PEČARIĆ, J.E. and STOLARSKY, K.B.:
[PS1] Carleman's inequality: history and new generalizations, *Aequationes Math.* **61** (2001) (1–2), 49–62. MR1820809

PERSSON, L.E.:
[Pe1] On a weak-type theorem with applications, *Proc. Lond. Math. Soc.* **38** (1979), 295-308. MR0739034
[Pe2] Lecture Notes, Collège de France, Paris, France, November 2015 (P.L. Lions seminar)

PERSSON, L.E. and KUFNER, A.:
[PeK1] Some difference inequalities with weights and interpolation, *Math. Inequal. Appl.* **1** (1998) (3), 437–444. MR1629416

PERSSON, L.E. and OGUNTUASE, J.:

[PeO1] Hardy-type inequalities via convexity-the journey so far, *Austral. J. Math. Anal. Appl.* (2010), No. **2**, Art. 18. MR2794513

PERSSON, L.E., POPOVA, O. and STEPANOV, V.D.:

[PPS1] Two-sided Hardy-type inequalities for monotone functions, *Complex Var. Elliptic. Equ.* **55** (2010), (8–10), 973–989. MR2681427

PERSSON, L.E. and SAMKO, N.:

[PeS1] Some remarks and new developments concerning Hardy-type inequalities, *Rend. Circ. Mat. Palermo*, Suppl. **82** (2010), 93–122. MR3307194

[PeS2] Weighted Hardy and potential operators in the generalized Morrey spaces. *J. Math. Anal. Appl.* **377** (2011) (2), 792–806. MR2769175

[PeS3] What should have happened if Hardy had discovered this?, *J. Inequal. Appl.* 2012, 2012: 29, 11 pp. MR2925639

PERSSON, L.E., SAMKO, N. and WALL, P.:

[PSW1] Quasi-monotone weight functions and their characteristics and applications, *Math. Inequal. Appl.* **15** (2012) (3), 685–705. MR2925639

PERSSON, L.E. and SAMKO. S.:

[PSa1] A note on the best constants in some Hardy inequalities, *J. Math. Inequal.* **9** (2015) (2), 437–447. MR3333874

PERSSON, L.E., SHAMBILOVA, G.E. and STEPANOV, V.D.:

[PSS1] Hardy-type inequalities on the weighted cones of quasi-concave functions, *Banach J. Math. Anal.* **9** (2015) (2), 21–34. MR3296103

PERSSON, L.E. and STEPANOV, V.D.:

[PeSt1] Weighted integral inequalities with the geometric mean operator, *J. Inequal. Appl.* **7** (2002) (5), 727–746 (an abbreviated version can also be found in *Dokl. Akad. Nauk.* **377** (2001) (4), 439–440). MR1931263

PERSSON, L.E., STEPANOV, V.D. and USHAKOVA, E.:

[PSU1] On integral operators with monotone kernels, *Russian Akad. Sci. Dokl. Math.* **403** (2005) (1), 11–14. MR2162464

[PSU2] Equivalence of Hardy-type inequalities with general measures on the cones of non-negative respective non-increasing functions, *Proc. Amer. Math. Soc.* **134** (2006) (8), 2363–2372. MR2213710

PERSSON, L.E., STEPANOV, V. and WALL, P.:

[PStW1] Some scales of equivalent weight characterizations of Hardy's inequality: the case q < p, *Math. Inequal. Appl.* **10** (2007) (2), 267–279. MR2312082

PICK, L. and OPIC, B.:

[PO1] On the geometric mean operator, *J. Math. Anal. Appl.* **183** (1994) (3), 652–662. MR1274866

POPOVA, O.:

[Po1] Weighted Hardy-type Inequalities on the Cones of Monotone and Quasi-concave Functions, Doctoral thesis, Department of Mathematics, Luleå University of Technology, Sweden, 2012.

PROKHOROV, D.V.:
  [Pr1] On the boundedness and compactness of a class of integral operators, *J. London Math. Soc.* **61** (2000) (2), 617–628. MR1760684
  [Pr2] Weighted Hardy inequalities for negative indices, *Publ. Mat.* **48** (2004) (2), 423–443. MR2091014
PROKHOROV, D.V., STEPANOV, V.D. and USHAKOVA, E.:
  [PrSU1] Hardy-Steklov Integral Operators, *Sovrem. Probl. Math.* 22, Steklov Math. Institute of RAS, Moscow, 2016 (in Russian).
RIESZ, F.:
  [R1] Sur un théorème de maximum de MM. Hardy et Littlewood, *J. London Math. Soc.* **7** (1932), 10–13.
SAMKO, N.:
  [S1] On non-equilibrated almost monotonic functions of the Zygmund-Bary-Stechkin class. *Real Anal. Exchange* **30** (2004/05) (2), 727–745. MR2177430
  [S2] Weighted Hardy and singular operators in Morrey spaces, *J. Math. Anal. Appl.* **350** (2009) (1), 56–72. MR2476892
  [S3] Weighted Hardy operators in the local generalized vanishing Morrey spaces. *Positivity* **17** (2013) (3), 683–706. MR3090687
  [S4] On two-weight estimates for the maximal operator in local Morrey spaces. *Internat. J. Math.* **25** (2014) (11), 1450099. MR3285299
  [S5] On a Muckenhoupt condition for Morrey spaces, *Mediterr. J. Math.* **10** (2013) (2), 941–951. MR3045688
SAMKO, S.:
  [Sam1] Proof of the Babenko-Stein theorem (Russian), *Izu. Vysch. Uchebn. Zaved. Matematika* **5** (1975), 47–51. MR0387979
SAWYER, E.:
  [Sa1] Weighted Lebesgue and Lorentz norm inequalities for the Hardy operator, *Trans. Amer. Math. Soc.* **281** (1984) (1), 329–337. MR0719673
  [Sa2] Weighted inequalities for the two-dimensional Hardy operator, *Studia Math.* **82** (1985) (1), 1–16. MR0809769
  [Sa3] Boundedness of classical operators on classical Lorentz spaces, *Studia Math.* **96** (1990) (2), 145–158. MR1052631
SBORDONE, C. and WIK, I.:
  [SW1] Maximal functions and related weight classes, *Publ. Mat.* **38** (1994), 127–155. MR1291957
SCHUR, I.:
  [Sh1] Bemerkungen zur Theorie der beschränkten Bilinearformen mit unendlich vielen Veränderlichen, *J. Reine Angew. Math.* **140** (1911), 1–28. MR1580823
SINNAMON, G.:
  [Si1] Operators on Lebesgue spaces with general measures, Doctoral thesis, McMaster University, 1987
  [Si2] A weighted gradient inequality, *Proc. Royal. Soc. Edinburgh Sect. A* **111** (1989) (3–4), 329–335. MR1007530

[Si3] Spaces defined by level function and their duals, *Studia Math.* **111** (1994) (1), 19–52. MR1292851

[Si4] Kufner's conjecture for higher order Hardy inequalities, *Real. Anal. Exchange* **21** (1995/96) (2), 590–603. MR1407271

[Si5] One-dimensional Hardy-type inequalities in many dimensions, *Proc. Royal. Soc. Edinburgh* **128A** (1998), 833–848. MR1635444

[Si6] Hardy-type inequalities for a new class of integral operators, in: *Analysis of divergence* (Orono, ME, 1997), *Appl. Numer. Harmon. Anal.*, Birkhäuser, Boston, 1999, 297–307. MR1731271

SINNAMON, G. and STEPANOV, V.D.:

[SS1] The weighted Hardy inequality: new proofs and the case $p = 1$, *J. London Math. Soc.* **54** (1996) (2), 89–101. MR1395069

SORIA, J.:

[So1] Lorentz spaces of weak-type, *Quart. J. Math. Oxford* **49** (1998), 93–103. MR1617343

STEIN, E.M. and WEISS, G:

[StW1] Fractional integrals on n-dimensional Euclidean space, *J. Math. Mech.* **7** (1958) (4), 503–514. MR0098285

[StW2] *Introduction to Fourier Analysis on Euclidean Spaces,* Princeton Mathematical Series, No. 32, Princeton University Press, Princeton, N.J., 1971. MR0304972

STEPANOV, V.D.:

[St1] Weighted inequalities of Hardy type for Riemann-Liouville fractional integrals (Russian), *Sibirsk. Mat. Zh.* **31** (1990) (3), 186–197; translation in *Sibirian Math. J.* **31** (1990) (3) 513–522. MR1084772

[St2] Weighted inequalities for a class of Volterra convolution operators, *J. London. Math. Soc.* **45** (1992) (2), 232–242. MR1171551

[St3] Integral operators on the cone of monotone functions, *J. London Math. Soc.* **48** (1993) (3), 465–487. MR1241782

[St4] The weighted Hardy's inequality for nonincreasing functions, *Trans. Amer. Math. Soc.* **338** (1993) (1), 173–186. MR1097171

[St5] Weighted norm inequalities of Hardy type for a class of integral operators, *J. London Math. Soc.* **50** (1994), 105–120. MR1277757

STEPANOV, V.D. and USHAKOVA, E.:

[SU1] On the geometric mean operator with variable limits of integration, *Proc. Steklov Inst. Math.* **260** (2008), 254–278. MR2489517

[SU2] Kernel operators with variable intervals of integration in Lebesgue spaces and applications, *Math. Inequal. Appl.* **13** (2010) (3), 449–510. MR2662834

[SU3] On boundedness of a certain class of Hardy-Steklov operators in Lebesgue spaces, *Banach J. Math. Anal.* **4** (2010) (1), 28–52. MR2593905

TALENTI, G.:

[T1] Osservazioni sopra una classe di disuguaglianze (Italian), *Rend. Scm. Mat. Fis. Milano* **39** (1969), 171–185. MR0280661

TEMIRKHANOVA, A.:
[Te1] Estimates for Discrete Hardy and Pólya-Knopp-type Operators in Weighted Lebesgue Spaces, Doctoral thesis, Department of Mathematics, Luleå University of Technology, Sweden, 2015.

TOMASELLI, G.:
[To1] A class of inequalities, *Boll. Un. Mat. Hal.* **2** (1969), 622–631. MR0255751

TRIEBEL, H.:
[Tr1] *Interpolation Theory, Function Spaces, Differential Operators,* 2nd edition, Johann Ambrosius Barth Verlag, Heidelberg-Leipzig, 1995. MR1328645
[Tr2] Hardy inequalities in function spaces, *Math. Bohem.* **124** (1999) (2), 123–130. MR1780686

USHAKOVA, E.:
[U1] Norm Inequalities of Hardy and Pólya-Knopp Types, Doctoral thesis, Department of Mathematics, Luleå University of Technology, Sweden, 2006.
[U2] On boundedness and compactness of a certain class of kernel operators, *J. Funct. Spaces Appl.* **9** (2011) (1), 67–107. MR2796726
[U3] Boundedness criteria for the Hardy-Steklov operator expressed in terms of a fairway function (Russian), *Dokl. Acad. Nauk.* **461** (2015) (4), 398–399; translation in *Dokl. Math.* **91** (2015) (2), 197–198. MR3410693

WANNEBO, A.:
[W1] Hardy inequalities, *Proc. Amer. Math. Soc.* **109** (1990) (1), 85–95. MR1010807
[W2] Hardy inequalities and imbeddings in domains generalizing $C^{o,\lambda}$ domains, *Proc. Amer. Math. Soc.* **122** (1994) (4), 1181–1190. MR1211593

WEDESTIG, A.:
[We1] Weighted Inequalities of Hardy type and their Limiting Inequalities, Doctoral thesis, Department of Mathematics, Luleå University of Technology, Sweden, 2003.
[We2] Some new Hardy-type inequalities and their limiting inequalities, *JIPAM. J. Inequal. Pure Appl. Math.* **4** (2003) (4), Article 68, 12 pp. MR2044410

WEYL, H.:
[Wey1] Singuläre *Integralgleichngen* mit besonderen Berucksichtigung des Fourierschen Integraltheorems, Doctoral thesis, Göttingen, 1908.

WIENER, F.:
[Wi1] Elementarer Beweis eines Reihensatzes von Herrn Hilbert, *Math. Ann.* **68** (1910) (3), 361–366. MR1511566

YAFAEV, D.:
[Y1] Sharp constants in the Hardy-Rellich inequalities, *J. Funct. Anal.* **168** (1999) (1), 121–144. MR1717839

YAN, P. and SUN, G.:

[YS1] A strengthened Carleman inequality,   *J. Math. Anal. Appl.* **240**
       (1999), 290–293. MR1728192

YANG, B. and DEBNATH, L.:

[YD1] Some inequalities involving the constant e and its applications to
       Carleman's inequality, *J. Math. Anal. Appl.* **223** (1998), 347–353.
       MR1627296

[YD2] Generalizations of Hardy's integral inequality, *Int. J. Math. Sci.* **22**
       (1999) (3), 535–542. MR1717174

YANG, B., ZENG, Z. and DEBNATH, L.:

[YZD1] On new generalizations of Hardy's integral inequality, *J. Math. Anal.
       Appl.* **217** (1998) (1), 321–327. MR1492091

ZHAROV, P.A.:

[Z1]   On a two-weight inequality. Generalization of inequalities of Hardy
       and Poincaré (Russian),   *Trudy Mat. Inst. Steklov* **194** (1992),
       97–110; translation in *Proc. Steklov Inst. Math.* **194** (1993) (4),
       101–114. MR1289650

ZHONG, W.:

[Zh1] The Hilbert-type integral inequalities with a homogeneous kernel of
       λ-degree, *J. Inequal. Appl.* **2008** (2008), Article ID 917392, 12 pp.
       MR2415411

Index

Printed in the United States
By Bookmasters